Algebraic Theory for True Concurrency

Algebraic Theory for True Concurrency

Yong Wang
Faculty of Information
Beijing University of Technology
Beijing, China

ACADEMIC PRESS

An imprint of Elsevier

ISBN: 978-0-443-18912-8

For information on all Academic Press publications
visit our website at https://www.elsevier.com/books-and-journals

Publisher: Mara E. Conner
Acquisitions Editor: Chris Katsaropoulos
Editorial Project Manager: Tom Mearns
Production Project Manager: Fahmida Sultana
Cover Designer: Matthew Limbert

Typeset by VTeX

Working together
to grow libraries in
developing countries

www.elsevier.com • www.bookaid.org

To my family

Contents

1

Introduction

Parallelism and concurrency [7] are the core concepts within computer science. There are mainly two camps in capturing concurrency: the interleaving concurrency and the true concurrency.

The situation for interleaving concurrency is as Fig. 1.1 illustrated. In this situation, there are m concurrent processes (programs) with each process consists of a set of atomic actions and their execution logic (modeled by sequential composition, alternative composition, and recursion or iteration), and each two actions within two processes may communicate. And there are only single thread within single core, single processor and single computer. All the actions (including the new communication actions caused by two actions within two processes) of m concurrent processes are laid on the single thread to be executed. The equivalent execution logic of all actions is the main research contents of interleaving concurrency.

The situation for true concurrency is as Fig. 1.2 illustrated. In this situation, there are also m concurrent processes with each process consists of a set of atomic actions and their execution logic (but modeled by causality and conflicts, these causality and conflicts may exist within two actions in the same process and between actions in two processes). All the m processes actually construct a graph by causality and conflict relations. And there are n threads within multi-cores, multi-processors or multi-computers. All the actions of m concurrent processes are laid on the n threads to be executed. The equivalent execution logic of all actions is the corresponding main research contents of true concurrency.

The representative of interleaving concurrency is bisimulation/rooted branching bisimulation equivalences. CCS (Calculus of Communicating Systems) [3] is a calculus based on bisimulation semantics model. Hennessy and Milner (HM) logic for bisimulation equivalence is also designed. Later, algebraic laws to capture computational properties modulo bisimulation equivalence was introduced in [1], and eventually founded the comprehensive axiomatization modulo bisimulation equivalence – ACP (Algebra of Communicating Processes) [4].

The other camp of concurrency is true concurrency. The researches on true concurrency are still active. Firstly, there are several truly concurrent bisimulation equivalences, the representatives are: pomset bisimulation equivalence, step bisimulation equivalence, history-preserving (hp-) bisimulation equivalence, and especially hereditary history-preserving (hhp-) bisimulation equivalence [8,9], the well-known finest truly concurrent bisimulation equivalence. These truly concurrent bisimulations are studied in different structures [5–7]: Petri nets, event structures, domains, and also a uniform form called TSI (Transition System with Independence) [13]. There are also several logics based on different truly concurrent bisimulation equivalences, for example, SFL (Separation Fixpoint

Algebraic Theory for True Concurrency. https://doi.org/10.1016/B978-0-44-318912-8.00006-1

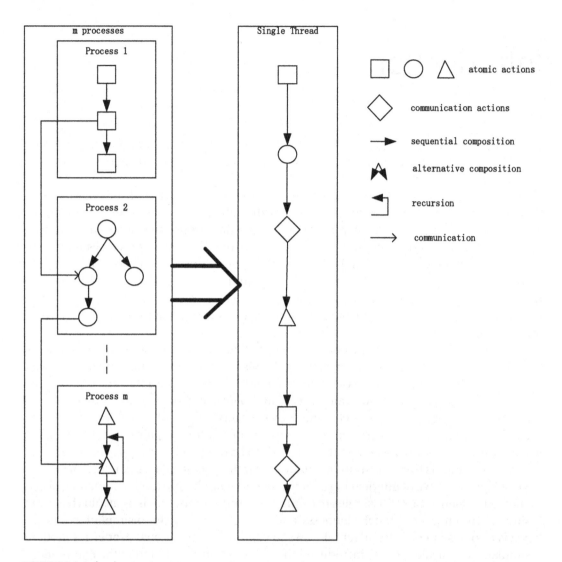

FIGURE 1.1 Interleaving concurrency.

Logic) and TFL (Trace Fixpoint Logic) [13] are extensions on true concurrency of mu-calculi [10] on bisimulation equivalence, and also a logic with reverse modalities [11,12] based on the so-called reverse bisimulations with a reverse flavor. It must be pointed out that, a uniform logic for true concurrency [14,15] was represented, which used a logical framework to unify several truly concurrent bisimulation equivalences, including pomset bisimulation, step bisimulation, hp-bisimulation and hhp-bisimulation.

There are simple comparisons between HM logic for bisimulation equivalence and the uniform logic [14,15] for truly concurrent bisimulation equivalences, the algebraic calculus

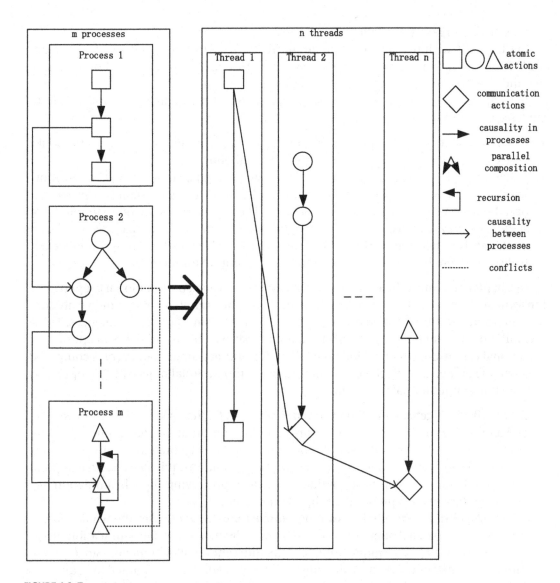

FIGURE 1.2 True concurrency.

CCS, algebraic laws [1], ACP [4], π calculus [23,24], guards [21] for bisimulation equivalence, and *those* for truly concurrent bisimulation equivalences, are still void.

Yes, we try to find the algebraic theory for true concurrency following the way paved by CCS, ACP, π, guards for bisimulation equivalence. And finally, we establish a whole algebraic theory for true concurrency called CTC (Calculus for True Concurrency), $APTC$ (Algebra for Processes in True Concurrency), π_{tc} (π for True Concurrency), $APTC_G$ ($APTC$ with Guards), respectively.

This book is organized as follows.

In Chapter 2, we introduce some preliminaries to make this book self-satisfied, including brief introduction to process algebra, preliminaries on structured operational semantics and proof techniques, and preliminaries on true concurrency.

In Chapter 3, we introduce CTC, which is a calculus of truly concurrent systems. It includes syntax and semantics:

1. Its syntax includes actions, process constant, and operators acting between actions, like Prefix, Summation, Composition, Restriction, Relabeling.
2. Its semantics is based on labeled transition systems, Prefix, Summation, Composition, Restriction, Relabeling have their transition rules. CTC has good semantic properties based on the truly concurrent bisimulations. These properties include monoid laws, static laws, new expansion law for strongly truly concurrent bisimulations, τ laws for weakly truly concurrent bisimulations, and full congruences for strongly and weakly truly concurrent bisimulations, and also unique solution for recursion.

In Chapter 4, we introduce $APTC$, which captures several computational properties in the form of algebraic laws, and proves the soundness and completeness modulo truly concurrent bisimulation/rooted branching truly concurrent bisimulation equivalences. These computational properties are organized in a modular way by use of the concept of conservational extension, which include the following modules, note that, every is composed of constants and operators, the constants are the computational objects, while operators capture the computational properties.

1. $BATC$ **(Basic Algebras for True Concurrency)**. $BATC$ has sequential composition $\cdot$ and alternative composition $+$ to capture causality computation and conflict. The constants are ranged over $\mathbb{E}$, the set of atomic events. The algebraic laws on $\cdot$ and $+$ are sound and complete modulo truly concurrent bisimulation equivalences, such as pomset bisimulation $\sim_p$, step bisimulation $\sim_s$, history-preserving (hp-) bisimulation $\sim_{hp}$ and hereditary history-preserving (hhp-) bisimulation $\sim_{hhp}$.
2. $APTC$ **(Algebra for Parallelism in True Concurrency)**. $APTC$ uses the whole parallel operator $\between$, the parallel operator $\parallel$ to model parallelism, and the communication merge $\mid$ to model causality (communication) among different parallel branches. Since a communication may be blocked, a new constant called deadlock δ is extended to $\mathbb{E}$, and also a new unary encapsulation operator ∂_H is introduced to eliminate δ, which may exist in the processes. And also a conflict elimination operator Θ to eliminate conflicts existing in different parallel branches. The algebraic laws on these operators are also sound and complete modulo truly concurrent bisimulation equivalences, such as pomset bisimulation $\sim_p$, step bisimulation $\sim_s$, history-preserving (hp-) bisimulation $\sim_{hp}$. Note that, these operators in a process except the parallel operator $\parallel$ can be eliminated by deductions on the process using axioms of $APTC$, and eventually be steadied by $\cdot$, $+$ and $\parallel$, this is also why bisimulations are called a *truly concurrent* semantics.
3. **Recursion**. To model infinite computation, recursion is introduced into $APTC$. In order to obtain a sound and complete theory, guarded recursion and linear recursion

are needed. The corresponding axioms are RSP (Recursive Specification Principle) and RDP (Recursive Definition Principle), RDP says the solutions of a recursive specification can represent the behaviors of the specification, while RSP says that a guarded recursive specification has only one solution, they are sound with respect to $APTC$ with guarded recursion modulo truly concurrent bisimulation equivalences, such as pomset bisimulation $\sim_p$, step bisimulation $\sim_s$, history-preserving (hp-) bisimulation $\sim_{hp}$, and they are complete with respect to $APTC$ with linear recursion modulo truly concurrent bisimulation equivalence, such as pomset bisimulation $\sim_p$, step bisimulation $\sim_s$, history-preserving (hp-) bisimulation $\sim_{hp}$.

4. **Abstraction**. To abstract away internal implementations from the external behaviors, a new constant τ called silent step is added to $\mathbb{E}$, and also a new unary abstraction operator τ_I is used to rename actions in I into τ (the resulted $APTC$ with silent step and abstraction operator is called $APTC_\tau$). The recursive specification is adapted to guarded linear recursion to prevent infinite τ-loops specifically. The axioms for τ and τ_I are sound modulo rooted branching truly concurrent bisimulation equivalences (a kind of weak truly concurrent bisimulation equivalence), such as rooted branching pomset bisimulation $\approx_p$, rooted branching step bisimulation $\approx_s$, rooted branching history-preserving (hp-) bisimulation $\approx_{hp}$. To eliminate infinite τ-loops caused by τ_I and obtain the completeness, $CFAR$ (Cluster Fair Abstraction Rule) is used to prevent infinite τ-loops in a constructible way.

In Chapter 5, we introduce π_{tc}, which is an extension of CTC and a generalization of π calculus.

1. It treats names, variables and substitutions more carefully, since names may be free or bound.
2. Names are mobile by references, rather than by values.
3. There are three kinds of prefixes, τ prefix $\tau.P$, output prefix $\overline{x}y.P$ and input prefix $x(y).P$, which are most distinctive to CTC.
4. π_{tc} has good semantic properties based on the truly concurrent bisimulations. These properties include summation laws, identity laws, restriction laws, parallel laws, new expansion law for strongly truly concurrent bisimulations, and full congruences for truly concurrent bisimulations, and also unique solution for recursion.

In Chapter 6, to support data manipulation, we introduce $APTC_G$, which is an extension of $APTC$ with guards.

1. It has a sound and complete theory of concurrency and parallelism with guards.
2. It has a sound and complete theory of recursion including concurrency with guards.
3. It has a sound and complete theory of abstraction with guards.
4. It has a sound Hoare logic [20] including concurrency and parallelism, recursion, and abstraction.

2

Backgrounds

To make this book self-satisfied, we introduce some preliminaries in this chapter, including some introductions on process algebra and true concurrency.

2.1 Process algebra

In this subsection, we introduce the preliminaries on process algebra CCS, ACP, mobility, guards which are based on the interleaving bisimulation semantics.

2.1.1 CCS

A crucial initial observation that is at the heart of the notion of process algebra is due to Milner, who noticed that concurrent processes have an algebraic structure. CCS [2,3] is a calculus of concurrent systems. It includes syntax and semantics:

1. Its syntax includes actions, process constant, and operators acting between actions, like Prefix, Summation, Composition, Restriction, Relabeling.
2. Its semantics is based on labeled transition systems, Prefix, Summation, Composition, Restriction, Relabeling have their transition rules. CCS has good semantic properties based on the interleaving bisimulation. These properties include monoid laws, static laws, expansion law for strongly interleaving bisimulation, τ laws for weakly interleaving bisimulation, and full congruences for strongly and weakly interleaving bisimulations, and also unique solution for recursion.

CCS can be used widely in verification of computer systems with an interleaving concurrent flavor.

2.1.2 ACP

ACP captures several computational properties in the form of algebraic laws, and proves the soundness and completeness modulo bisimulation/rooted branching bisimulation equivalences. These computational properties are organized in a modular way by use of the concept of conservational extension, which include the following modules, note that, every algebra is composed of constants and operators, the constants are the computational objects, while operators capture the computational properties.

1. BPA **(Basic Process Algebras)**. BPA has sequential composition $\cdot$ and alternative composition $+$ to capture sequential computation and nondeterminacy. The constants are

Algebraic Theory for True Concurrency. https://doi.org/10.1016/B978-0-44-318912-8.00007-3

ranged over A, the set of atomic actions. The algebraic laws on $\cdot$ and $+$ are sound and complete modulo bisimulation equivalence.

2. ACP **(Algebra of Communicating Processes)**. ACP uses the parallel operator $\parallel$, the auxiliary binary left merge $\lfloor\!\lfloor$ to model parallelism, and the communication merge $\mid$ to model communications among different parallel branches. Since a communication may be blocked, a new constant called deadlock δ is extended to A, and also a new unary encapsulation operator ∂_H is introduced to eliminate δ, which may exist in the processes. The algebraic laws on these operators are also sound and complete modulo bisimulation equivalence. Note that, these operators in a process can be eliminated by deductions on the process using axioms of ACP, and eventually be steadied by $\cdot$ and $+$, this is also why bisimulation is called an *interleaving* semantics.

3. **Recursion**. To model infinite computation, recursion is introduced into ACP. In order to obtain a sound and complete theory, guarded recursion and linear recursion are needed. The corresponding axioms are RSP (Recursive Specification Principle) and RDP (Recursive Definition Principle), RDP says the solutions of a recursive specification can represent the behaviors of the specification, while RSP says that a guarded recursive specification has only one solution, they are sound with respect to ACP with guarded recursion modulo bisimulation equivalence, and they are complete with respect to ACP with linear recursion modulo bisimulation equivalence.

4. **Abstraction**. To abstract away internal implementations from the external behaviors, a new constant τ called silent step is added to A, and also a new unary abstraction operator τ_I is used to rename actions in I into τ (the resulted ACP with silent step and abstraction operator is called ACP_τ). The recursive specification is adapted to guarded linear recursion to prevent infinite τ-loops specifically. The axioms for τ and τ_I are sound modulo rooted branching bisimulation equivalence (a kind of weak bisimulation equivalence). To eliminate infinite τ-loops caused by τ_I and obtain the completeness, $CFAR$ (Cluster Fair Abstraction Rule) is used to prevent infinite τ-loops in a constructible way.

ACP can be used to verify the correctness of system behaviors, by deduction on the description of the system using the axioms of ACP. Base on the modularity of ACP, it can be extended easily and elegantly. For more details, please refer to the book of ACP [4].

2.1.3 π-calculus

π-calculus [23,24] is a calculus for mobile processes, which have changing structure. The component processes not only can be arbitrarily linked, but also can change the linkages by communications among them. π-calculus is an extension of the process algebra CCS [3]:

1. It treats names, variables and substitutions more carefully, since names may be free or bound.
2. Names are mobile by references, rather than by values.
3. There are three kinds of prefixes, τ prefix $\tau.P$, output prefix $\overline{x}y.P$ and input prefix $x(y).P$, which are most distinctive to CCS.

4. Since strong bisimilarity is not preserved by substitutions because of its interleaving nature, π-calculus develops several kinds of strong bisimulations, and discusses their laws modulo these bisimulations.

2.1.4 Guards

Traditional process algebras, such as CCS [2,3] and ACP [4], are well-known for capturing concurrency based on the interleaving bisimulation semantics. These algebras do not involve anything about data and conditionals, because data are hidden behind of actions. There are some efforts to extend these algebras with conditionals, for example, [21] extended ACP with guards. But these works did not form a perfect solution, just because it is difficult to deal conditionals with the interleaving concurrency.

2.2 Operational semantics

The semantics of ACP is based on bisimulation/rooted branching bisimulation equivalences, and the modularity of ACP relies on the concept of conservative extension, for the conveniences, we introduce some concepts and conclusions on them.

Definition 2.1 (Bisimulation). A bisimulation relation R is a binary relation on processes p, p', q, q' such that: (1) if pRq and $p \xrightarrow{a} p'$ then $q \xrightarrow{a} q'$ with $p'Rq'$; (2) if pRq and $q \xrightarrow{a} q'$ then $p \xrightarrow{a} p'$ with $p'Rq'$; (3) if pRq and pP, then qP; (4) if pRq and qP, then pP. Two processes p and q are bisimilar, denoted by $p \sim_{HM} q$, if there is a bisimulation relation R such that pRq. Note that p, p', q, q' are processes, a is an atomic action, and P is a predicate.

Definition 2.2 (Congruence). Let Σ be a signature. An equivalence relation R on terms $\mathcal{T}(\Sigma)$ is a congruence if for each $f \in \Sigma$, if $s_i R t_i$ for $i \in \{1, \cdots, ar(f)\}$, then $f(s_1, \cdots, s_{ar(f)}) R f(t_1, \cdots, t_{ar(f)})$.

Definition 2.3 (Branching bisimulation). A branching bisimulation relation R is a binary relation on the collection of processes p, p', q, q' such that: (1) if pRq and $p \xrightarrow{a} p'$ then either $a \equiv \tau$ and $p'Rq$ or there is a sequence of (zero or more) τ-transitions $q \xrightarrow{\tau} \cdots \xrightarrow{\tau} q_0$ such that pRq_0 and $q_0 \xrightarrow{a} q'$ with $p'Rq'$; (2) if pRq and $q \xrightarrow{a} q'$ then either $a \equiv \tau$ and pRq' or there is a sequence of (zero or more) τ-transitions $p \xrightarrow{\tau} \cdots \xrightarrow{\tau} p_0$ such that p_0Rq and $p_0 \xrightarrow{a} p'$ with $p'Rq'$; (3) if pRq and pP, then there is a sequence of (zero or more) τ-transitions $q \xrightarrow{\tau} \cdots \xrightarrow{\tau} q_0$ such that pRq_0 and q_0P; (4) if pRq and qP, then there is a sequence of (zero or more) τ-transitions $p \xrightarrow{\tau} \cdots \xrightarrow{\tau} p_0$ such that p_0Rq and p_0P. Two processes p and q are branching bisimilar, denoted by $p \approx_{bHM} q$, if there is a branching bisimulation relation R such that pRq.

Definition 2.4 (Rooted branching bisimulation). A rooted branching bisimulation relation R is a binary relation on processes p, p', q, q' such that: (1) if pRq and $p \xrightarrow{a} p'$ then $q \xrightarrow{a} q'$ with $p' \approx_{bHM} q'$; (2) if pRq and $q \xrightarrow{a} q'$ then $p \xrightarrow{a} p'$ with $p' \approx_{bHM} q'$; (3) if pRq and pP, then qP; (4) if pRq and qP, then pP. Two processes p and q are rooted branching bisimilar, denoted by $p \approx_{rbHM} q$, if there is a rooted branching bisimulation relation R such that pRq.

Definition 2.5 (Conservative extension). Let T_0 and T_1 be TSSs (transition system specifications) over signatures Σ_0 and Σ_1, respectively. The TSS $T_0 \oplus T_1$ is a conservative extension of T_0 if the LTSs (labeled transition systems) generated by T_0 and $T_0 \oplus T_1$ contain exactly the same transitions $t \xrightarrow{a} t'$ and $t P$ with $t \in \mathcal{T}(\Sigma_0)$.

Definition 2.6 (Source-dependency). The source-dependent variables in a transition rule of ρ are defined inductively as follows: (1) all variables in the source of ρ are source-dependent; (2) if $t \xrightarrow{a} t'$ is a premise of ρ and all variables in t are source-dependent, then all variables in t' are source-dependent. A transition rule is source-dependent if all its variables are. A TSS is source-dependent if all its rules are.

Definition 2.7 (Freshness). Let T_0 and T_1 be TSSs over signatures Σ_0 and Σ_1, respectively. A term in $\mathbb{T}(T_0 \oplus T_1)$ is said to be fresh if it contains a function symbol from $\Sigma_1 \setminus \Sigma_0$. Similarly, a transition label or predicate symbol in T_1 is fresh if it does not occur in T_0.

Theorem 2.8 (Conservative extension). *Let T_0 and T_1 be TSSs over signatures Σ_0 and Σ_1, respectively, where T_0 and $T_0 \oplus T_1$ are positive after reduction. Under the following conditions, $T_0 \oplus T_1$ is a conservative extension of T_0. (1) T_0 is source-dependent. (2) For each $\rho \in T_1$, either the source of ρ is fresh, or ρ has a premise of the form $t \xrightarrow{a} t'$ or $t P$, where $t \in \mathbb{T}(\Sigma_0)$, all variables in t occur in the source of ρ and t', a or P is fresh.*

2.3 Proof techniques

In this subsection, we introduce the concepts and conclusions about elimination, which is very important in the proof of completeness theorem.

Definition 2.9 (Elimination property). Let a process algebra with a defined set of basic terms as a subset of the set of closed terms over the process algebra. Then the process algebra has the elimination to basic terms property if for every closed term s of the algebra, there exists a basic term t of the algebra such that the algebra $\vdash s = t$.

Definition 2.10 (Strongly normalizing). A term s_0 is called strongly normalizing if does not have an infinite series of reductions beginning in s_0.

Definition 2.11. We write $s >_{lpo} t$ if $s \rightarrow^+ t$ where $\rightarrow^+$ is the transitive closure of the reduction relation defined by the transition rules of an algebra.

Theorem 2.12 (Strong normalization). *Let a term rewriting system (TRS) with finitely many rewriting rules and let $>$ be a well-founded ordering on the signature of the corresponding algebra. If $s >_{lpo} t$ for each rewriting rule $s \rightarrow t$ in the TRS, then the term rewriting system is strongly normalizing.*

2.4 True concurrency

In this subsection, we introduce the concepts of prime event structure, and also concurrent behavior equivalences [5–7], and also we extend prime event structure with silent event τ, and explain the concept of weakly true concurrency, i.e., concurrent behavioral equivalences with considering silent event τ [16].

2.4.1 Event structure

We give the definition of prime event structure (PES) [5–7] extended with the silent event τ as follows.

Definition 2.13 (Prime event structure with silent event). Let Λ be a fixed set of labels, ranged over $a, b, c, \cdots$ and τ. A (Λ-labeled) prime event structure with silent event τ is a tuple $\mathcal{E} = \langle \mathbb{E}, \leq, \sharp, \lambda \rangle$, where $\mathbb{E}$ is a denumerable set of events, including the silent event τ. Let $\hat{\mathbb{E}} = \mathbb{E} \backslash \{\tau\}$, exactly excluding τ, it is obvious that $\hat{\tau^*} = \epsilon$, where ϵ is the empty event. Let $\lambda : \mathbb{E} \to \Lambda$ be a labeling function and let $\lambda(\tau) = \tau$. And $\leq, \sharp$ are binary relations on $\mathbb{E}$, called causality and conflict respectively, such that:

1. $\leq$ is a partial order and $\lceil e \rceil = \{e' \in \mathbb{E} | e' \leq e\}$ is finite for all $e \in \mathbb{E}$. It is easy to see that $e \leq \tau^* \leq e' = e \leq \tau \leq \cdots \leq \tau \leq e'$, then $e \leq e'$.
2. $\sharp$ is irreflexive, symmetric and hereditary with respect to $\leq$, that is, for all $e, e', e'' \in \mathbb{E}$, if $e \sharp e' \leq e''$, then $e \sharp e''$.

Then, the concepts of consistency and concurrency can be drawn from the above definition:

1. $e, e' \in \mathbb{E}$ are consistent, denoted as $e \frown e'$, if $\neg(e \sharp e')$. A subset $X \subseteq \mathbb{E}$ is called consistent, if $e \frown e'$ for all $e, e' \in X$.
2. $e, e' \in \mathbb{E}$ are concurrent, denoted as $e \parallel e'$, if $\neg(e \leq e')$, $\neg(e' \leq e)$, and $\neg(e \sharp e')$.

The prime event structure without considering silent event τ is the original one in [5–7].

Definition 2.14 (Configuration). Let $\mathcal{E}$ be a PES. A (finite) configuration in $\mathcal{E}$ is a (finite) consistent subset of events $C \subseteq \mathcal{E}$, closed with respect to causality (i.e. $\lceil C \rceil = C$). The set of finite configurations of $\mathcal{E}$ is denoted by $\mathcal{C}(\mathcal{E})$. We let $\hat{C} = C \backslash \{\tau\}$.

A consistent subset of $X \subseteq \mathbb{E}$ of events can be seen as a pomset. Given $X, Y \subseteq \mathbb{E}$, $\hat{X} \sim \hat{Y}$ if $\hat{X}$ and $\hat{Y}$ are isomorphic as pomsets. In the following of the paper, we say $C_1 \sim C_2$, we mean $\hat{C}_1 \sim \hat{C}_2$.

Definition 2.15 (Pomset transitions and step). Let $\mathcal{E}$ be a PES and let $C \in \mathcal{C}(\mathcal{E})$, and $\emptyset \neq X \subseteq \mathbb{E}$, if $C \cap X = \emptyset$ and $C' = C \cup X \in \mathcal{C}(\mathcal{E})$, then $C \xrightarrow{X} C'$ is called a pomset transition from C to C'. When the events in X are pairwise concurrent, we say that $C \xrightarrow{X} C'$ is a step.

Definition 2.16 (Weak pomset transitions and weak step). Let $\mathcal{E}$ be a PES and let $C \in \mathcal{C}(\mathcal{E})$, and $\emptyset \neq X \subseteq \hat{\mathbb{E}}$, if $C \cap X = \emptyset$ and $\hat{C}' = \hat{C} \cup X \in \mathcal{C}(\mathcal{E})$, then $C \xRightarrow{X} C'$ is called a weak pomset transition from C to C', where we define $\overset{e}{\Rightarrow} \triangleq \overset{\tau^*}{\rightarrow} \overset{e}{\rightarrow} \overset{\tau^*}{\rightarrow}$. And $\overset{X}{\Rightarrow} \triangleq \overset{\tau^*}{\rightarrow} \overset{e}{\rightarrow} \overset{\tau^*}{\rightarrow}$, for every $e \in X$. When the events in X are pairwise concurrent, we say that $C \overset{X}{\Rightarrow} C'$ is a weak step.

We will also suppose that all the PESs in this paper are image finite, that is, for any PES $\mathcal{E}$ and $C \in \mathcal{C}(\mathcal{E})$ and $a \in \Lambda$, $\{e \in \mathbb{E} | C \overset{e}{\rightarrow} C' \wedge \lambda(e) = a\}$ and $\{e \in \hat{\mathbb{E}} | C \overset{e}{\Rightarrow} C' \wedge \lambda(e) = a\}$ is finite.

2.4.2 Concurrent behavioral equivalences

Definition 2.17 (Pomset, step bisimulation). Let $\mathcal{E}_1$, $\mathcal{E}_2$ be PESs. A pomset bisimulation is a relation $R \subseteq \mathcal{C}(\mathcal{E}_1) \times \mathcal{C}(\mathcal{E}_2)$, such that if $(C_1, C_2) \in R$, and $C_1 \overset{X_1}{\rightarrow} C_1'$ then $C_2 \overset{X_2}{\rightarrow} C_2'$, with $X_1 \subseteq \mathbb{E}_1$, $X_2 \subseteq \mathbb{E}_2$, $X_1 \sim X_2$ and $(C_1', C_2') \in R$, and vice-versa. We say that $\mathcal{E}_1$, $\mathcal{E}_2$ are pomset bisimilar, written $\mathcal{E}_1 \sim_p \mathcal{E}_2$, if there exists a pomset bisimulation R, such that $(\emptyset, \emptyset) \in R$. By replacing pomset transitions with steps, we can get the definition of step bisimulation. When PESs $\mathcal{E}_1$ and $\mathcal{E}_2$ are step bisimilar, we write $\mathcal{E}_1 \sim_s \mathcal{E}_2$.

Definition 2.18 (Weak pomset, step bisimulation). Let $\mathcal{E}_1$, $\mathcal{E}_2$ be PESs. A weak pomset bisimulation is a relation $R \subseteq \mathcal{C}(\mathcal{E}_1) \times \mathcal{C}(\mathcal{E}_2)$, such that if $(C_1, C_2) \in R$, and $C_1 \overset{X_1}{\Rightarrow} C_1'$ then $C_2 \overset{X_2}{\Rightarrow} C_2'$, with $X_1 \subseteq \hat{\mathbb{E}}_1$, $X_2 \subseteq \hat{\mathbb{E}}_2$, $X_1 \sim X_2$ and $(C_1', C_2') \in R$, and vice-versa. We say that $\mathcal{E}_1$, $\mathcal{E}_2$ are weak pomset bisimilar, written $\mathcal{E}_1 \approx_p \mathcal{E}_2$, if there exists a weak pomset bisimulation R, such that $(\emptyset, \emptyset) \in R$. By replacing weak pomset transitions with weak steps, we can get the definition of weak step bisimulation. When PESs $\mathcal{E}_1$ and $\mathcal{E}_2$ are weak step bisimilar, we write $\mathcal{E}_1 \approx_s \mathcal{E}_2$.

Definition 2.19 (Posetal product). Given two PESs $\mathcal{E}_1$, $\mathcal{E}_2$, the posetal product of their configurations, denoted $\mathcal{C}(\mathcal{E}_1) \overline{\times} \mathcal{C}(\mathcal{E}_2)$, is defined as

$$\{(C_1, f, C_2) | C_1 \in \mathcal{C}(\mathcal{E}_1), C_2 \in \mathcal{C}(\mathcal{E}_2), f : C_1 \rightarrow C_2 \text{ isomorphism}\}.$$

A subset $R \subseteq \mathcal{C}(\mathcal{E}_1) \overline{\times} \mathcal{C}(\mathcal{E}_2)$ is called a posetal relation. We say that R is downward closed when for any $(C_1, f, C_2), (C_1', f', C_2') \in \mathcal{C}(\mathcal{E}_1) \overline{\times} \mathcal{C}(\mathcal{E}_2)$, if $(C_1, f, C_2) \subseteq (C_1', f', C_2')$ pointwise and $(C_1', f', C_2') \in R$, then $(C_1, f, C_2) \in R$.

For $f : X_1 \rightarrow X_2$, we define $f[x_1 \mapsto x_2] : X_1 \cup \{x_1\} \rightarrow X_2 \cup \{x_2\}$, $z \in X_1 \cup \{x_1\}$, (1) $f[x_1 \mapsto x_2](z) = x_2$, if $z = x_1$; (2) $f[x_1 \mapsto x_2](z) = f(z)$, otherwise. Where $X_1 \subseteq \mathbb{E}_1$, $X_2 \subseteq \mathbb{E}_2$, $x_1 \in \mathbb{E}_1$, $x_2 \in \mathbb{E}_2$.

Definition 2.20 (Weakly posetal product). Given two PESs $\mathcal{E}_1$, $\mathcal{E}_2$, the weakly posetal product of their configurations, denoted $\mathcal{C}(\mathcal{E}_1) \overline{\times} \mathcal{C}(\mathcal{E}_2)$, is defined as

$$\{(C_1, f, C_2) | C_1 \in \mathcal{C}(\mathcal{E}_1), C_2 \in \mathcal{C}(\mathcal{E}_2), f : \hat{C}_1 \rightarrow \hat{C}_2 \text{ isomorphism}\}.$$

A subset $R \subseteq \mathcal{C}(\mathcal{E}_1) \overline{\times} \mathcal{C}(\mathcal{E}_2)$ is called a weakly posetal relation. We say that R is downward closed when for any $(C_1, f, C_2), (C_1', f', C_2') \in \mathcal{C}(\mathcal{E}_1) \overline{\times} \mathcal{C}(\mathcal{E}_2)$, if $(C_1, f, C_2) \subseteq (C_1', f', C_2')$ pointwise and $(C_1', f', C_2') \in R$, then $(C_1, f, C_2) \in R$.

For $f : X_1 \rightarrow X_2$, we define $f[x_1 \mapsto x_2] : X_1 \cup \{x_1\} \rightarrow X_2 \cup \{x_2\}$, $z \in X_1 \cup \{x_1\}$, (1) $f[x_1 \mapsto x_2](z) = x_2$, if $z = x_1$; (2) $f[x_1 \mapsto x_2](z) = f(z)$, otherwise. Where $X_1 \subseteq \hat{\mathbb{E}}_1$, $X_2 \subseteq \hat{\mathbb{E}}_2$, $x_1 \in \hat{\mathbb{E}}_1$, $x_2 \in \hat{\mathbb{E}}_2$. Also, we define $f(\tau^*) = f(\tau^*)$.

Definition 2.21 ((Hereditary) history-preserving bisimulation). A history-preserving (hp-) bisimulation is a posetal relation $R \subseteq \mathcal{C}(\mathcal{E}_1)\overline{\times}\mathcal{C}(\mathcal{E}_2)$ such that if $(C_1, f, C_2) \in R$, and $C_1 \xrightarrow{e_1} C_1'$, then $C_2 \xrightarrow{e_2} C_2'$, with $(C_1', f[e_1 \mapsto e_2], C_2') \in R$, and vice-versa. $\mathcal{E}_1, \mathcal{E}_2$ are history-preserving (hp-)bisimilar and are written $\mathcal{E}_1 \sim_{hp} \mathcal{E}_2$ if there exists a hp-bisimulation R such that $(\emptyset, \emptyset, \emptyset) \in R$.

A hereditary history-preserving (hhp-)bisimulation is a downward closed hp-bisimulation. $\mathcal{E}_1, \mathcal{E}_2$ are hereditary history-preserving (hhp-)bisimilar and are written $\mathcal{E}_1 \sim_{hhp} \mathcal{E}_2$.

Definition 2.22 (Weak (hereditary) history-preserving bisimulation). A weak history-preserving (hp-) bisimulation is a weakly posetal relation $R \subseteq \mathcal{C}(\mathcal{E}_1)\overline{\times}\mathcal{C}(\mathcal{E}_2)$ such that if $(C_1, f, C_2) \in R$, and $C_1 \xRightarrow{e_1} C_1'$, then $C_2 \xRightarrow{e_2} C_2'$, with $(C_1', f[e_1 \mapsto e_2], C_2') \in R$, and vice-versa. $\mathcal{E}_1, \mathcal{E}_2$ are weak history-preserving (hp-)bisimilar and are written $\mathcal{E}_1 \approx_{hp} \mathcal{E}_2$ if there exists a weak hp-bisimulation R such that $(\emptyset, \emptyset, \emptyset) \in R$.

A weakly hereditary history-preserving (hhp-)bisimulation is a downward closed weak hp-bisimulation. $\mathcal{E}_1, \mathcal{E}_2$ are weakly hereditary history-preserving (hhp-)bisimilar and are written $\mathcal{E}_1 \approx_{hhp} \mathcal{E}_2$.

Proposition 2.23 (Weakly concurrent behavioral equivalences). *(Strongly) concurrent behavioral equivalences imply weakly concurrent behavioral equivalences. That is, $\sim_p$ implies $\approx_p$, $\sim_s$ implies $\approx_s$, $\sim_{hp}$ implies $\approx_{hp}$, $\sim_{hhp}$ implies $\approx_{hhp}$.*

Proof. From the definition of weak pomset transition, weak step transition, weakly posetal product and weakly concurrent behavioral equivalence, it is easy to see that $\xrightarrow{e} = \xrightarrow{\epsilon}\xrightarrow{e}\xrightarrow{\epsilon}$ for $e \in \mathbb{E}$, where ϵ is the empty event. □

Note that in the above definitions, truly concurrent behavioral equivalences are defined by events $e \in \mathcal{E}$ and prime event structure $\mathcal{E}$, in contrast to interleaving behavioral equivalences by actions $a, b \in \mathcal{P}$ and process (graph) $\mathcal{P}$. Indeed, they have correspondences, in [13], models of concurrency, including Petri nets, transition systems and event structures, are unified in a uniform representation – TSI (Transition System with Independence). If x is a process, let $C(x)$ denote the corresponding configuration (the already executed part of the process x, of course, it is free of conflicts), when $x \xrightarrow{e} x'$, the corresponding configuration $C(x) \xrightarrow{e} C(x')$ with $C(x') = C(x) \cup \{e\}$, where e may be caused by some events in $C(x)$ and concurrent with the other events in $C(x)$, or entirely concurrent with all events in $C(x)$, or entirely caused by all events in $C(x)$. Though the concurrent behavioral equivalences (Definition 2.17, 2.18, 2.21, and 2.22) are defined based on configurations (pasts of processes), they can also be defined based on processes (futures of configurations), we omit the concrete definitions. One key difference between definitions based on configurations and processes is that, the definitions based on configurations are stressing the structures of two equivalent configurations and the concrete atomic events may be different, but the definitions based on processes require not only the equivalent structures, but also the same atomic events by their labels, since we try to establish the algebraic equations modulo the corresponding concurrent behavior equivalences.

With a little abuse of concepts, in the following of the paper, we will not distinguish actions and events, prime event structures and processes, also concurrent behavior equivalences based on configurations and processes, and use them freely, unless they have specific meanings. Usually, in congruence theorem and soundness, we show them in a structure only flavor (equivalences based on configuration); but in proof of the completeness theorem, we must require not only the equivalent structure, but also the same set of atomic events.

3

A calculus for true concurrency

In this chapter, we design a calculus for true concurrency (CTC) following the way paved by CCS for bisimulation equivalence. This chapter is organized as follows. We introduce strongly truly concurrent bisimulations in section 3.2, its properties for weakly truly concurrent bisimulations in section 3.3. In section 3.4, we show the applications of CTC by an example called alternating-bit protocol. Finally, in section 3.5, we conclude this chapter.

3.1 Syntax and operational semantics

We assume an infinite set $\mathcal{N}$ of (action or event) names, and use $a, b, c, \cdots$ to range over $\mathcal{N}$. We denote by $\overline{\mathcal{N}}$ the set of co-names and let $\overline{a}, \overline{b}, \overline{c}, \cdots$ range over $\overline{\mathcal{N}}$. Then we set $\mathcal{L} = \mathcal{N} \cup \overline{\mathcal{N}}$ as the set of labels, and use $l, \overline{l}$ to range over $\mathcal{L}$. We extend complementation to $\mathcal{L}$ such that $\overline{\overline{a}} = a$. Let τ denote the silent step (internal action or event) and define $Act = \mathcal{L} \cup \{\tau\}$ to be the set of actions, α, β range over Act. And K, L are used to stand for subsets of $\mathcal{L}$ and $\overline{L}$ is used for the set of complements of labels in L. A relabeling function f is a function from $\mathcal{L}$ to $\mathcal{L}$ such that $f(\overline{l}) = \overline{f(l)}$. By defining $f(\tau) = \tau$, we extend f to Act.

Further, we introduce a set $\mathcal{X}$ of process variables, and a set $\mathcal{K}$ of process constants, and let $X, Y, \cdots$ range over $\mathcal{X}$, and $A, B, \cdots$ range over $\mathcal{K}$, $\widetilde{X}$ is a tuple of distinct process variables, and also $E, F, \cdots$ range over the recursive expressions. We write $\mathcal{P}$ for the set of processes. Sometimes, we use I, J to stand for an indexing set, and we write $E_i : i \in I$ for a family of expressions indexed by I. Id_D is the identity function or relation over set D.

For each process constant schema A, a defining equation of the form

$$A \overset{\text{def}}{=} P$$

is assumed, where P is a process.

3.1.1 Syntax

We use the Prefix . to model the causality relation $\leq$ in true concurrency, the Summation $+$ to model the conflict relation $\sharp$ in true concurrency, and the Composition $\parallel$ to explicitly model concurrent relation in true concurrency. And we follow the conventions of process algebra.

Definition 3.1 (Syntax). Truly concurrent processes are defined inductively by the following formation rules:

1. $A \in \mathcal{P}$;
2. $\mathbf{nil} \in \mathcal{P}$;

Algebraic Theory for True Concurrency. https://doi.org/10.1016/B978-0-44-318912-8.00008-5

3. if $P \in \mathcal{P}$, then the Prefix $\alpha.P \in \mathcal{P}$, for $\alpha \in Act$;
4. if $P, Q \in \mathcal{P}$, then the Summation $P + Q \in \mathcal{P}$;
5. if $P, Q \in \mathcal{P}$, then the Composition $P \parallel Q \in \mathcal{P}$;
6. if $P \in \mathcal{P}$, then the Prefix $(\alpha_1 \parallel \cdots \parallel \alpha_n).P \in \mathcal{P}$ $(n \in I)$, for $\alpha, \cdots, \alpha_n \in Act$;
7. if $P \in \mathcal{P}$, then the Restriction $P \setminus L \in \mathcal{P}$ with $L \in \mathcal{L}$;
8. if $P \in \mathcal{P}$, then the Relabeling $P[f] \in \mathcal{P}$.

The standard BNF grammar of syntax of CTC can be summarized as follows:

$$P ::= A \quad | \quad \mathbf{nil} \quad | \quad \alpha.P \quad | \quad P+P \quad | \quad P \parallel P \quad | \quad (\alpha_1 \parallel \cdots \parallel \alpha_n).P \quad | \quad P \setminus L \quad | \quad P[f]$$

3.1.2 Operational semantics

The operational semantics is defined by LTSs (labeled transition systems), and it is detailed by the following definition.

Definition 3.2 (Semantics). The operational semantics of CTC corresponding to the syntax in Definition 3.1 is defined by a series of transition rules, named **Act**, **Sum**, **Com**, **Res**, **Rel**, and **Con** indicate that the rules are associated respectively with Prefix, Summation, Composition, Restriction, Relabeling, and Constants in Definition 3.1. They are shown in Table 3.1.

3.1.3 Properties of transitions

Definition 3.3 (Sorts). Given the sorts $\mathcal{L}(A)$ and $\mathcal{L}(X)$ of constants and variables, we define $\mathcal{L}(P)$ inductively as follows.

1. $\mathcal{L}(l.P) = \{l\} \cup \mathcal{L}(P)$;
2. $\mathcal{L}((l_1 \parallel \cdots \parallel l_n).P) = \{l_1, \cdots, l_n\} \cup \mathcal{L}(P)$;
3. $\mathcal{L}(\tau.P) = \mathcal{L}(P)$;
4. $\mathcal{L}(P + Q) = \mathcal{L}(P) \cup \mathcal{L}(Q)$;
5. $\mathcal{L}(P \parallel Q) = \mathcal{L}(P) \cup \mathcal{L}(Q)$;
6. $\mathcal{L}(P \setminus L) = \mathcal{L}(P) - (L \cup \overline{L})$;
7. $\mathcal{L}(P[f]) = \{f(l) : l \in \mathcal{L}(P)\}$;
8. for $A \stackrel{\text{def}}{=} P, \mathcal{L}(P) \subseteq \mathcal{L}(A)$.

Now, we present some properties of the transition rules defined in Table 3.1.

Proposition 3.4. *If $P \xrightarrow{\alpha} P'$, then*

1. $\alpha \in \mathcal{L}(P) \cup \{\tau\}$;
2. $\mathcal{L}(P') \subseteq \mathcal{L}(P)$.

If $P \xrightarrow{\{\alpha_1, \cdots, \alpha_n\}} P'$, then

1. $\alpha_1, \cdots, \alpha_n \in \mathcal{L}(P) \cup \{\tau\}$;
2. $\mathcal{L}(P') \subseteq \mathcal{L}(P)$.

Table 3.1   Transition rules of CTC.

$$\textbf{Act}_1 \quad \frac{}{\alpha.P \xrightarrow{\alpha} P}$$

$$\textbf{Sum}_1 \quad \frac{P \xrightarrow{\alpha} P'}{P + Q \xrightarrow{\alpha} P'}$$

$$\textbf{Com}_1 \quad \frac{P \xrightarrow{\alpha} P' \quad Q \nrightarrow}{P \parallel Q \xrightarrow{\alpha} P' \parallel Q}$$

$$\textbf{Com}_2 \quad \frac{Q \xrightarrow{\alpha} Q' \quad P \nrightarrow}{P \parallel Q \xrightarrow{\alpha} P \parallel Q'}$$

$$\textbf{Com}_3 \quad \frac{P \xrightarrow{\alpha} P' \quad Q \xrightarrow{\beta} Q'}{P \parallel Q \xrightarrow{\{\alpha,\beta\}} P' \parallel Q'} \quad (\beta \neq \overline{\alpha})$$

$$\textbf{Com}_4 \quad \frac{P \xrightarrow{l} P' \quad Q \xrightarrow{\overline{l}} Q'}{P \parallel Q \xrightarrow{\tau} P' \parallel Q'}$$

$$\textbf{Act}_2 \quad \frac{}{(\alpha_1 \parallel \cdots \parallel \alpha_n).P \xrightarrow{\{\alpha_1,\cdots,\alpha_n\}} P} \quad (\alpha_i \neq \overline{\alpha_j} \quad i, j \in \{1,\cdots,n\})$$

$$\textbf{Sum}_2 \quad \frac{P \xrightarrow{\{\alpha_1,\cdots,\alpha_n\}} P'}{P + Q \xrightarrow{\{\alpha_1,\cdots,\alpha_n\}} P'}$$

$$\textbf{Res}_1 \quad \frac{P \xrightarrow{\alpha} P'}{P \setminus L \xrightarrow{\alpha} P' \setminus L} \quad (\alpha, \overline{\alpha} \notin L)$$

$$\textbf{Res}_2 \quad \frac{P \xrightarrow{\{\alpha_1,\cdots,\alpha_n\}} P'}{P \setminus L \xrightarrow{\{\alpha_1,\cdots,\alpha_n\}} P' \setminus L} \quad (\alpha_1, \overline{\alpha_1}, \cdots, \alpha_n, \overline{\alpha_n} \notin L)$$

$$\textbf{Rel}_1 \quad \frac{P \xrightarrow{\alpha} P'}{P[f] \xrightarrow{f(\alpha)} P'[f]}$$

$$\textbf{Rel}_2 \quad \frac{P \xrightarrow{\{\alpha_1,\cdots,\alpha_n\}} P'}{P[f] \xrightarrow{\{f(\alpha_1),\cdots,f(\alpha_n)\}} P'[f]}$$

continued on next page

Table 3.1 (continued)

$$\textbf{Con}_1 \quad \frac{P \xrightarrow{\alpha} P'}{A \xrightarrow{\alpha} P'} \quad (A \overset{\text{def}}{=} P)$$

$$\textbf{Con}_2 \quad \frac{P \xrightarrow{\{\alpha_1, \cdots, \alpha_n\}} P'}{A \xrightarrow{\{\alpha_1, \cdots, \alpha_n\}} P'} \quad (A \overset{\text{def}}{=} P)$$

Proof. By induction on the inference of $P \xrightarrow{\alpha} P'$ and $P \xrightarrow{\{\alpha_1, \cdots, \alpha_n\}} P'$, there are fourteen cases corresponding to the transition rules named $\textbf{Act}_{1,2}$, $\textbf{Sum}_{1,2}$, $\textbf{Com}_{1,2,3,4}$, $\textbf{Res}_{1,2}$, $\textbf{Rel}_{1,2}$, and $\textbf{Con}_{1,2}$ in Table 3.1, we just prove the one case $\textbf{Act}_1$ and $\textbf{Act}_2$, and omit the others.

Case $\textbf{Act}_1$: by $\textbf{Act}_1$, with $P \equiv \alpha.P'$. Then by Definition 3.3, we have (1) $\mathcal{L}(P) = \{\alpha\} \cup \mathcal{L}(P')$ if $\alpha \neq \tau$; (2) $\mathcal{L}(P) = \mathcal{L}(P')$ if $\alpha = \tau$. So, $\alpha \in \mathcal{L}(P) \cup \{\tau\}$, and $\mathcal{L}(P') \subseteq \mathcal{L}(P)$, as desired.

Case $\textbf{Act}_2$: by $\textbf{Act}_2$, with $P \equiv (\alpha_1 \parallel \cdots \parallel \alpha_n).P'$. Then by Definition 3.3, we have (1) $\mathcal{L}(P) = \{\alpha_1, \cdots, \alpha_n\} \cup \mathcal{L}(P')$ if $\alpha_i \neq \tau$ for $i \leq n$; (2) $\mathcal{L}(P) = \mathcal{L}(P')$ if $\alpha_1, \cdots, \alpha_n = \tau$. So, $\alpha_1, \cdots, \alpha_n \in \mathcal{L}(P) \cup \{\tau\}$, and $\mathcal{L}(P') \subseteq \mathcal{L}(P)$, as desired. $\square$

3.2 Strongly truly concurrent bisimulations

3.2.1 Laws and congruence

Based on the concepts of strongly truly concurrent bisimulation equivalences, we get the following laws.

Proposition 3.5 (Monoid laws for strong pomset bisimulation). *The monoid laws for strong pomset bisimulation are as follows.*

1. $P + Q \sim_p Q + P$;
2. $P + (Q + R) \sim_p (P + Q) + R$;
3. $P + P \sim_p P$;
4. $P + \textbf{nil} \sim_p P$.

Proof. 1. $P + Q \sim_p Q + P$. By the transition rules $\textbf{Sum}_{1,2}$ in Table 3.1, we get

$$\frac{P \xrightarrow{p} P'}{P + Q \xrightarrow{p} P'}(p \subseteq P) \quad \frac{P \xrightarrow{p} P'}{Q + P \xrightarrow{p} P'}(p \subseteq P)$$

$$\frac{Q \xrightarrow{q} Q'}{P + Q \xrightarrow{q} Q'}(q \subseteq Q) \quad \frac{Q \xrightarrow{q} Q'}{Q + P \xrightarrow{q} Q'}(q \subseteq Q)$$

Since $P' \sim_p P'$ and $Q' \sim_p Q'$, $P + Q \sim_p Q + P$, as desired.

2. $P + (Q + R) \sim_p (P + Q) + R$. By the transition rules **Sum**$_{1,2}$ in Table 3.1, we get

$$\frac{P \xrightarrow{p} P'}{P + (Q + R) \xrightarrow{p} P'}(p \subseteq P) \qquad \frac{P \xrightarrow{p} P'}{(P + Q) + R \xrightarrow{p} P'}(p \subseteq P)$$

$$\frac{Q \xrightarrow{q} Q'}{P + (Q + R) \xrightarrow{q} Q'}(q \subseteq Q) \qquad \frac{Q \xrightarrow{q} Q'}{(P + Q) + R \xrightarrow{q} Q'}(q \subseteq Q)$$

$$\frac{R \xrightarrow{r} R'}{P + (Q + R) \xrightarrow{r} R'}(r \subseteq R) \qquad \frac{R \xrightarrow{r} R'}{(P + Q) + R \xrightarrow{r} R'}(r \subseteq R)$$

Since $P' \sim_p P'$, $Q' \sim_p Q'$, and $R' \sim_p R'$, $P + (Q + R) \sim_p (P + Q) + R$, as desired.

3. $P + P \sim_p P$. By the transition rules **Sum**$_{1,2}$ in Table 3.1, we get

$$\frac{P \xrightarrow{p} P'}{P + P \xrightarrow{p} P'}(p \subseteq P) \qquad \frac{P \xrightarrow{p} P'}{P \xrightarrow{p} P'}(p \subseteq P)$$

Since $P' \sim_p P'$, $P + P \sim_p P$, as desired.

4. $P + \mathbf{nil} \sim_p P$. By the transition rules **Sum**$_{1,2}$ in Table 3.1, we get

$$\frac{P \xrightarrow{p} P'}{P + \mathbf{nil} \xrightarrow{p} P'}(p \subseteq P) \qquad \frac{P \xrightarrow{p} P'}{P \xrightarrow{p} P'}(p \subseteq P)$$

Since $P' \sim_p P'$, $P + \mathbf{nil} \sim_p P$, as desired. $\qquad\qquad\square$

Proposition 3.6 (Monoid laws for strong step bisimulation). *The monoid laws for strong step bisimulation are as follows.*

1. $P + Q \sim_s Q + P$;
2. $P + (Q + R) \sim_s (P + Q) + R$;
3. $P + P \sim_s P$;
4. $P + \mathbf{nil} \sim_s P$.

Proof. **1.** $P + Q \sim_s Q + P$. By the transition rules **Sum**$_{1,2}$ in Table 3.1, we get

$$\frac{P \xrightarrow{p} P'}{P + Q \xrightarrow{p} P'}(p \subseteq P, \forall \alpha, \beta \in p, \text{ are pairwise concurrent})$$

$$\frac{P \xrightarrow{p} P'}{Q + P \xrightarrow{p} P'}(p \subseteq P, \forall \alpha, \beta \in p, \text{ are pairwise concurrent})$$

$$\frac{Q \xrightarrow{q} Q'}{P + Q \xrightarrow{q} Q'}(q \subseteq Q, \forall \alpha, \beta \in q, \text{ are pairwise concurrent})$$

$$\frac{Q \xrightarrow{q} Q'}{Q + P \xrightarrow{q} Q'}(q \subseteq Q, \forall \alpha, \beta \in q, \text{ are pairwise concurrent})$$

Since $P' \sim_s P'$ and $Q' \sim_s Q'$, $P + Q \sim_s Q + P$, as desired.

2. $P + (Q + R) \sim_s (P + Q) + R$. By the transition rules $\mathbf{Sum}_{1,2}$ in Table 3.1, we get

$$\frac{P \xrightarrow{p} P'}{P + (Q + R) \xrightarrow{p} P'}(p \subseteq P, \forall \alpha, \beta \in p, \text{ are pairwise concurrent})$$

$$\frac{P \xrightarrow{p} P'}{(P + Q) + R \xrightarrow{p} P'}(p \subseteq P, \forall \alpha, \beta \in p, \text{ are pairwise concurrent})$$

$$\frac{Q \xrightarrow{q} Q'}{P + (Q + R) \xrightarrow{q} Q'}(q \subseteq Q, \forall \alpha, \beta \in q, \text{ are pairwise concurrent})$$

$$\frac{Q \xrightarrow{q} Q'}{(P + Q) + R \xrightarrow{q} Q'}(q \subseteq Q, \forall \alpha, \beta \in q, \text{ are pairwise concurrent})$$

$$\frac{R \xrightarrow{r} R'}{P + (Q + R) \xrightarrow{r} R'}(r \subseteq R, \forall \alpha, \beta \in r, \text{ are pairwise concurrent})$$

$$\frac{R \xrightarrow{r} R'}{(P + Q) + R \xrightarrow{r} R'}(r \subseteq R, \forall \alpha, \beta \in r, \text{ are pairwise concurrent})$$

Since $P' \sim_s P'$, $Q' \sim_s Q'$, and $R' \sim_s R'$, $P + (Q + R) \sim_s (P + Q) + R$, as desired.

3. $P + P \sim_s P$. By the transition rules $\mathbf{Sum}_{1,2}$ in Table 3.1, we get

$$\frac{P \xrightarrow{p} P'}{P + P \xrightarrow{p} P'}(p \subseteq P, \forall \alpha, \beta \in p, \text{ are pairwise concurrent})$$

$$\frac{P \xrightarrow{p} P'}{P \xrightarrow{p} P'}(p \subseteq P, \forall \alpha, \beta \in p, \text{ are pairwise concurrent})$$

Since $P' \sim_s P'$, $P + P \sim_s P$, as desired.

4. $P + \mathbf{nil} \sim_s P$. By the transition rules $\mathbf{Sum}_{1,2}$ in Table 3.1, we get

$$\frac{P \xrightarrow{p} P'}{P + \mathbf{nil} \xrightarrow{p} P'}(p \subseteq P, \forall \alpha, \beta \in p, \text{ are pairwise concurrent})$$

$$\frac{P \xrightarrow{p} P'}{P \xrightarrow{p} P'}(p \subseteq P, \forall \alpha, \beta \in p, \text{ are pairwise concurrent})$$

Since $P' \sim_s P'$, $P + \mathbf{nil} \sim_s P$, as desired. $\square$

Proposition 3.7 (Monoid laws for strong hp-bisimulation). *The monoid laws for strong hp-bisimulation are as follows.*

1. $P + Q \sim_{hp} Q + P$;
2. $P + (Q + R) \sim_{hp} (P + Q) + R$;
3. $P + P \sim_{hp} P$;
4. $P + \mathbf{nil} \sim_{hp} P$.

Proof. **1.** $P + Q \sim_{hp} Q + P$. By the transition rules **Sum**$_{1,2}$ in Table 3.1, we get

$$\frac{P \xrightarrow{\alpha} P'}{P + Q \xrightarrow{\alpha} P'} \quad \frac{P \xrightarrow{\alpha} P'}{Q + P \xrightarrow{\alpha} P'}$$

$$\frac{Q \xrightarrow{\beta} Q'}{P + Q \xrightarrow{\beta} Q'} \quad \frac{Q \xrightarrow{\beta} Q'}{Q + P \xrightarrow{\beta} Q'}$$

Since $(C(P + Q), f, C(Q + P)) \in \sim_{hp}$, $(C((P + Q)'), f[\alpha \mapsto \alpha], C((Q + P)')) \in \sim_{hp}$, and $(C((P + Q)'), f[\beta \mapsto \beta], C((Q + P)')) \in \sim_{hp}$, $P + Q \sim_{hp} Q + P$, as desired.

2. $P + (Q + R) \sim_{hp} (P + Q) + R$. By the transition rules **Sum**$_{1,2}$ in Table 3.1, we get

$$\frac{P \xrightarrow{\alpha} P'}{P + (Q + R) \xrightarrow{\alpha} P'} \quad \frac{P \xrightarrow{\alpha} P'}{(P + Q) + R \xrightarrow{\alpha} P'}$$

$$\frac{Q \xrightarrow{\beta} Q'}{P + (Q + R) \xrightarrow{\beta} Q'} \quad \frac{Q \xrightarrow{\beta} Q'}{(P + Q) + R \xrightarrow{\beta} Q'}$$

$$\frac{R \xrightarrow{\gamma} R'}{P + (Q + R) \xrightarrow{\gamma} R'} \quad \frac{R \xrightarrow{\gamma} R'}{(P + Q) + R \xrightarrow{\gamma} R'}$$

Since $(C(P + (Q + R)), f, C((P + Q) + R)) \in \sim_{hp}$, $(C((P + (Q + R))'), f[\alpha \mapsto \alpha], C(((P + Q) + R)')) \in \sim_{hp}$, $(C((P + (Q + R))'), f[\beta \mapsto \beta], C(((P + Q) + R)')) \in \sim_{hp}$, and $(C((P + (Q + R))'), f[\gamma \mapsto \gamma], C(((P + Q) + R)')) \in \sim_{hp}$, $P + (Q + R) \sim_{hp} (P + Q) + R$, as desired.

3. $P + P \sim_{hp} P$. By the transition rules **Sum**$_{1,2}$ in Table 3.1, we get

$$\frac{P \xrightarrow{\alpha} P'}{P + P \xrightarrow{\alpha} P'} \quad \frac{P \xrightarrow{\alpha} P'}{P \xrightarrow{\alpha} P'}$$

Since $(C(P + P), f, C(P)) \in \sim_{hp}$, $(C((P + P)'), f[\alpha \mapsto \alpha], C((P)')) \in \sim_{hp}$, $P + P \sim_{hp} P$, as desired.

4. $P + \mathbf{nil} \sim_{hp} P$. By the transition rules **Sum**$_{1,2}$ in Table 3.1, we get

$$\frac{P \xrightarrow{\alpha} P'}{P + \mathbf{nil} \xrightarrow{\alpha} P'} \quad \frac{P \xrightarrow{\alpha} P'}{P \xrightarrow{\alpha} P'}$$

Since $(C(P + \textbf{nil}), f, C(P)) \in \sim_{hp}$, $(C((P + \textbf{nil})'), f[\alpha \mapsto \alpha], C((P)')) \in \sim_{hp}$, $P + \textbf{nil} \sim_{hp} P$, as desired. $\qquad\square$

Proposition 3.8 (Monoid laws for strongly hhp-bisimulation). *The monoid laws for strongly hhp-bisimulation are as follows.*

1. $P + Q \sim_{hhp} Q + P$;
2. $P + (Q + R) \sim_{hhp} (P + Q) + R$;
3. $P + P \sim_{hhp} P$;
4. $P + \textbf{nil} \sim_{hhp} P$.

Proof. **1.** $P + Q \sim_{hhp} Q + P$. By the transition rules $\textbf{Sum}_{1,2}$ in Table 3.1, we get

$$\frac{P \xrightarrow{\alpha} P'}{P + Q \xrightarrow{\alpha} P'} \quad \frac{P \xrightarrow{\alpha} P'}{Q + P \xrightarrow{\alpha} P'}$$

$$\frac{Q \xrightarrow{\beta} Q'}{P + Q \xrightarrow{\beta} Q'} \quad \frac{Q \xrightarrow{\beta} Q'}{Q + P \xrightarrow{\beta} Q'}$$

Since $(C(P + Q), f, C(Q + P)) \in \sim_{hhp}$, $(C((P + Q)'), f[\alpha \mapsto \alpha], C((Q + P)')) \in \sim_{hhp}$ and $(C((P + Q)'), f[\beta \mapsto \beta], C((Q + P)')) \in \sim_{hhp}$, $P + Q \sim_{hhp} Q + P$, as desired.

2. $P + (Q + R) \sim_{hhp} (P + Q) + R$. By the transition rules $\textbf{Sum}_{1,2}$ in Table 3.1, we get

$$\frac{P \xrightarrow{\alpha} P'}{P + (Q + R) \xrightarrow{\alpha} P'} \quad \frac{P \xrightarrow{\alpha} P'}{(P + Q) + R \xrightarrow{\alpha} P'}$$

$$\frac{Q \xrightarrow{\beta} Q'}{P + (Q + R) \xrightarrow{\beta} Q'} \quad \frac{Q \xrightarrow{\beta} Q'}{(P + Q) + R \xrightarrow{\beta} Q'}$$

$$\frac{R \xrightarrow{\gamma} R'}{P + (Q + R) \xrightarrow{\gamma} R'} \quad \frac{R \xrightarrow{\gamma} R'}{(P + Q) + R \xrightarrow{\gamma} R'}$$

Since $(C(P + (Q + R)), f, C((P + Q) + R)) \in \sim_{hhp}$, $(C((P + (Q + R))'), f[\alpha \mapsto \alpha], C(((P + Q) + R)')) \in \sim_{hhp}$, $(C((P + (Q + R))'), f[\beta \mapsto \beta], C(((P + Q) + R)')) \in \sim_{hhp}$ and $(C((P + (Q + R))'), f[\gamma \mapsto \gamma], C(((P + Q) + R)')) \in \sim_{hhp}$, $P + (Q + R) \sim_{hhp} (P + Q) + R$, as desired.

3. $P + P \sim_{hhp} P$. By the transition rules $\textbf{Sum}_{1,2}$ in Table 3.1, we get

$$\frac{P \xrightarrow{\alpha} P'}{P + P \xrightarrow{\alpha} P'} \quad \frac{P \xrightarrow{\alpha} P'}{P \xrightarrow{\alpha} P'}$$

Since $(C(P + P), f, C(P)) \in \sim_{hhp}$, $(C((P + P)'), f[\alpha \mapsto \alpha], C((P)')) \in \sim_{hhp}$, $P + P \sim_{hhp} P$, as desired.

4. $P + \mathbf{nil} \sim_{hhp} P$. By the transition rules $\mathbf{Sum}_{1,2}$ in Table 3.1, we get

$$\frac{P \xrightarrow{\alpha} P'}{P + \mathbf{nil} \xrightarrow{\alpha} P'} \quad \frac{P \xrightarrow{\alpha} P'}{P \xrightarrow{\alpha} P'}$$

Since $(C(P + \mathbf{nil}), f, C(P)) \in \sim_{hhp}$, $(C((P + \mathbf{nil})'), f[\alpha \mapsto \alpha], C((P)')) \in \sim_{hhp}$, $P + \mathbf{nil} \sim_{hhp} P$, as desired. $\qquad\square$

Proposition 3.9 (Static laws for strong step bisimulation). *The static laws for strong step bisimulation are as follows.*

1. $P \parallel Q \sim_s Q \parallel P$;
2. $P \parallel (Q \parallel R) \sim_s (P \parallel Q) \parallel R$;
3. $P \parallel \mathbf{nil} \sim_s P$;
4. $P \setminus L \sim_s P$, if $\mathcal{L}(P) \cap (L \cup \overline{L}) = \emptyset$;
5. $P \setminus K \setminus L \sim_s P \setminus (K \cup L)$;
6. $P[f] \setminus L \sim_s P \setminus f^{-1}(L)[f]$;
7. $(P \parallel Q) \setminus L \sim_s P \setminus L \parallel Q \setminus L$, if $\mathcal{L}(P) \cap \overline{\mathcal{L}(Q)} \cap (L \cup \overline{L}) = \emptyset$;
8. $P[Id] \sim_s P$;
9. $P[f] \sim_s P[f']$, if $f \upharpoonright \mathcal{L}(P) = f' \upharpoonright \mathcal{L}(P)$;
10. $P[f][f'] \sim_s P[f' \circ f]$;
11. $(P \parallel Q)[f] \sim_s P[f] \parallel Q[f]$, if $f \upharpoonright (L \cup \overline{L})$ is one-to-one, where $L = \mathcal{L}(P) \cup \mathcal{L}(Q)$.

Proof. Though transition rules in Table 3.1 are defined in the flavor of single event, they can be modified into a step (a set of events within which each event is pairwise concurrent), we omit them. If we treat a single event as a step containing just one event, the proof of the static laws does not exist any problem, so we use this way and still use the transition rules in Table 3.1.

1. $P \parallel Q \sim_s Q \parallel P$. By the transition rules $\mathbf{Com}_{1,2,3,4}$ in Table 3.1, we get

$$\frac{P \xrightarrow{\alpha} P' \quad Q \nrightarrow}{P \parallel Q \xrightarrow{\alpha} P' \parallel Q} \quad \frac{P \xrightarrow{\alpha} P' \quad Q \nrightarrow}{Q \parallel P \xrightarrow{\alpha} Q \parallel P'}$$

$$\frac{Q \xrightarrow{\beta} Q' \quad P \nrightarrow}{P \parallel Q \xrightarrow{\beta} P \parallel Q'} \quad \frac{Q \xrightarrow{\beta} Q' \quad P \nrightarrow}{Q \parallel P \xrightarrow{\beta} Q' \parallel P}$$

$$\frac{P \xrightarrow{\alpha} P' \quad Q \xrightarrow{\beta} Q'}{P \parallel Q \xrightarrow{\{\alpha,\beta\}} P' \parallel Q'}(\beta \neq \overline{\alpha}) \quad \frac{P \xrightarrow{\alpha} P' \quad Q \xrightarrow{\beta} Q'}{Q \parallel P \xrightarrow{\{\alpha,\beta\}} Q' \parallel P'}(\beta \neq \overline{\alpha})$$

$$\frac{P \xrightarrow{l} P' \quad Q \xrightarrow{\bar{l}} Q'}{P \parallel Q \xrightarrow{\tau} P' \parallel Q'} \quad \frac{P \xrightarrow{l} P' \quad Q \xrightarrow{\bar{l}} Q'}{Q \parallel P \xrightarrow{\tau} Q' \parallel P'}$$

So, with the assumptions $P' \parallel Q \sim_s Q \parallel P'$, $P \parallel Q' \sim_s Q' \parallel P$, and $P' \parallel Q' \sim_s Q' \parallel P'$, $P \parallel Q \sim_s Q \parallel P$, as desired.

2. $P \parallel (Q \parallel R) \sim_s (P \parallel Q) \parallel R$. By the transition rules $\mathbf{Com}_{1,2,3,4}$ in Table 3.1, we get

$$\frac{P \xrightarrow{\alpha} P' \quad Q \nrightarrow \quad R \nrightarrow}{P \parallel (Q \parallel R) \xrightarrow{\alpha} P' \parallel (Q \parallel R)} \qquad \frac{P \xrightarrow{\alpha} P' \quad Q \nrightarrow \quad R \nrightarrow}{(P \parallel Q) \parallel R \xrightarrow{\alpha} (P' \parallel Q) \parallel R}$$

$$\frac{Q \xrightarrow{\beta} Q' \quad P \nrightarrow \quad R \nrightarrow}{P \parallel (Q \parallel R) \xrightarrow{\beta} P \parallel (Q' \parallel R)} \qquad \frac{Q \xrightarrow{\beta} Q' \quad P \nrightarrow \quad R \nrightarrow}{(P \parallel Q) \parallel R \xrightarrow{\beta} (P \parallel Q') \parallel R}$$

$$\frac{R \xrightarrow{\gamma} R' \quad P \nrightarrow \quad Q \nrightarrow}{P \parallel (Q \parallel R) \xrightarrow{\gamma} P \parallel (Q \parallel R')} \qquad \frac{R \xrightarrow{\gamma} R' \quad P \nrightarrow \quad Q \nrightarrow}{(P \parallel Q) \parallel R \xrightarrow{\gamma} (P \parallel Q) \parallel R'}$$

$$\frac{P \xrightarrow{\alpha} P' \quad Q \xrightarrow{\beta} Q' \quad R \nrightarrow}{P \parallel (Q \parallel R) \xrightarrow{\{\alpha,\beta\}} P' \parallel (Q' \parallel R)}(\beta \neq \overline{\alpha}) \qquad \frac{P \xrightarrow{\alpha} P' \quad Q \xrightarrow{\beta} Q' \quad R \nrightarrow}{(P \parallel Q) \parallel R \xrightarrow{\{\alpha,\beta\}} (P' \parallel Q') \parallel R}(\beta \neq \overline{\alpha})$$

$$\frac{P \xrightarrow{\alpha} P' \quad R \xrightarrow{\gamma} R' \quad Q \nrightarrow}{P \parallel (Q \parallel R) \xrightarrow{\{\alpha,\gamma\}} P' \parallel (Q \parallel R')}(\gamma \neq \overline{\alpha}) \qquad \frac{P \xrightarrow{\alpha} P' \quad R \xrightarrow{\gamma} R' \quad Q \nrightarrow}{(P \parallel Q) \parallel R \xrightarrow{\{\alpha,\gamma\}} (P' \parallel Q) \parallel R]}(\gamma \neq \overline{\alpha})$$

$$\frac{Q \xrightarrow{\beta} P' \quad R \xrightarrow{\gamma} R' \quad P \nrightarrow}{P \parallel (Q \parallel R) \xrightarrow{\{\beta,\gamma\}} P \parallel (Q' \parallel R')}(\gamma \neq \overline{\beta}) \qquad \frac{Q \xrightarrow{\beta} Q' \quad R \xrightarrow{\gamma} R' \quad P \nrightarrow}{(P \parallel Q) \parallel R \xrightarrow{\{\beta,\gamma\}} (P \parallel Q') \parallel R'}(\gamma \neq \overline{\beta})$$

$$\frac{P \xrightarrow{\alpha} P' \quad Q \xrightarrow{\beta} Q' \quad R \xrightarrow{\gamma} R'}{P \parallel (Q \parallel R) \xrightarrow{\{\alpha,\beta,\gamma\}} P' \parallel (Q' \parallel R')}(\beta \neq \overline{\alpha}, \gamma \neq \overline{\alpha}, \gamma \neq \overline{\beta})$$

$$\frac{P \xrightarrow{\alpha} P' \quad Q \xrightarrow{\beta} Q' \quad R \xrightarrow{\gamma} R'}{(P \parallel Q) \parallel R \xrightarrow{\{\alpha,\beta,\gamma\}} (P' \parallel Q') \parallel R'}(\beta \neq \overline{\alpha}, \gamma \neq \overline{\alpha}, \gamma \neq \overline{\beta})$$

$$\frac{P \xrightarrow{l} P' \quad Q \xrightarrow{\overline{l}} Q' \quad R \nrightarrow}{P \parallel (Q \parallel R) \xrightarrow{\tau} P' \parallel (Q' \parallel R)} \qquad \frac{P \xrightarrow{l} P' \quad Q \xrightarrow{\overline{l}} Q' \quad R \nrightarrow}{(P \parallel Q) \parallel R \xrightarrow{\tau} (P' \parallel Q') \parallel R}$$

$$\frac{P \xrightarrow{l} P' \quad R \xrightarrow{\overline{l}} R' \quad Q \nrightarrow}{P \parallel (Q \parallel R) \xrightarrow{\tau} P' \parallel (Q \parallel R')} \qquad \frac{P \xrightarrow{l} P' \quad R \xrightarrow{\overline{l}} R' \quad Q \nrightarrow}{(P \parallel Q) \parallel R \xrightarrow{\tau} (P' \parallel Q) \parallel R]}$$

$$\frac{Q \xrightarrow{l} P' \quad R \xrightarrow{\overline{l}} R' \quad P \nrightarrow}{P \parallel (Q \parallel R) \xrightarrow{\tau} P \parallel (Q' \parallel R')} \qquad \frac{Q \xrightarrow{l} Q' \quad R \xrightarrow{\overline{l}} R' \quad P \nrightarrow}{(P \parallel Q) \parallel R \xrightarrow{\tau} (P \parallel Q') \parallel R'}$$

$$\frac{P \xrightarrow{l} P' \quad Q \xrightarrow{\bar{l}} Q' \quad R \xrightarrow{\gamma} R'}{P \parallel (Q \parallel R) \xrightarrow{\{\tau,\gamma\}} P' \parallel (Q' \parallel R')} \qquad \frac{P \xrightarrow{l} P' \quad Q \xrightarrow{\bar{l}} Q' \quad R \xrightarrow{\gamma} R'}{(P \parallel Q) \parallel R \xrightarrow{\{\tau,\gamma\}} (P' \parallel Q') \parallel R'}$$

$$\frac{P \xrightarrow{l} P' \quad R \xrightarrow{\bar{l}} R' \quad Q \xrightarrow{\beta} Q'}{P \parallel (Q \parallel R) \xrightarrow{\{\tau,\beta\}} P' \parallel (Q' \parallel R')} \qquad \frac{P \xrightarrow{l} P' \quad R \xrightarrow{\bar{l}} R' \quad Q \xrightarrow{\beta} Q'}{(P \parallel Q) \parallel R \xrightarrow{\{\tau,\beta\}} (P' \parallel Q') \parallel R]}$$

$$\frac{Q \xrightarrow{l} Q' \quad R \xrightarrow{\bar{l}} R' \quad P \xrightarrow{\alpha} P'}{P \parallel (Q \parallel R) \xrightarrow{\{\tau,\alpha\}} P' \parallel (Q' \parallel R')} \qquad \frac{Q \xrightarrow{l} Q' \quad R \xrightarrow{\bar{l}} R' \quad P \xrightarrow{\alpha} P'}{(P \parallel Q) \parallel R \xrightarrow{\{\tau,\alpha\}} (P' \parallel Q') \parallel R'}$$

So, with the assumptions $P' \parallel (Q \parallel R) \sim_s (P' \parallel Q) \parallel R$, $P \parallel (Q' \parallel R) \sim_s (P \parallel Q') \parallel R$, $P \parallel (Q \parallel R') \sim_s (P \parallel Q) \parallel R'$, $P' \parallel (Q' \parallel R) \sim_s (P' \parallel Q') \parallel R$, $P' \parallel (Q \parallel R') \sim_s (P' \parallel Q) \parallel R'$, $P \parallel (Q' \parallel R') \sim_s (P \parallel Q') \parallel R'$, $P' \parallel (Q' \parallel R') \sim_s (P' \parallel Q') \parallel R'$, $P \parallel (Q \parallel R) \sim_s (P \parallel Q) \parallel R$, as desired.

3. $P \parallel \mathbf{nil} \sim_s P$. By the transition rules $\mathbf{Com}_{1,2,3,4}$ in Table 3.1, we get

$$\frac{P \xrightarrow{\alpha} P'}{P \parallel \mathbf{nil} \xrightarrow{\alpha} P'} \qquad \frac{P \xrightarrow{\alpha} P'}{P \xrightarrow{\alpha} P'}$$

Since $P' \sim_s P'$, $P \parallel \mathbf{nil} \sim_s P$, as desired.

4. $P \setminus L \sim_s P$, if $\mathcal{L}(P) \cap (L \cup \bar{L}) = \emptyset$. By the transition rules $\mathbf{Res}_{1,2}$ in Table 3.1, we get

$$\frac{P \xrightarrow{\alpha} P'}{P \setminus L \xrightarrow{\alpha} P' \setminus L}(\mathcal{L}(P) \cap (L \cup \bar{L}) = \emptyset) \qquad \frac{P \xrightarrow{\alpha} P'}{P \xrightarrow{\alpha} P'}$$

Since $P' \sim_s P'$, and with the assumption $P' \setminus L \sim_s P'$, $P \setminus L \sim_s P$, if $\mathcal{L}(P) \cap (L \cup \bar{L}) = \emptyset$, as desired.

5. $P \setminus K \setminus L \sim_s P \setminus (K \cup L)$. By the transition rules $\mathbf{Res}_{1,2}$ in Table 3.1, we get

$$\frac{P \xrightarrow{\alpha} P'}{P \setminus K \setminus L \xrightarrow{\alpha} P' \setminus K \setminus L} \qquad \frac{P \xrightarrow{\alpha} P'}{P \setminus (K \cup L) \xrightarrow{\alpha} P' \setminus (K \cup L)}$$

Since $P' \sim_s P'$, and with the assumption $P' \setminus K \setminus L \sim_s P' \setminus (K \cup L)$, $P \setminus K \setminus L \sim_s P \setminus (K \cup L)$, as desired.

6. $P[f] \setminus L \sim_s P \setminus f^{-1}(L)[f]$. By the transition rules $\mathbf{Res}_{1,2}$ and $\mathbf{Rel}_{1,2}$ in Table 3.1, we get

$$\frac{P \xrightarrow{\alpha} P'}{P[f] \setminus L \xrightarrow{f(\alpha)} P'[f] \setminus L} \qquad \frac{P \xrightarrow{\alpha} P'}{P \setminus f^{-1}(L)[f] \xrightarrow{f(\alpha)} P' \setminus f^{-1}(L)[f]}$$

So, with the assumption $P'[f] \setminus L \sim_s P' \setminus f^{-1}(L)[f]$, $P[f] \setminus L \sim_s P \setminus f^{-1}(L)[f]$, as desired.

7. $(P \parallel Q) \setminus L \sim_s P \setminus L \parallel Q \setminus L$, if $\mathcal{L}(P) \cap \overline{\mathcal{L}(Q)} \cap (L \cup \overline{L}) = \emptyset$. By the transition rules $\mathbf{Com}_{1,2,3,4}$ and $\mathbf{Res}_{1,2}$ in Table 3.1, we get

$$\frac{P \xrightarrow{\alpha} P' \quad Q \nrightarrow}{(P \parallel Q) \setminus L \xrightarrow{\alpha} (P' \parallel Q) \setminus L}(\mathcal{L}(P) \cap \overline{\mathcal{L}(Q)} \cap (L \cup \overline{L}) = \emptyset)$$

$$\frac{P \xrightarrow{\alpha} P' \quad Q \nrightarrow}{P \setminus L \parallel Q \setminus L \xrightarrow{\alpha} P' \setminus L \parallel Q \setminus L}(\mathcal{L}(P) \cap \overline{\mathcal{L}(Q)} \cap (L \cup \overline{L}) = \emptyset)$$

$$\frac{Q \xrightarrow{\beta} Q' \quad P \nrightarrow}{(P \parallel Q) \setminus L \xrightarrow{\beta} (P \parallel Q') \setminus L}(\mathcal{L}(P) \cap \overline{\mathcal{L}(Q)} \cap (L \cup \overline{L}) = \emptyset)$$

$$\frac{Q \xrightarrow{\beta} Q' \quad P \nrightarrow}{P \setminus L \parallel Q \setminus L \xrightarrow{\beta} P \setminus L \parallel Q' \setminus L}(\mathcal{L}(P) \cap \overline{\mathcal{L}(Q)} \cap (L \cup \overline{L}) = \emptyset)$$

$$\frac{P \xrightarrow{\alpha} P' \quad Q \xrightarrow{\beta} Q'}{(P \parallel Q) \setminus L \xrightarrow{\{\alpha,\beta\}} (P' \parallel Q') \setminus L}(\mathcal{L}(P) \cap \overline{\mathcal{L}(Q)} \cap (L \cup \overline{L}) = \emptyset)$$

$$\frac{P \xrightarrow{\alpha} P' \quad Q \xrightarrow{\beta} Q'}{P \setminus L \parallel Q \setminus L \xrightarrow{\{\alpha,\beta\}} (P' \parallel Q') \setminus L}(\mathcal{L}(P) \cap \overline{\mathcal{L}(Q)} \cap (L \cup \overline{L}) = \emptyset)$$

$$\frac{P \xrightarrow{l} P' \quad Q \xrightarrow{\overline{l}} Q'}{(P \parallel Q) \setminus L \xrightarrow{\tau} (P' \parallel Q') \setminus L}(\mathcal{L}(P) \cap \overline{\mathcal{L}(Q)} \cap (L \cup \overline{L}) = \emptyset)$$

$$\frac{P \xrightarrow{l} P' \quad Q \xrightarrow{\overline{l}} Q'}{(P \setminus L \parallel Q \setminus L \xrightarrow{\tau} P' \setminus L \parallel Q' \setminus L}(\mathcal{L}(P) \cap \overline{\mathcal{L}(Q)} \cap (L \cup \overline{L}) = \emptyset)$$

Since $(P' \parallel Q) \setminus L \sim_s P' \setminus L \parallel Q \setminus L$, $(P \parallel Q') \setminus L \sim_s P \setminus L \parallel Q' \setminus L$ and $(P' \parallel Q') \setminus L \sim_s P' \setminus L \parallel Q' \setminus L$, $(P \parallel Q) \setminus L \sim_s P \setminus L \parallel Q \setminus L$, if $\mathcal{L}(P) \cap \overline{\mathcal{L}(Q)} \cap (L \cup \overline{L}) = \emptyset$, as desired.

8. $P[Id] \sim_s P$. By the transition rules $\mathbf{Rel}_{1,2}$ in Table 3.1, we get

$$\frac{P \xrightarrow{\alpha} P'}{P[Id] \xrightarrow{Id(\alpha)} P'[Id]} \qquad \frac{P \xrightarrow{\alpha} P'}{P \xrightarrow{\alpha} P'}$$

So, with the assumption $P'[Id] \sim_s P'$ and $Id(\alpha) = \alpha$, $P[Id] \sim_s P$, as desired.

9. $P[f] \sim_s P[f']$, if $f \upharpoonright \mathcal{L}(P) = f' \upharpoonright \mathcal{L}(P)$. By the transition rules $\mathbf{Rel}_{1,2}$ in Table 3.1, we get

$$\frac{P \xrightarrow{\alpha} P'}{P[f] \xrightarrow{f(\alpha)} P'[f]} \qquad \frac{P \xrightarrow{\alpha} P'}{P[f'] \xrightarrow{f'(\alpha)} P'[f']}$$

So, with the assumption $P'[f] \sim_s P'[f']$ and $f(\alpha) = f'(\alpha)$, if $f \upharpoonright \mathcal{L}(P) = f' \upharpoonright \mathcal{L}(P)$, $P[f] \sim_s P[f']$, as desired.

10. $P[f][f'] \sim_s P[f' \circ f]$. By the transition rules **Rel**$_{1,2}$ in Table 3.1, we get

$$\frac{P \xrightarrow{\alpha} P'}{P[f][f'] \xrightarrow{f'(f(\alpha))} P'[f][f']} \qquad \frac{P \xrightarrow{\alpha} P'}{P[f' \circ f] \xrightarrow{f'(f(\alpha))} P'[f' \circ f]}$$

So, with the assumption $P'[f][f'] \sim_s P'[f' \circ f]$, $P[f][f'] \sim_s P[f' \circ f]$, as desired.

11. $(P \parallel Q)[f] \sim_s P[f] \parallel Q[f]$, if $f \upharpoonright (L \cup \overline{L})$ is one-to-one, where $L = \mathcal{L}(P) \cup \mathcal{L}(Q)$. By the transition rules **Com**$_{1,2,3,4}$ and **Rel**$_{1,2}$ in Table 3.1, we get

$$\frac{P \xrightarrow{\alpha} P' \quad Q \nrightarrow}{(P \parallel Q)[f] \xrightarrow{f(\alpha)} (P' \parallel Q)[f]} \text{(if } f \upharpoonright (L \cup \overline{L}) \text{ is one-to-one, where } L = \mathcal{L}(P) \cup \mathcal{L}(Q))$$

$$\frac{P \xrightarrow{\alpha} P' \quad Q \nrightarrow}{P[f] \parallel Q[f] \xrightarrow{f(\alpha)} P'[f] \parallel Q[f]} \text{(if } f \upharpoonright (L \cup \overline{L}) \text{ is one-to-one, where } L = \mathcal{L}(P) \cup \mathcal{L}(Q))$$

$$\frac{Q \xrightarrow{\beta} Q' \quad P \nrightarrow}{(P \parallel Q)[f] \xrightarrow{f(\beta)} (P \parallel Q')[f]} \text{(if } f \upharpoonright (L \cup \overline{L}) \text{ is one-to-one, where } L = \mathcal{L}(P) \cup \mathcal{L}(Q))$$

$$\frac{Q \xrightarrow{\beta} Q' \quad P \nrightarrow}{P[f] \parallel Q[f] \xrightarrow{f(\beta)} P[f] \parallel Q'[f]} \text{(if } f \upharpoonright (L \cup \overline{L}) \text{ is one-to-one, where } L = \mathcal{L}(P) \cup \mathcal{L}(Q))$$

$$\frac{P \xrightarrow{\alpha} P' \quad Q \xrightarrow{\beta} Q'}{(P \parallel Q)[f] \xrightarrow{\{f(\alpha), f(\beta)\}} (P' \parallel Q')[f]} \text{(if } f \upharpoonright (L \cup \overline{L}) \text{ is one-to-one, where } L = \mathcal{L}(P) \cup \mathcal{L}(Q))$$

$$\frac{P \xrightarrow{\alpha} P' \quad Q \xrightarrow{\beta} Q'}{P[f] \parallel Q[f] \xrightarrow{\{f(\alpha), f(\beta)\}} P'[f] \parallel Q'[f]} \text{(if } f \upharpoonright (L \cup \overline{L}) \text{ is one-to-one, where}$$

$$L = \mathcal{L}(P) \cup \mathcal{L}(Q))$$

$$\frac{P \xrightarrow{l} P' \quad Q \xrightarrow{\bar{l}} Q'}{(P \parallel Q)[f] \xrightarrow{\tau} (P' \parallel Q')[f]} \text{(if } f \upharpoonright (L \cup \overline{L}) \text{ is one-to-one, where } L = \mathcal{L}(P) \cup \mathcal{L}(Q))$$

$$\frac{P \xrightarrow{l} P' \quad Q \xrightarrow{\bar{l}} Q'}{(P[f] \parallel Q[f] \xrightarrow{\tau} P'[f] \parallel Q'[f]} \text{(if } f \upharpoonright (L \cup \overline{L}) \text{ is one-to-one, where } L = \mathcal{L}(P) \cup \mathcal{L}(Q))$$

So, with the assumptions $(P' \parallel Q)[f] \sim_s P'[f] \parallel Q[f]$, $(P \parallel Q')[f] \sim_s P[f] \parallel Q'[f]$, and $(P' \parallel Q')[f] \sim_s P'[f] \parallel Q'[f]$, $(P \parallel Q)[f] \sim_s P[f] \parallel Q[f]$, if $f \upharpoonright (L \cup \overline{L})$ is one-to-one, where $L = \mathcal{L}(P) \cup \mathcal{L}(Q)$, as desired. $\square$

Proposition 3.10 (Static laws for strong pomset bisimulation). *The static laws for strong pomset bisimulation are as follows.*

1. $P \parallel Q \sim_p Q \parallel P$;
2. $P \parallel (Q \parallel R) \sim_p (P \parallel Q) \parallel R$;
3. $P \parallel \textbf{nil} \sim_p P$;
4. $P \setminus L \sim_p P$, if $\mathcal{L}(P) \cap (L \cup \overline{L}) = \emptyset$;
5. $P \setminus K \setminus L \sim_p P \setminus (K \cup L)$;
6. $P[f] \setminus L \sim_p P \setminus f^{-1}(L)[f]$;
7. $(P \parallel Q) \setminus L \sim_p P \setminus L \parallel Q \setminus L$, if $\mathcal{L}(P) \cap \overline{\mathcal{L}(Q)} \cap (L \cup \overline{L}) = \emptyset$;
8. $P[Id] \sim_p P$;
9. $P[f] \sim_p P[f']$, if $f \upharpoonright \mathcal{L}(P) = f' \upharpoonright \mathcal{L}(P)$;
10. $P[f][f'] \sim_p P[f' \circ f]$;
11. $(P \parallel Q)[f] \sim_p P[f] \parallel Q[f]$, if $f \upharpoonright (L \cup \overline{L})$ is one-to-one, where $L = \mathcal{L}(P) \cup \mathcal{L}(Q)$.

Proof. From the definition of strong pomset bisimulation (see Definition 2.17), we know that strong pomset bisimulation is defined by pomset transitions, which are labeled by pomsets. In a pomset transition, the events in the pomset are either within causality relations (defined by the prefix .) or in concurrency (implicitly defined by . and +, and explicitly defined by $\parallel$), of course, they are pairwise consistent (without conflicts). In Proposition 3.9, we have already proven the case that all events are pairwise concurrent, so, we only need to prove the case of events in causality. Without loss of generality, we take a pomset of $p = \{\alpha, \beta : \alpha.\beta\}$. Then the pomset transition labeled by the above p is just composed of one single event transition labeled by α succeeded by another single event transition labeled by β, that is, $\xrightarrow{p} = \xrightarrow{\alpha} \xrightarrow{\beta}$.

Similarly to the proof of static laws for strong step bisimulation (see Proposition 3.9), we can prove that the static laws hold for strong pomset bisimulation, we omit them. $\square$

Proposition 3.11 (Static laws for strong hp-bisimulation). *The static laws for strong hp-bisimulation are as follows.*

1. $P \parallel Q \sim_{hp} Q \parallel P$;
2. $P \parallel (Q \parallel R) \sim_{hp} (P \parallel Q) \parallel R$;
3. $P \parallel \textbf{nil} \sim_{hp} P$;
4. $P \setminus L \sim_{hp} P$, if $\mathcal{L}(P) \cap (L \cup \overline{L}) = \emptyset$;
5. $P \setminus K \setminus L \sim_{hp} P \setminus (K \cup L)$;
6. $P[f] \setminus L \sim_{hp} P \setminus f^{-1}(L)[f]$;
7. $(P \parallel Q) \setminus L \sim_{hp} P \setminus L \parallel Q \setminus L$, if $\mathcal{L}(P) \cap \overline{\mathcal{L}(Q)} \cap (L \cup \overline{L}) = \emptyset$;
8. $P[Id] \sim_{hp} P$;
9. $P[f] \sim_{hp} P[f']$, if $f \upharpoonright \mathcal{L}(P) = f' \upharpoonright \mathcal{L}(P)$;

10. $P[f][f'] \sim_{hp} P[f' \circ f]$;
11. $(P \parallel Q)[f] \sim_{hp} P[f] \parallel Q[f]$, *if* $f \upharpoonright (L \cup \overline{L})$ *is one-to-one, where* $L = \mathcal{L}(P) \cup \mathcal{L}(Q)$.

Proof. From the definition of strong hp-bisimulation (see Definition 2.21), we know that strong hp-bisimulation is defined on the posetal product $(C_1, f, C_2), f : C_1 \to C_2$ isomorphism. Two processes P related to C_1 and Q related to C_2, and $f : C_1 \to C_2$ isomorphism. Initially, $(C_1, f, C_2) = (\emptyset, \emptyset, \emptyset)$, and $(\emptyset, \emptyset, \emptyset) \in \sim_{hp}$. When $P \xrightarrow{\alpha} P'$ $(C_1 \xrightarrow{\alpha} C_1')$, there will be $Q \xrightarrow{\alpha} Q'$ $(C_2 \xrightarrow{\alpha} C_2')$, and we define $f' = f[\alpha \mapsto \alpha]$. Then, if $(C_1, f, C_2) \in \sim_{hp}$, then $(C_1', f', C_2') \in \sim_{hp}$.

Similarly to the proof of static laws for strong pomset bisimulation (see Proposition 3.10), we can prove that static laws hold for strong hp-bisimulation, we just need additionally to check the above conditions on hp-bisimulation, we omit them. □

Proposition 3.12 (Static laws for strongly hhp-bisimulation). *The static laws for strongly hhp-bisimulation are as follows.*

1. $P \parallel Q \sim_{hhp} Q \parallel P$;
2. $P \parallel (Q \parallel R) \sim_{hhp} (P \parallel Q) \parallel R$;
3. $P \parallel \mathbf{nil} \sim_{hhp} P$;
4. $P \setminus L \sim_{hhp} P$, *if* $\mathcal{L}(P) \cap (L \cup \overline{L}) = \emptyset$;
5. $P \setminus K \setminus L \sim_{hhp} P \setminus (K \cup L)$;
6. $P[f] \setminus L \sim_{hhp} P \setminus f^{-1}(L)[f]$;
7. $(P \parallel Q) \setminus L \sim_{hhp} P \setminus L \parallel Q \setminus L$, *if* $\mathcal{L}(P) \cap \overline{\mathcal{L}(Q)} \cap (L \cup \overline{L}) = \emptyset$;
8. $P[Id] \sim_{hhp} P$;
9. $P[f] \sim_{hhp} P[f']$, *if* $f \upharpoonright \mathcal{L}(P) = f' \upharpoonright \mathcal{L}(P)$;
10. $P[f][f'] \sim_{hhp} P[f' \circ f]$;
11. $(P \parallel Q)[f] \sim_{hhp} P[f] \parallel Q[f]$, *if* $f \upharpoonright (L \cup \overline{L})$ *is one-to-one, where* $L = \mathcal{L}(P) \cup \mathcal{L}(Q)$.

Proof. From the definition of strongly hhp-bisimulation (see Definition 2.21), we know that strongly hhp-bisimulation is downward closed for strong hp-bisimulation.

Similarly to the proof of static laws for strong hp-bisimulation (see Proposition 3.11), we can prove that static laws hold for strongly hhp-bisimulation, that is, they are downward closed for strong hp-bisimulation, we omit them. □

Proposition 3.13 (Milner's expansion law for strongly truly concurrent bisimulations). *Milner's expansion law does not hold any more for any strongly truly concurrent bisimulation, that is,*

1. $\alpha \parallel \beta \nsim_p \alpha.\beta + \beta.\alpha$;
2. $\alpha \parallel \beta \nsim_s \alpha.\beta + \beta.\alpha$;
3. $\alpha \parallel \beta \nsim_{hp} \alpha.\beta + \beta.\alpha$;
4. $\alpha \parallel \beta \nsim_{hhp} \alpha.\beta + \beta.\alpha$.

Proof. In nature, it is caused by $\alpha \parallel \beta$ and $\alpha.\beta + \beta.\alpha$ having different causality structure. By the transition rules for $\mathbf{Com}_{1,2,3,4}$, $\mathbf{Sum}_{1,2}$, and $\mathbf{Act}_{1,2}$, we have

$$\alpha \parallel \beta \xrightarrow{\{\alpha,\beta\}} \mathbf{nil}$$

while

$$\alpha.\beta + \beta.\alpha \not\xrightarrow{\{\alpha,\beta\}} . \qquad \square$$

Proposition 3.14 (New expansion law for strong step bisimulation). *Let* $P \equiv (P_1[f_1] \parallel \cdots \parallel P_n[f_n]) \setminus L$, *with* $n \geq 1$. *Then*

$$P \sim_s \{(f_1(\alpha_1) \parallel \cdots \parallel f_n(\alpha_n)).(P_1'[f_1] \parallel \cdots \parallel P_n'[f_n]) \setminus L :$$

$$P_i \xrightarrow{\alpha_i} P_i', i \in \{1, \cdots, n\}, f_i(\alpha_i) \notin L \cup \overline{L}\}$$

$$+ \sum \{\tau.(P_1[f_1] \parallel \cdots \parallel P_i'[f_i] \parallel \cdots \parallel P_j'[f_j] \parallel \cdots \parallel P_n[f_n]) \setminus L :$$

$$P_i \xrightarrow{l_1} P_i', P_j \xrightarrow{l_2} P_j', f_i(l_1) = \overline{f_j(l_2)}, i < j\}$$

Proof. Though transition rules in Table 3.1 are defined in the flavor of single event, they can be modified into a step (a set of events within which each event is pairwise concurrent), we omit them. If we treat a single event as a step containing just one event, the proof of the new expansion law has not any problem, so we use this way and still use the transition rules in Table 3.1.

Firstly, we consider the case without Restriction and Relabeling. That is, we suffice to prove the following case by induction on the size n.

For $P \equiv P_1 \parallel \cdots \parallel P_n$, with $n \geq 1$, we need to prove

$$P \sim_s \{(\alpha_1 \parallel \cdots \parallel \alpha_n).(P_1' \parallel \cdots \parallel P_n') : P_i \xrightarrow{\alpha_i} P_i', i \in \{1, \cdots, n\}$$

$$+ \sum \{\tau.(P_1 \parallel \cdots \parallel P_i' \parallel \cdots \parallel P_j' \parallel \cdots \parallel P_n) : P_i \xrightarrow{l} P_i', P_j \xrightarrow{\overline{l}} P_j', i < j\}$$

For $n = 1$, $P_1 \sim_s \alpha_1.P_1' : P_1 \xrightarrow{\alpha_1} P_1'$ is obvious. Then with a hypothesis n, we consider $R \equiv P \parallel P_{n+1}$. By the transition rules $\mathbf{Com}_{1,2,3,4}$, we can get

$$R \sim_s \{(p \parallel \alpha_{n+1}).(P' \parallel P_{n+1}') : P \xrightarrow{p} P', P_{n+1} \xrightarrow{\alpha_{n+1}} P_{n+1}', p \subseteq P\}$$

$$+ \sum \{\tau.(P' \parallel P_{n+1}') : P \xrightarrow{l} P', P_{n+1} \xrightarrow{\overline{l}} P_{n+1}'\}$$

Now with the induction assumption $P \equiv P_1 \parallel \cdots \parallel P_n$, the right-hand side can be reformulated as follows.

$$\{(\alpha_1 \parallel \cdots \parallel \alpha_n \parallel \alpha_{n+1}).(P_1' \parallel \cdots \parallel P_n' \parallel P_{n+1}') :$$

$$P_i \xrightarrow{\alpha_i} P_i', i \in \{1, \cdots, n+1\}$$

$$+ \sum \{\tau.(P_1 \parallel \cdots \parallel P_i' \parallel \cdots \parallel P_j' \parallel \cdots \parallel P_n \parallel P_{n+1}):$$

$$P_i \xrightarrow{l} P_i', P_j \xrightarrow{\bar{l}} P_j', i < j\}$$

$$+ \sum \{\tau.(P_1 \parallel \cdots \parallel P_i' \parallel \cdots \parallel P_j \parallel \cdots \parallel P_n \parallel P_{n+1}'):$$

$$P_i \xrightarrow{l} P_i', P_{n+1} \xrightarrow{\bar{l}} P_{n+1}', i \in \{1, \cdots, n\}\}$$

So,

$$R \sim_s \{(\alpha_1 \parallel \cdots \parallel \alpha_n \parallel \alpha_{n+1}).(P_1' \parallel \cdots \parallel P_n' \parallel P_{n+1}'):$$

$$P_i \xrightarrow{\alpha_i} P_i', i \in \{1, \cdots, n+1\}$$

$$+ \sum \{\tau.(P_1 \parallel \cdots \parallel P_i' \parallel \cdots \parallel P_j' \parallel \cdots \parallel P_n):$$

$$P_i \xrightarrow{l} P_i', P_j \xrightarrow{\bar{l}} P_j', 1 \le i < j \ge n+1\}$$

Then, we can easily add the full conditions with Restriction and Relabeling. □

Proposition 3.15 (New expansion law for strong pomset bisimulation). *Let* $P \equiv (P_1[f_1] \parallel \cdots \parallel P_n[f_n]) \setminus L$, *with* $n \ge 1$. *Then*

$$P \sim_p \{(f_1(\alpha_1) \parallel \cdots \parallel f_n(\alpha_n)).(P_1'[f_1] \parallel \cdots \parallel P_n'[f_n]) \setminus L:$$

$$P_i \xrightarrow{\alpha_i} P_i', i \in \{1, \cdots, n\}, f_i(\alpha_i) \notin L \cup \overline{L}\}$$

$$+ \sum \{\tau.(P_1[f_1] \parallel \cdots \parallel P_i'[f_i] \parallel \cdots \parallel P_j'[f_j] \parallel \cdots \parallel P_n[f_n]) \setminus L:$$

$$P_i \xrightarrow{l_1} P_i', P_j \xrightarrow{l_2} P_j', f_i(l_1) = \overline{f_j(l_2)}, i < j\}$$

Proof. From the definition of strong pomset bisimulation (see Definition 2.17), we know that strong pomset bisimulation is defined by pomset transitions, which are labeled by pomsets. In a pomset transition, the events in the pomset are either within causality relations (defined by the prefix .) or in concurrency (implicitly defined by . and +, and explicitly defined by $\parallel$), of course, they are pairwise consistent (without conflicts). In Proposition 3.14, we have already proven the case that all events are pairwise concurrent, so, we only need to prove the case of events in causality. Without loss of generality, we take a pomset of $p = \{\alpha, \beta : \alpha.\beta\}$. Then the pomset transition labeled by the above p is just composed of one single event transition labeled by α succeeded by another single event transition labeled by β, that is, $\xrightarrow{p} = \xrightarrow{\alpha} \xrightarrow{\beta}$.

Similarly to the proof of new expansion law for strong step bisimulation (see Proposition 3.14), we can prove that the new expansion law holds for strong pomset bisimulation, we omit them. □

Proposition 3.16 (New expansion law for strong hp-bisimulation). *Let* $P \equiv (P_1[f_1] \parallel \cdots \parallel P_n[f_n]) \setminus L$, *with* $n \ge 1$. *Then*

$$P \sim_{hp} \{(f_1(\alpha_1) \parallel \cdots \parallel f_n(\alpha_n)).(P_1'[f_1] \parallel \cdots \parallel P_n'[f_n]) \setminus L :$$

$$P_i \xrightarrow{\alpha_i} P_i', i \in \{1, \cdots, n\}, f_i(\alpha_i) \notin L \cup \overline{L}\}$$

$$+ \sum \{\tau.(P_1[f_1] \parallel \cdots \parallel P_i'[f_i] \parallel \cdots \parallel P_j'[f_j] \parallel \cdots \parallel P_n[f_n]) \setminus L :$$

$$P_i \xrightarrow{l_1} P_i', P_j \xrightarrow{l_2} P_j', f_i(l_1) = \overline{f_j(l_2)}, i < j\}$$

Proof. From the definition of strong hp-bisimulation (see Definition 2.21), we know that strong hp-bisimulation is defined on the posetal product (C_1, f, C_2), $f : C_1 \rightarrow C_2$ isomorphism. Two processes P related to C_1 and Q related to C_2, and $f : C_1 \rightarrow C_2$ isomorphism. Initially, $(C_1, f, C_2) = (\emptyset, \emptyset, \emptyset)$, and $(\emptyset, \emptyset, \emptyset) \in \sim_{hp}$. When $P \xrightarrow{\alpha} P'$ $(C_1 \xrightarrow{\alpha} C_1')$, there will be $Q \xrightarrow{\alpha} Q'$ $(C_2 \xrightarrow{\alpha} C_2')$, and we define $f' = f[\alpha \mapsto \alpha]$. Then, if $(C_1, f, C_2) \in \sim_{hp}$, then $(C_1', f', C_2') \in \sim_{hp}$.

Similarly to the proof of new expansion law for strong pomset bisimulation (see Proposition 3.15), we can prove that the new expansion law holds for strong hp-bisimulation, we just need additionally to check the above conditions on hp-bisimulation, we omit them. □

Proposition 3.17 (New expansion law for strongly hhp-bisimulation). *Let $P \equiv (P_1[f_1] \parallel \cdots \parallel P_n[f_n]) \setminus L$, with $n \geq 1$. Then*

$$P \sim_{hhp} \{(f_1(\alpha_1) \parallel \cdots \parallel f_n(\alpha_n)).(P_1'[f_1] \parallel \cdots \parallel P_n'[f_n]) \setminus L :$$

$$P_i \xrightarrow{\alpha_i} P_i', i \in \{1, \cdots, n\}, f_i(\alpha_i) \notin L \cup \overline{L}\}$$

$$+ \sum \{\tau.(P_1[f_1] \parallel \cdots \parallel P_i'[f_i] \parallel \cdots \parallel P_j'[f_j] \parallel \cdots \parallel P_n[f_n]) \setminus L :$$

$$P_i \xrightarrow{l_1} P_i', P_j \xrightarrow{l_2} P_j', f_i(l_1) = \overline{f_j(l_2)}, i < j\}$$

Proof. From the definition of strongly hhp-bisimulation (see Definition 2.21), we know that strongly hhp-bisimulation is downward closed for strong hp-bisimulation.

Similarly to the proof of the new expansion law for strong hp-bisimulation (see Proposition 3.16), we can prove that the new expansion law holds for strongly hhp-bisimulation, that is, they are downward closed for strong hp-bisimulation, we omit them. □

Theorem 3.18 (Congruence for strong step bisimulation). *We can enjoy the full congruence for strong step bisimulation as follows.*

1. *If $A \stackrel{def}{=} P$, then $A \sim_s P$;*

2. *Let $P_1 \sim_s P_2$. Then*

    a. $\alpha.P_1 \sim_s \alpha.P_2$;

    b. $(\alpha_1 \parallel \cdots \parallel \alpha_n).P_1 \sim_s (\alpha_1 \parallel \cdots \parallel \alpha_n).P_2$;

    c. $P_1 + Q \sim_s P_2 + Q$;

    d. $P_1 \parallel Q \sim_s P_2 \parallel Q$;

    e. $P_1 \setminus L \sim_s P_2 \setminus L$;

    f. $P_1[f] \sim_s P_2[f]$.

Proof. Though transition rules in Table 3.1 are defined in the flavor of single event, they can be modified into a step (a set of events within which each event is pairwise concurrent), we omit them. If we treat a single event as a step containing just one event, the proof of the congruence does not exist any problem, so we use this way and still use the transition rules in Table 3.1.

1. If $A \overset{\text{def}}{=} P$, then $A \sim_s P$. It is obvious.
2. Let $P_1 \sim_s P_2$. Then
   a. $\alpha.P_1 \sim_s \alpha.P_2$. By the transition rules of **Act**$_{1,2}$ in Table 3.1, we can get

   $$\alpha.P_1 \overset{\alpha}{\to} P_1$$

   $$\alpha.P_2 \overset{\alpha}{\to} P_2$$

   Since $P_1 \sim_s P_2$, we get $\alpha.P_1 \sim_s \alpha.P_2$, as desired.
   b. $(\alpha_1 \parallel \cdots \parallel \alpha_n).P_1 \sim_s (\alpha_1 \parallel \cdots \parallel \alpha_n).P_2$. By the transition rules of **Act**$_{1,2}$ in Table 3.1, we can get

   $$(\alpha_1 \parallel \cdots \parallel \alpha_n).P_1 \xrightarrow{\{\alpha_1,\cdots,\alpha_n\}} P_1$$

   $$(\alpha_1 \parallel \cdots \parallel \alpha_n).P_2 \xrightarrow{\{\alpha_1,\cdots,\alpha_n\}} P_2$$

   Since $P_1 \sim_s P_2$, we get $(\alpha_1 \parallel \cdots \parallel \alpha_n).P_1 \sim_s (\alpha_1 \parallel \cdots \parallel \alpha_n).P_2$, as desired.
   c. $P_1 + Q \sim_s P_2 + Q$. By the transition rules of **Sum**$_{1,2}$ in Table 3.1, we can get

   $$\frac{P_1 \overset{\alpha}{\to} P_1'}{P_2 \overset{\alpha}{\to} P_2'}(P_1' \sim_s P_2')$$

   $$\frac{P_1 \overset{\alpha}{\to} P_1'}{P_1 + Q \overset{\alpha}{\to} P_1'} \qquad \frac{P_2 \overset{\alpha}{\to} P_2'}{P_2 + Q \overset{\alpha}{\to} P_2'}$$

   $$\frac{Q \overset{\beta}{\to} Q'}{P_1 + Q \overset{\beta}{\to} Q'} \qquad \frac{Q \overset{\beta}{\to} Q'}{P_2 + Q \overset{\beta}{\to} Q'}$$

   Since $P_1' \sim_s P_2'$ and $Q' \sim_s Q'$, we get $P_1 + Q \sim_s P_2 + Q$, as desired.
   d. $P_1 \parallel Q \sim_s P_2 \parallel Q$. By the transition rules of **Com**$_{1,2,3,4}$ in Table 3.1, we can get

   $$\frac{P_1 \overset{\alpha}{\to} P_1'}{P_2 \overset{\alpha}{\to} P_2'}(P_1' \sim_s P_2')$$

   $$\frac{P_1 \overset{\alpha}{\to} P_1' \quad Q \nrightarrow}{P_1 \parallel Q \overset{\alpha}{\to} P_1' \parallel Q} \qquad \frac{P_2 \overset{\alpha}{\to} P_2' \quad Q \nrightarrow}{P_2 \parallel Q \overset{\alpha}{\to} P_2' \parallel Q}$$

$$\frac{Q \xrightarrow{\beta} Q' \quad P_1 \not\rightarrow}{P_1 \parallel Q \xrightarrow{\beta} P_1 \parallel Q'} \qquad \frac{Q \xrightarrow{\beta} P_2' \quad P_2 \not\rightarrow}{P_2 \parallel Q \xrightarrow{\beta} P_2 \parallel Q'}$$

$$\frac{P_1 \xrightarrow{\alpha} P_1' \quad Q \xrightarrow{\beta} Q'}{P_1 \parallel Q \xrightarrow{\{\alpha,\beta\}} P_1' \parallel Q'}(\beta \neq \overline{\alpha}) \qquad \frac{P_2 \xrightarrow{\alpha} P_2' \quad Q \xrightarrow{\beta} Q'}{P_2 \parallel Q \xrightarrow{\{\alpha,\beta\}} P_2' \parallel Q'}(\beta \neq \overline{\alpha})$$

$$\frac{P_1 \xrightarrow{l} P_1' \quad Q \xrightarrow{\overline{l}} Q'}{P_1 \parallel Q \xrightarrow{\tau} P_1' \parallel Q'} \qquad \frac{P_2 \xrightarrow{l} P_2' \quad Q \xrightarrow{\overline{l}} Q'}{P_2 \parallel Q \xrightarrow{\tau} P_2' \parallel Q'}$$

Since $P_1' \sim_s P_2'$ and $Q' \sim_s Q'$, and with the assumptions $P_1' \parallel Q \sim_s P_2' \parallel Q$, $P_1 \parallel Q' \sim_s P_2 \parallel Q'$, and $P_1' \parallel Q' \sim_s P_2' \parallel Q'$, we get $P_1 \parallel Q \sim_s P_2 \parallel Q$, as desired.

e. $P_1 \setminus L \sim_s P_2 \setminus L$. By the transition rules of $\mathbf{Res}_{1,2}$ in Table 3.1, we get

$$\frac{P_1 \xrightarrow{\alpha} P_1'}{P_2 \xrightarrow{\alpha} P_2'}(P_1' \sim_s P_2')$$

$$\frac{P_1 \xrightarrow{\alpha} P_1'}{P_1 \setminus L \xrightarrow{\alpha} P_1' \setminus L}$$

$$\frac{P_2 \xrightarrow{\alpha} P_2'}{P_2 \setminus L \xrightarrow{\alpha} P_2' \setminus L}$$

Since $P_1' \sim_s P_2'$, and with the assumption $P_1' \setminus L \sim_s P_2' \setminus L$, we get $P_1 \setminus L \sim_s P_2 \setminus L$, as desired.

f. $P_1[f] \sim_s P_2[f]$. By the transition rules of $\mathbf{Rel}_{1,2}$ in Table 3.1, we get

$$\frac{P_1 \xrightarrow{\alpha} P_1'}{P_2 \xrightarrow{\alpha} P_2'}(P_1' \sim_s P_2')$$

$$\frac{P_1 \xrightarrow{\alpha} P_1'}{P_1[f] \xrightarrow{f(\alpha)} P_1'[f]}$$

$$\frac{P_2 \xrightarrow{\alpha} P_2'}{P_2[f] \xrightarrow{f(\alpha)} P_2'[f]}$$

Since $P_1' \sim_s P_2'$, and with the assumption $P_1'[f] \sim_s P_2'[f]$, we get $P_1[f] \sim_s P_2[f]$, as desired.   □

Theorem 3.19 (Congruence for strong pomset bisimulation). *We can enjoy the full congruence for strong pomset bisimulation as follows.*

1. *If $A \stackrel{def}{=} P$, then $A \sim_p P$;*
2. *Let $P_1 \sim_p P_2$. Then*
   a. $\alpha.P_1 \sim_p \alpha.P_2$;
   b. $(\alpha_1 \parallel \cdots \parallel \alpha_n).P_1 \sim_p (\alpha_1 \parallel \cdots \parallel \alpha_n).P_2$;
   c. $P_1 + Q \sim_p P_2 + Q$;
   d. $P_1 \parallel Q \sim_p P_2 \parallel Q$;
   e. $P_1 \setminus L \sim_p P_2 \setminus L$;
   f. $P_1[f] \sim_p P_2[f]$.

Proof. From the definition of strong pomset bisimulation (see Definition 2.17), we know that strong pomset bisimulation is defined by pomset transitions, which are labeled by pomsets. In a pomset transition, the events in the pomset are either within causality relations (defined by the prefix .) or in concurrency (implicitly defined by . and +, and explicitly defined by $\parallel$), of course, they are pairwise consistent (without conflicts). In Theorem 3.18, we have already proven the case that all events are pairwise concurrent, so, we only need to prove the case of events in causality. Without loss of generality, we take a pomset of $p = \{\alpha, \beta : \alpha.\beta\}$. Then the pomset transition labeled by the above p is just composed of one single event transition labeled by α succeeded by another single event transition labeled by β, that is, $\stackrel{p}{\to} = \stackrel{\alpha}{\to}\stackrel{\beta}{\to}$.

Similarly to the proof of congruence for strong step bisimulation (see Theorem 3.18), we can prove that the congruence holds for strong pomset bisimulation, we omit them. □

Theorem 3.20 (Congruence for strong hp-bisimulation). *We can enjoy the full congruence for strong hp-bisimulation as follows.*

1. *If $A \stackrel{def}{=} P$, then $A \sim_{hp} P$;*
2. *Let $P_1 \sim_{hp} P_2$. Then*
   a. $\alpha.P_1 \sim_{hp} \alpha.P_2$;
   b. $(\alpha_1 \parallel \cdots \parallel \alpha_n).P_1 \sim_{hp} (\alpha_1 \parallel \cdots \parallel \alpha_n).P_2$;
   c. $P_1 + Q \sim_{hp} P_2 + Q$;
   d. $P_1 \parallel Q \sim_{hp} P_2 \parallel Q$;
   e. $P_1 \setminus L \sim_{hp} P_2 \setminus L$;
   f. $P_1[f] \sim_{hp} P_2[f]$.

Proof. From the definition of strong hp-bisimulation (see Definition 2.21), we know that strong hp-bisimulation is defined on the posetal product (C_1, f, C_2), $f : C_1 \to C_2$ isomorphism. Two processes P related to C_1 and Q related to C_2, and $f : C_1 \to C_2$ isomorphism. Initially, $(C_1, f, C_2) = (\emptyset, \emptyset, \emptyset)$, and $(\emptyset, \emptyset, \emptyset) \in \sim_{hp}$. When $P \stackrel{\alpha}{\to} P'$ $(C_1 \stackrel{\alpha}{\to} C_1')$, there will be $Q \stackrel{\alpha}{\to} Q'$ $(C_2 \stackrel{\alpha}{\to} C_2')$, and we define $f' = f[\alpha \mapsto \alpha]$. Then, if $(C_1, f, C_2) \in \sim_{hp}$, then $(C_1', f', C_2') \in \sim_{hp}$.

Similarly to the proof of congruence for strong pomset bisimulation (see Theorem 3.19), we can prove that the congruence holds for strong hp-bisimulation, we just need additionally to check the above conditions on hp-bisimulation, we omit them. □

Theorem 3.21 (Congruence for strongly hhp-bisimulation). *We can enjoy the full congruence for strongly hhp-bisimulation as follows.*

1. *If $A \overset{def}{=} P$, then $A \sim_{hhp} P$;*
2. *Let $P_1 \sim_{hhp} P_2$. Then*
   a. $\alpha.P_1 \sim_{hhp} \alpha.P_2$;
   b. $(\alpha_1 \parallel \cdots \parallel \alpha_n).P_1 \sim_{hhp} (\alpha_1 \parallel \cdots \parallel \alpha_n).P_2$;
   c. $P_1 + Q \sim_{hhp} P_2 + Q$;
   d. $P_1 \parallel Q \sim_{hhp} P_2 \parallel Q$;
   e. $P_1 \setminus L \sim_{hhp} P_2 \setminus L$;
   f. $P_1[f] \sim_{hhp} P_2[f]$.

Proof. From the definition of strongly hhp-bisimulation (see Definition 2.21), we know that strongly hhp-bisimulation is downward closed for strong hp-bisimulation.

Similarly to the proof of congruence for strong hp-bisimulation (see Theorem 3.20), we can prove that the congruence holds for strongly hhp-bisimulation, we omit them.   □

3.2.2 Recursion

Definition 3.22 (Weakly guarded recursive expression). *X is weakly guarded in E if each occurrence of X is with some subexpression $\alpha.F$ or $(\alpha_1 \parallel \cdots \parallel \alpha_n).F$ of E.*

Lemma 3.23. *If the variables $\widetilde{X}$ are weakly guarded in E, and $E\{\widetilde{P}/\widetilde{X}\} \xrightarrow{\{\alpha_1,\cdots,\alpha_n\}} P'$, then P' takes the form $E'\{\widetilde{P}/\widetilde{X}\}$ for some expression E', and moreover, for any $\widetilde{Q}$, $E\{\widetilde{Q}/\widetilde{X}\} \xrightarrow{\{\alpha_1,\cdots,\alpha_n\}} E'\{\widetilde{Q}/\widetilde{X}\}$.*

Proof. It needs to induct on the depth of the inference of $E\{\widetilde{P}/\widetilde{X}\} \xrightarrow{\{\alpha_1,\cdots,\alpha_n\}} P'$.

1. Case $E \equiv Y$, a variable. Then $Y \notin \widetilde{X}$. Since $\widetilde{X}$ are weakly guarded, $Y\{\widetilde{P}/\widetilde{X} \equiv Y\} \nrightarrow$, this case is impossible.
2. Case $E \equiv \beta.F$. Then we must have $\alpha = \beta$, and $P' \equiv F\{\widetilde{P}/\widetilde{X}\}$, and $E\{\widetilde{Q}/\widetilde{X}\} \equiv \beta.F\{\widetilde{Q}/\widetilde{X}\} \xrightarrow{\beta} F\{\widetilde{Q}/\widetilde{X}\}$, then, let E' be F, as desired.
3. Case $E \equiv (\beta_1 \parallel \cdots \parallel \beta_n).F$. Then we must have $\alpha_i = \beta_i$ for $1 \le i \le n$, and $P' \equiv F\{\widetilde{P}/\widetilde{X}\}$, and $E\{\widetilde{Q}/\widetilde{X}\} \equiv (\beta_1 \parallel \cdots \parallel \beta_n).F\{\widetilde{Q}/\widetilde{X}\} \xrightarrow{\{\beta_1,\cdots,\beta_n\}} F\{\widetilde{Q}/\widetilde{X}\}$, then, let E' be F, as desired.
4. Case $E \equiv E_1 + E_2$. Then either $E_1\{\widetilde{P}/\widetilde{X}\} \xrightarrow{\{\alpha_1,\cdots,\alpha_n\}} P'$ or $E_2\{\widetilde{P}/\widetilde{X}\} \xrightarrow{\{\alpha_1,\cdots,\alpha_n\}} P'$, then, we can apply this lemma in either case, as desired.
5. Case $E \equiv E_1 \parallel E_2$. There are four possibilities.
   a. We may have $E_1\{\widetilde{P}/\widetilde{X}\} \xrightarrow{\alpha} P_1'$ and $E_2\{\widetilde{P}/\widetilde{X}\} \nrightarrow$ with $P' \equiv P_1' \parallel (E_2\{\widetilde{P}/\widetilde{X}\})$, then by applying this lemma, P_1' is of the form $E_1'\{\widetilde{P}/\widetilde{X}\}$, and for any Q, $E_1\{\widetilde{Q}/\widetilde{X}\} \xrightarrow{\alpha} E_1'\{\widetilde{Q}/\widetilde{X}\}$. So, P' is of the form $E_1' \parallel E_2\{\widetilde{P}/\widetilde{X}\}$, and for any Q, $E\{\widetilde{Q}/\widetilde{X}\} \equiv E_1\{\widetilde{Q}/\widetilde{X}\} \parallel E_2\{\widetilde{Q}/\widetilde{X}\} \xrightarrow{\alpha} (E_1' \parallel E_2)\{\widetilde{Q}/\widetilde{X}\}$, then, let E' be $E_1' \parallel E_2$, as desired.
   b. We may have $E_2\{\widetilde{P}/\widetilde{X}\} \xrightarrow{\alpha} P_2'$ and $E_1\{\widetilde{P}/\widetilde{X}\} \nrightarrow$ with $P' \equiv P_2' \parallel (E_1\{\widetilde{P}/\widetilde{X}\})$, this case can be proven similarly to the above subcase, as desired.

c. We may have $E_1\{\widetilde{P}/\widetilde{X}\} \xrightarrow{\alpha} P_1'$ and $E_2\{\widetilde{P}/\widetilde{X}\} \xrightarrow{\beta} P_2'$ with $\alpha \neq \overline{\beta}$ and $P' \equiv P_1' \parallel P_2'$, then by applying this lemma, P_1' is of the form $E_1'\{\widetilde{P}/\widetilde{X}\}$, and for any Q, $E_1\{\widetilde{Q}/\widetilde{X}\} \xrightarrow{\alpha} E_1'\{\widetilde{Q}/\widetilde{X}\}$; P_2' is of the form $E_2'\{\widetilde{P}/\widetilde{X}\}$, and for any Q, $E_2\{\widetilde{Q}/\widetilde{X}\} \xrightarrow{\alpha} E_2'\{\widetilde{Q}/\widetilde{X}\}$. So, P' is of the form $E_1' \parallel E_2'\{\widetilde{P}/\widetilde{X}\}$, and for any Q, $E\{\widetilde{Q}/\widetilde{X}\} \equiv E_1\{\widetilde{Q}/\widetilde{X}\} \parallel E_2\{\widetilde{Q}/\widetilde{X}\} \xrightarrow{\{\alpha,\beta\}} (E_1' \parallel E_2')\{\widetilde{Q}/\widetilde{X}\}$, then, let E' be $E_1' \parallel E_2'$, as desired.

d. We may have $E_1\{\widetilde{P}/\widetilde{X}\} \xrightarrow{l} P_1'$ and $E_2\{\widetilde{P}/\widetilde{X}\} \xrightarrow{\bar{l}} P_2'$ with $P' \equiv P_1' \parallel P_2'$, then by applying this lemma, P_1' is of the form $E_1'\{\widetilde{P}/\widetilde{X}\}$, and for any Q, $E_1\{\widetilde{Q}/\widetilde{X}\} \xrightarrow{l} E_1'\{\widetilde{Q}/\widetilde{X}\}$; P_2' is of the form $E_2'\{\widetilde{P}/\widetilde{X}\}$, and for any Q, $E_2\{\widetilde{Q}/\widetilde{X}\} \xrightarrow{\bar{l}} E_2'\{\widetilde{Q}/\widetilde{X}\}$. So, P' is of the form $E_1' \parallel E_2'\{\widetilde{P}/\widetilde{X}\}$, and for any Q, $E\{\widetilde{Q}/\widetilde{X}\} \equiv E_1\{\widetilde{Q}/\widetilde{X}\} \parallel E_2\{\widetilde{Q}/\widetilde{X}\} \xrightarrow{\tau} (E_1' \parallel E_2')\{\widetilde{Q}/\widetilde{X}\}$, then, let E' be $E_1' \parallel E_2'$, as desired.

6. Case $E \equiv F[R]$ and $E \equiv F \setminus L$. These cases can be proven similarly to the above case.

7. Case $E \equiv C$, an agent constant defined by $C \overset{\text{def}}{=} R$. Then there is no $X \in \widetilde{X}$ occurring in E, so $C \xrightarrow{\{\alpha_1,\cdots,\alpha_n\}} P'$, let E' be P', as desired. □

Theorem 3.24 (Unique solution of equations for strong step bisimulation). *Let the recursive expressions $E_i (i \in I)$ contain at most the variables $X_i (i \in I)$, and let each $X_j (j \in I)$ be weakly guarded in each E_i. Then,*
   If $\widetilde{P} \sim_s \widetilde{E}\{\widetilde{P}/\widetilde{X}\}$ and $\widetilde{Q} \sim_s \widetilde{E}\{\widetilde{Q}/\widetilde{X}\}$, then $\widetilde{P} \sim_s \widetilde{Q}$.

Proof. It is sufficient to induct on the depth of the inference of $E\{\widetilde{P}/\widetilde{X}\} \xrightarrow{\{\alpha_1,\cdots,\alpha_n\}} P'$.

1. Case $E \equiv X_i$. Then we have $E\{\widetilde{P}/\widetilde{X}\} \equiv P_i \xrightarrow{\{\alpha_1,\cdots,\alpha_n\}} P'$, since $P_i \sim_s E_i\{\widetilde{P}/\widetilde{X}\}$, we have $E_i\{\widetilde{P}/\widetilde{X}\} \xrightarrow{\{\alpha_1,\cdots,\alpha_n\}} P'' \sim_s P'$. Since $\widetilde{X}$ are weakly guarded in E_i, by Lemma 3.23, $P'' \equiv E'\{\widetilde{P}/\widetilde{X}\}$ and $E_i\{\widetilde{P}/\widetilde{X}\} \xrightarrow{\{\alpha_1,\cdots,\alpha_n\}} E'\{\widetilde{P}/\widetilde{X}\}$. Since $E\{\widetilde{Q}/\widetilde{X}\} \equiv X_i\{\widetilde{Q}/\widetilde{X}\} \equiv Q_i \sim_s E_i\{\widetilde{Q}/\widetilde{X}\}$, $E\{\widetilde{Q}/\widetilde{X}\} \xrightarrow{\{\alpha_1,\cdots,\alpha_n\}} Q' \sim_s E'\{\widetilde{Q}/\widetilde{X}\}$. So, $P' \sim_s Q'$, as desired.

2. Case $E \equiv \alpha.F$. This case can be proven similarly.

3. Case $E \equiv (\alpha_1 \parallel \cdots \parallel \alpha_n).F$. This case can be proven similarly.

4. Case $E \equiv E_1 + E_2$. We have $E_i\{\widetilde{P}/\widetilde{X}\} \xrightarrow{\{\alpha_1,\cdots,\alpha_n\}} P'$, $E_i\{\widetilde{Q}/\widetilde{X}\} \xrightarrow{\{\alpha_1,\cdots,\alpha_n\}} Q'$, then, $P' \sim_s Q'$, as desired.

5. Case $E \equiv E_1 \parallel E_2$, $E \equiv F[R]$ and $E \equiv F \setminus L$, $E \equiv C$. These cases can be proven similarly to the above case. □

Theorem 3.25 (Unique solution of equations for strong pomset bisimulation). *Let the recursive expressions $E_i (i \in I)$ contain at most the variables $X_i (i \in I)$, and let each $X_j (j \in I)$ be weakly guarded in each E_i. Then,*
   If $\widetilde{P} \sim_p \widetilde{E}\{\widetilde{P}/\widetilde{X}\}$ and $\widetilde{Q} \sim_p \widetilde{E}\{\widetilde{Q}/\widetilde{X}\}$, then $\widetilde{P} \sim_p \widetilde{Q}$.

Proof. From the definition of strong pomset bisimulation (see Definition 2.17), we know that strong pomset bisimulation is defined by pomset transitions, which are labeled by pomsets. In a pomset transition, the events in the pomset are either within causality relations (defined by the prefix .) or in concurrency (implicitly defined by . and $+$, and explicitly defined by $\parallel$), of course, they are pairwise consistent (without conflicts). In Theorem 3.24, we have already proven the case that all events are pairwise concurrent, so, we only need to prove the case of events in causality. Without loss of generality, we take a pomset of $p = \{\alpha, \beta : \alpha.\beta\}$. Then the pomset transition labeled by the above p is just composed of one single event transition labeled by α succeeded by another single event transition labeled by β, that is, $\xrightarrow{p} = \xrightarrow{\alpha}\xrightarrow{\beta}$.

Similarly to the proof of unique solution of equations for strong step bisimulation (see Theorem 3.24), we can prove that the unique solution of equations holds for strong pomset bisimulation, we omit them. $\qquad\square$

Theorem 3.26 (Unique solution of equations for strong hp-bisimulation). *Let the recursive expressions $E_i (i \in I)$ contain at most the variables $X_i (i \in I)$, and let each $X_j (j \in I)$ be weakly guarded in each E_i. Then,*

If $\widetilde{P} \sim_{hp} \widetilde{E}\{\widetilde{P}/\widetilde{X}\}$ and $\widetilde{Q} \sim_{hp} \widetilde{E}\{\widetilde{Q}/\widetilde{X}\}$, then $\widetilde{P} \sim_{hp} \widetilde{Q}$.

Proof. From the definition of strong hp-bisimulation (see Definition 2.21), we know that strong hp-bisimulation is defined on the posetal product (C_1, f, C_2), $f : C_1 \to C_2$ isomorphism. Two processes P related to C_1 and Q related to C_2, and $f : C_1 \to C_2$ isomorphism. Initially, $(C_1, f, C_2) = (\emptyset, \emptyset, \emptyset)$, and $(\emptyset, \emptyset, \emptyset) \in \sim_{hp}$. When $P \xrightarrow{\alpha} P'$ $(C_1 \xrightarrow{\alpha} C_1')$, there will be $Q \xrightarrow{\alpha} Q'$ $(C_2 \xrightarrow{\alpha} C_2')$, and we define $f' = f[\alpha \mapsto \alpha]$. Then, if $(C_1, f, C_2) \in \sim_{hp}$, then $(C_1', f', C_2') \in \sim_{hp}$.

Similarly to the proof of unique solution of equations for strong pomset bisimulation (see Theorem 3.25), we can prove that the unique solution of equations holds for strong hp-bisimulation, we just need additionally to check the above conditions on hp-bisimulation, we omit them. $\qquad\square$

Theorem 3.27 (Unique solution of equations for strongly hhp-bisimulation). *Let the recursive expressions $E_i (i \in I)$ contain at most the variables $X_i (i \in I)$, and let each $X_j (j \in I)$ be weakly guarded in each E_i. Then,*

If $\widetilde{P} \sim_{hhp} \widetilde{E}\{\widetilde{P}/\widetilde{X}\}$ and $\widetilde{Q} \sim_{hhp} \widetilde{E}\{\widetilde{Q}/\widetilde{X}\}$, then $\widetilde{P} \sim_{hhp} \widetilde{Q}$.

Proof. From the definition of strongly hhp-bisimulation (see Definition 2.21), we know that strongly hhp-bisimulation is downward closed for strong hp-bisimulation.

Similarly to the proof of unique solution of equations for strong hp-bisimulation (see Theorem 3.26), we can prove that the unique solution of equations holds for strongly hhp-bisimulation, we omit them. $\qquad\square$

3.3 Weakly truly concurrent bisimulations

The weak transition rules for CTC are listed in Table 3.2.

Table 3.2   Weak transition rules of CTC.

$$\textbf{WAct}_1 \quad \frac{}{\alpha.P \overset{\alpha}{\Rightarrow} P}$$

$$\textbf{WSum}_1 \quad \frac{P \overset{\alpha}{\Rightarrow} P'}{P + Q \overset{\alpha}{\Rightarrow} P'}$$

$$\textbf{WCom}_1 \quad \frac{P \overset{\alpha}{\Rightarrow} P' \quad Q \nrightarrow}{P \parallel Q \overset{\alpha}{\Rightarrow} P' \parallel Q}$$

$$\textbf{WCom}_2 \quad \frac{Q \overset{\alpha}{\Rightarrow} Q' \quad P \nrightarrow}{P \parallel Q \overset{\alpha}{\Rightarrow} P \parallel Q'}$$

$$\textbf{WCom}_3 \quad \frac{P \overset{\alpha}{\Rightarrow} P' \quad Q \overset{\beta}{\Rightarrow} Q'}{P \parallel Q \overset{\{\alpha,\beta\}}{\Longrightarrow} P' \parallel Q'} \quad (\beta \neq \overline{\alpha})$$

$$\textbf{WCom}_4 \quad \frac{P \overset{l}{\Rightarrow} P' \quad Q \overset{\overline{l}}{\Rightarrow} Q'}{P \parallel Q \overset{\tau}{\Rightarrow} P' \parallel Q'}$$

$$\textbf{WAct}_2 \quad \frac{}{(\alpha_1 \parallel \cdots \parallel \alpha_n).P \overset{\{\alpha_1,\cdots,\alpha_n\}}{\Longrightarrow} P} \quad (\alpha_i \neq \overline{\alpha_j} \quad i, j \in \{1, \cdots, n\})$$

$$\textbf{WSum}_2 \quad \frac{P \overset{\{\alpha_1,\cdots,\alpha_n\}}{\Longrightarrow} P'}{P + Q \overset{\{\alpha_1,\cdots,\alpha_n\}}{\Longrightarrow} P'}$$

$$\textbf{WRes}_1 \quad \frac{P \overset{\alpha}{\Rightarrow} P'}{P \setminus L \overset{\alpha}{\Rightarrow} P' \setminus L} \quad (\alpha, \overline{\alpha} \notin L)$$

$$\textbf{WRes}_2 \quad \frac{P \overset{\{\alpha_1,\cdots,\alpha_n\}}{\Longrightarrow} P'}{P \setminus L \overset{\{\alpha_1,\cdots,\alpha_n\}}{\Longrightarrow} P' \setminus L} \quad (\alpha_1, \overline{\alpha_1}, \cdots, \alpha_n, \overline{\alpha_n} \notin L)$$

$$\textbf{WRel}_1 \quad \frac{P \overset{\alpha}{\Rightarrow} P'}{P[f] \overset{f(\alpha)}{\Longrightarrow} P'[f]}$$

$$\textbf{WRel}_2 \quad \frac{P \overset{\{\alpha_1,\cdots,\alpha_n\}}{\Longrightarrow} P'}{P[f] \overset{\{f(\alpha_1),\cdots,f(\alpha_n)\}}{\Longrightarrow} P'[f]}$$

continued on next page

Table 3.2   (continued)

$$\textbf{WCon}_1 \quad \frac{P \overset{\alpha}{\Rightarrow} P'}{A \overset{\alpha}{\Rightarrow} P'} \quad (A \overset{\text{def}}{=} P)$$

$$\textbf{WCon}_2 \quad \frac{P \xRightarrow{\{\alpha_1, \cdots, \alpha_n\}} P'}{A \xRightarrow{\{\alpha_1, \cdots, \alpha_n\}} P'} \quad (A \overset{\text{def}}{=} P)$$

3.3.1 Laws and congruence

Remembering that τ can neither be restricted nor relabeled, by Proposition 2.23, we know that the monoid laws, the static laws and the new expansion law in section 3.2 still hold with respect to the corresponding weakly truly concurrent bisimulations. And also, we can enjoy the full congruence of Prefix, Summation, Composition, Restriction, Relabeling, and Constants with respect to corresponding weakly truly concurrent bisimulations. We will not retype these laws, and just give the τ-specific laws.

Proposition 3.28 (τ laws for weak step bisimulation). *The τ laws for weak step bisimulation are as follows.*

1. $P \approx_s \tau.P$;
2. $\alpha.\tau.P \approx_s \alpha.P$;
3. $(\alpha_1 \parallel \cdots \parallel \alpha_n).\tau.P \approx_s (\alpha_1 \parallel \cdots \parallel \alpha_n).P$;
4. $P + \tau.P \approx_s \tau.P$;
5. $\alpha.(P + \tau.Q) + \alpha.Q \approx_s \alpha.(P + \tau.Q)$;
6. $(\alpha_1 \parallel \cdots \parallel \alpha_n).(P + \tau.Q) + (\alpha_1 \parallel \cdots \parallel \alpha_n).Q \approx_s (\alpha_1 \parallel \cdots \parallel \alpha_n).(P + \tau.Q)$;
7. $P \approx_s \tau \parallel P$.

Proof. Though transition rules in Table 3.1 are defined in the flavor of single event, they can be modified into a step (a set of events within which each event is pairwise concurrent), we omit them. If we treat a single event as a step containing just one event, the proof of τ laws does not exist any problem, so we use this way and still use the transition rules in Table 3.1.

1. $P \approx_s \tau.P$. By the weak transition rules **WAct**$_{1,2}$ of CTC in Table 3.2, we get

$$\frac{P \overset{\alpha}{\Rightarrow} P'}{P \overset{\alpha}{\Rightarrow} P'} \qquad \frac{P \overset{\alpha}{\Rightarrow} P'}{\tau.P \overset{\alpha}{\Rightarrow} P'}$$

Since $P' \approx_s P'$, we get $P \approx_s \tau.P$, as desired.

2. $\alpha.\tau.P \approx_s \alpha.P$. By the weak transition rules **WAct**$_{1,2}$ in Table 3.2, we get

$$\frac{}{\alpha.\tau.P \overset{\alpha}{\Rightarrow} P} \qquad \frac{}{\alpha.P \overset{\alpha}{\Rightarrow} P}$$

Since $P \approx_s P$, we get $\alpha.\tau.P \approx_s \alpha.P$, as desired.

3. $(\alpha_1 \parallel \cdots \parallel \alpha_n).\tau.P \approx_s (\alpha_1 \parallel \cdots \parallel \alpha_n).P$. By the weak transition rules **WAct**$_{1,2}$ in Table 3.2, we get

$$\frac{}{(\alpha_1 \parallel \cdots \parallel \alpha_n).\tau.P \xrightarrow{\{\alpha_1,\cdots,\alpha_n\}} P} \qquad \frac{}{(\alpha_1 \parallel \cdots \parallel \alpha_n).P \xrightarrow{\{\alpha_1,\cdots,\alpha_n\}} P}$$

Since $P \approx_s P$, we get $(\alpha_1 \parallel \cdots \parallel \alpha_n).\tau.P \approx_s (\alpha_1 \parallel \cdots \parallel \alpha_n).P$, as desired.

4. $P + \tau.P \approx_s \tau.P$. By the weak transition rules **WSum**$_{1,2}$ of CTC in Table 3.2, we get

$$\frac{P \xRightarrow{\alpha} P'}{P + \tau.P \xRightarrow{\alpha} P'} \qquad \frac{P \xRightarrow{\alpha} P'}{\tau.P \xRightarrow{\alpha} P'}$$

Since $P' \approx_s P'$, we get $P + \tau.P \approx_s \tau.P$, as desired.

5. $\alpha.(P + \tau.Q) + \alpha.Q \approx_s \alpha.(P + \tau.Q)$. By the weak transition rules **WAct**$_{1,2}$ and **WSum**$_{1,2}$ of CTC in Table 3.2, we get

$$\frac{}{\alpha.(P + \tau.Q) + \alpha.Q \xRightarrow{\alpha} Q} \qquad \frac{}{\alpha.(P + \tau.Q) \xRightarrow{\alpha} Q}$$

Since $Q \approx_s Q$, we get $\alpha.(P + \tau.Q) + \alpha.Q \approx_s \alpha.(P + \tau.Q)$, as desired.

6. $(\alpha_1 \parallel \cdots \parallel \alpha_n).(P + \tau.Q) + (\alpha_1 \parallel \cdots \parallel \alpha_n).Q \approx_s (\alpha_1 \parallel \cdots \parallel \alpha_n).(P + \tau.Q)$. By the weak transition rules **WAct**$_{1,2}$ and **WSum**$_{1,2}$ of CTC in Table 3.2, we get

$$\frac{}{(\alpha_1 \parallel \cdots \parallel \alpha_n).(P + \tau.Q) + (\alpha_1 \parallel \cdots \parallel \alpha_n).Q \xrightarrow{\{\alpha_1,\cdots,\alpha_n\}} Q} \qquad \frac{}{(\alpha_1 \parallel \cdots \parallel \alpha_n).(P + \tau.Q) \xrightarrow{\{\alpha_1,\cdots,\alpha_n\}} Q}$$

Since $Q \approx_s Q$, we get $(\alpha_1 \parallel \cdots \parallel \alpha_n).(P + \tau.Q) + (\alpha_1 \parallel \cdots \parallel \alpha_n).Q \approx_s (\alpha_1 \parallel \cdots \parallel \alpha_n).(P + \tau.Q)$, as desired.

7. $P \approx_s \tau \parallel P$. By the weak transition rules **WCom**$_{1,2,3,4}$ of CTC in Table 3.2, we get

$$\frac{P \xRightarrow{\alpha} P'}{P \xRightarrow{\alpha} P'} \qquad \frac{P \xRightarrow{\alpha} P'}{\tau \parallel P \xRightarrow{\alpha} P'}$$

Since $P' \approx_s P'$, we get $P \approx_s \tau \parallel P$, as desired. $\qquad\qquad\square$

Proposition 3.29 (τ laws for weak pomset bisimulation). *The τ laws for weak pomset bisimulation are as follows.*

1. $P \approx_p \tau.P$;

2. $\alpha.\tau.P \approx_p \alpha.P$;

3. $(\alpha_1 \parallel \cdots \parallel \alpha_n).\tau.P \approx_p (\alpha_1 \parallel \cdots \parallel \alpha_n).P$;

4. $P + \tau.P \approx_p \tau.P$;

5. $\alpha.(P + \tau.Q) + \alpha.Q \approx_p \alpha.(P + \tau.Q)$;

6. $(\alpha_1 \parallel \cdots \parallel \alpha_n).(P + \tau.Q) + (\alpha_1 \parallel \cdots \parallel \alpha_n).Q \approx_p (\alpha_1 \parallel \cdots \parallel \alpha_n).(P + \tau.Q)$;

7. $P \approx_p \tau \parallel P$.

Proof. From the definition of weak pomset bisimulation $\approx_p$ (see Definition 2.18), we know that weak pomset bisimulation $\approx_p$ is defined by weak pomset transitions, which are labeled by pomsets with τ. In a weak pomset transition, the events in the pomset are either within causality relations (defined by .) or in concurrency (implicitly defined by . and +, and explicitly defined by ∥), of course, they are pairwise consistent (without conflicts). In Proposition 3.28, we have already proven the case that all events are pairwise concurrent, so, we only need to prove the case of events in causality. Without loss of generality, we take a pomset of $p = \{\alpha, \beta : \alpha.\beta\}$. Then the weak pomset transition labeled by the above p is just composed of one single event transition labeled by α succeeded by another single event transition labeled by β, that is, $\xRightarrow{p} = \xRightarrow{\alpha}\xRightarrow{\beta}$.

Similarly to the proof of τ laws for weak step bisimulation $\approx_s$ (Proposition 3.28), we can prove that τ laws hold for weak pomset bisimulation $\approx_p$, we omit them.   □

Proposition 3.30 (τ laws for weak hp-bisimulation). *The τ laws for weak hp-bisimulation are as follows.*

1. $P \approx_{hp} \tau.P$;
2. $\alpha.\tau.P \approx_{hp} \alpha.P$;
3. $(\alpha_1 \parallel \cdots \parallel \alpha_n).\tau.P \approx_{hp} (\alpha_1 \parallel \cdots \parallel \alpha_n).P$;
4. $P + \tau.P \approx_{hp} \tau.P$;
5. $\alpha.(P + \tau.Q) + \alpha.Q \approx_{hp} \alpha.(P + \tau.Q)$;
6. $(\alpha_1 \parallel \cdots \parallel \alpha_n).(P + \tau.Q) + (\alpha_1 \parallel \cdots \parallel \alpha_n).Q \approx_{hp} (\alpha_1 \parallel \cdots \parallel \alpha_n).(P + \tau.Q)$;
7. $P \approx_{hp} \tau \parallel P$.

Proof. From the definition of weak hp-bisimulation $\approx_{hp}$ (see Definition 2.22), we know that weak hp-bisimulation $\approx_{hp}$ is defined on the weakly posetal product (C_1, f, C_2), $f : \hat{C}_1 \to \hat{C}_2$ isomorphism. Two processes P related to C_1 and Q related to C_2, and $f : \hat{C}_1 \to \hat{C}_2$ isomorphism. Initially, $(C_1, f, C_2) = (\emptyset, \emptyset, \emptyset)$, and $(\emptyset, \emptyset, \emptyset) \in \approx_{hp}$. When $P \xrightarrow{\alpha} P'$ ($C_1 \xrightarrow{\alpha} C_1'$), there will be $Q \xRightarrow{\alpha} Q'$ ($C_2 \xRightarrow{\alpha} C_2'$), and we define $f' = f[\alpha \mapsto \alpha]$. Then, if $(C_1, f, C_2) \in \approx_{hp}$, then $(C_1', f', C_2') \in \approx_{hp}$.

Similarly to the proof of τ laws for weak pomset bisimulation (Proposition 3.29), we can prove that τ laws hold for weak hp-bisimulation, we just need additionally to check the above conditions on weak hp-bisimulation, we omit them.   □

Proposition 3.31 (τ laws for weakly hhp-bisimulation). *The τ laws for weakly hhp-bisimulation are as follows.*

1. $P \approx_{hhp} \tau.P$;
2. $\alpha.\tau.P \approx_{hhp} \alpha.P$;
3. $(\alpha_1 \parallel \cdots \parallel \alpha_n).\tau.P \approx_{hhp} (\alpha_1 \parallel \cdots \parallel \alpha_n).P$;
4. $P + \tau.P \approx_{hhp} \tau.P$;
5. $\alpha.(P + \tau.Q) + \alpha.Q \approx_{hhp} \alpha.(P + \tau.Q)$;
6. $(\alpha_1 \parallel \cdots \parallel \alpha_n).(P + \tau.Q) + (\alpha_1 \parallel \cdots \parallel \alpha_n).Q \approx_{hhp} (\alpha_1 \parallel \cdots \parallel \alpha_n).(P + \tau.Q)$;
7. $P \approx_{hhp} \tau \parallel P$.

Proof. From the definition of weakly hhp-bisimulation (see Definition 2.22), we know that weakly hhp-bisimulation is downward closed for weak hp-bisimulation.

Similarly to the proof of τ laws for weak hp-bisimulation (see Proposition 3.30), we can prove that the τ laws hold for weakly hhp-bisimulation, we omit them.   □

3.3.2 Recursion

Definition 3.32 (Sequential). *X is sequential in E if every subexpression of E which contains X, apart from X itself, is of the form $\alpha.F$, or $(\alpha_1 \parallel \cdots \parallel \alpha_n).F$, or $\sum \widetilde{F}$.*

Definition 3.33 (Guarded recursive expression). *X is guarded in E if each occurrence of X is with some subexpression $l.F$ or $(l_1 \parallel \cdots \parallel l_n).F$ of E.*

Lemma 3.34. *Let G be guarded and sequential, $Vars(G) \subseteq \widetilde{X}$, and let $G\{\widetilde{P}/\widetilde{X}\} \xrightarrow{\{\alpha_1,\cdots,\alpha_n\}} P'$. Then there is an expression H such that $G \xrightarrow{\{\alpha_1,\cdots,\alpha_n\}} H$, $P' \equiv H\{\widetilde{P}/\widetilde{X}\}$, and for any $\widetilde{Q}$, $G\{\widetilde{Q}/\widetilde{X}\} \xrightarrow{\{\alpha_1,\cdots,\alpha_n\}} H\{\widetilde{Q}/\widetilde{X}\}$. Moreover H is sequential, $Vars(H) \subseteq \widetilde{X}$, and if $\alpha_1 = \cdots = \alpha_n = \tau$, then H is also guarded.*

Proof. We need to induct on the structure of G.

If G is a Constant, a Composition, a Restriction or a Relabeling then it contains no variables, since it is sequential and guarded, then $G \xrightarrow{\{\alpha_1,\cdots,\alpha_n\}} P'$, then let $H \equiv P'$, as desired.

G cannot be a variable, since it is guarded.

If $G \equiv G_1 + G_2$. Then either $G_1\{\widetilde{P}/\widetilde{X}\} \xrightarrow{\{\alpha_1,\cdots,\alpha_n\}} P'$ or $G_2\{\widetilde{P}/\widetilde{X}\} \xrightarrow{\{\alpha_1,\cdots,\alpha_n\}} P'$, then, we can apply this lemma in either case, as desired.

If $G \equiv \beta.H$. Then we must have $\alpha = \beta$, and $P' \equiv H\{\widetilde{P}/\widetilde{X}\}$, and $G\{\widetilde{Q}/\widetilde{X}\} \equiv \beta.H\{\widetilde{Q}/\widetilde{X}\} \xrightarrow{\beta} H\{\widetilde{Q}/\widetilde{X}\}$, then, let G' be H, as desired.

If $G \equiv (\beta_1 \parallel \cdots \parallel \beta_n).H$. Then we must have $\alpha_i = \beta_i$ for $1 \leq i \leq n$, and $P' \equiv H\{\widetilde{P}/\widetilde{X}\}$, and $G\{\widetilde{Q}/\widetilde{X}\} \equiv (\beta_1 \parallel \cdots \parallel \beta_n).H\{\widetilde{Q}/\widetilde{X}\} \xrightarrow{\{\beta_1,\cdots,\beta_n\}} H\{\widetilde{Q}/\widetilde{X}\}$, then, let G' be H, as desired.

If $G \equiv \tau.H$. Then we must have $\tau = \tau$, and $P' \equiv H\{\widetilde{P}/\widetilde{X}\}$, and $G\{\widetilde{Q}/\widetilde{X}\} \equiv \tau.H\{\widetilde{Q}/\widetilde{X}\} \xrightarrow{\tau} H\{\widetilde{Q}/\widetilde{X}\}$, then, let G' be H, as desired.   □

Theorem 3.35 (Unique solution of equations for weak step bisimulation). *Let the guarded and sequential expressions $\widetilde{E}$ contain free variables $\subseteq \widetilde{X}$, then,*
If $\widetilde{P} \approx_s \widetilde{E}\{\widetilde{P}/\widetilde{X}\}$ and $\widetilde{Q} \approx_s \widetilde{E}\{\widetilde{Q}/\widetilde{X}\}$, then $\widetilde{P} \approx_s \widetilde{Q}$.

Proof. Like the corresponding theorem in CCS, without loss of generality, we only consider a single equation $X = E$. So we assume $P \approx_s E(P)$, $Q \approx_s E(Q)$, then $P \approx_s Q$.

We will prove $\{(H(P), H(Q)) : H\}$ sequential, if $H(P) \xrightarrow{\{\alpha_1,\cdots,\alpha_n\}} P'$, then, for some Q', $H(Q) \xRightarrow{\{\alpha_1,\cdots,\alpha_n\}} Q'$ and $P' \approx_s Q'$.

Let $H(P) \xrightarrow{\{\alpha_1,\cdot,\alpha_n\}} P'$, then $H(E(P)) \xRightarrow{\{\alpha_1,\cdots,\alpha_n\}} P''$ and $P' \approx_s P''$.

By Lemma 3.34, we know there is a sequential H' such that $H(E(P)) \xRightarrow{\{\alpha_1,\cdots,\alpha_n\}} H'(P) \Rightarrow P'' \approx_s P'$.

And, $H(E(Q)) \xRightarrow{\{\alpha_1,\cdots,\alpha_n\}} H'(Q) \Rightarrow Q''$ and $P'' \approx_s Q''$. And $H(Q) \xrightarrow{\{\alpha_1,\cdots,\alpha_n\}} Q' \approx_s Q''$. Hence, $P' \approx_s Q'$, as desired.   □

Theorem 3.36 (Unique solution of equations for weak pomset bisimulation). *Let the guarded and sequential expressions $\widetilde{E}$ contain free variables $\subseteq \widetilde{X}$, then,*
    If $\widetilde{P} \approx_p \widetilde{E}\{\widetilde{P}/\widetilde{X}\}$ and $\widetilde{Q} \approx_p \widetilde{E}\{\widetilde{Q}/\widetilde{X}\}$, then $\widetilde{P} \approx_p \widetilde{Q}$.

Proof. From the definition of weak pomset bisimulation $\approx_p$ (see Definition 2.18), we know that weak pomset bisimulation $\approx_p$ is defined by weak pomset transitions, which are labeled by pomsets with τ. In a weak pomset transition, the events in the pomset are either within causality relations (defined by .) or in concurrency (implicitly defined by . and +, and explicitly defined by $\parallel$), of course, they are pairwise consistent (without conflicts). In Theorem 3.35, we have already proven the case that all events are pairwise concurrent, so, we only need to prove the case of events in causality. Without loss of generality, we take a pomset of $p = \{\alpha, \beta : \alpha.\beta\}$. Then the weak pomset transition labeled by the above p is just composed of one single event transition labeled by α succeeded by another single event transition labeled by β, that is, $\xrightarrow{p} = \xrightarrow{\alpha}\xrightarrow{\beta}$.

Similarly to the proof of unique solution of equations for weak step bisimulation $\approx_s$ (Theorem 3.35), we can prove that unique solution of equations holds for weak pomset bisimulation $\approx_p$, we omit them. □

Theorem 3.37 (Unique solution of equations for weak hp-bisimulation). *Let the guarded and sequential expressions $\widetilde{E}$ contain free variables $\subseteq \widetilde{X}$, then,*
    If $\widetilde{P} \approx_{hp} \widetilde{E}\{\widetilde{P}/\widetilde{X}\}$ and $\widetilde{Q} \approx_{hp} \widetilde{E}\{\widetilde{Q}/\widetilde{X}\}$, then $\widetilde{P} \approx_{hp} \widetilde{Q}$.

Proof. From the definition of weak hp-bisimulation $\approx_{hp}$ (see Definition 2.22), we know that weak hp-bisimulation $\approx_{hp}$ is defined on the weakly posetal product (C_1, f, C_2), $f : \hat{C}_1 \to \hat{C}_2$ isomorphism. Two processes P related to C_1 and Q related to C_2, and $f : \hat{C}_1 \to \hat{C}_2$ isomorphism. Initially, $(C_1, f, C_2) = (\emptyset, \emptyset, \emptyset)$, and $(\emptyset, \emptyset, \emptyset) \in \approx_{hp}$. When $P \xrightarrow{\alpha} P'$ $(C_1 \xrightarrow{\alpha} C_1')$, there will be $Q \xrightarrow{\alpha} Q'$ $(C_2 \xrightarrow{\alpha} C_2')$, and we define $f' = f[\alpha \mapsto \alpha]$. Then, if $(C_1, f, C_2) \in \approx_{hp}$, then $(C_1', f', C_2') \in \approx_{hp}$.

Similarly to the proof of unique solution of equations for weak pomset bisimulation (Theorem 3.36), we can prove that unique solution of equations holds for weak hp-bisimulation, we just need additionally to check the above conditions on weak hp-bisimulation, we omit them. □

Theorem 3.38 (Unique solution of equations for weakly hhp-bisimulation). *Let the guarded and sequential expressions $\widetilde{E}$ contain free variables $\subseteq \widetilde{X}$, then,*
    If $\widetilde{P} \approx_{hhp} \widetilde{E}\{\widetilde{P}/\widetilde{X}\}$ and $\widetilde{Q} \approx_{hhp} \widetilde{E}\{\widetilde{Q}/\widetilde{X}\}$, then $\widetilde{P} \approx_{hhp} \widetilde{Q}$.

Proof. From the definition of weakly hhp-bisimulation (see Definition 2.22), we know that weakly hhp-bisimulation is downward closed for weak hp-bisimulation.

Similarly to the proof of unique solution of equations for weak hp-bisimulation (see Theorem 3.37), we can prove that the unique solution of equations holds for weakly hhp-bisimulation, we omit them. □

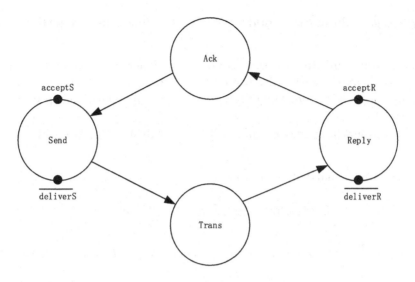

FIGURE 3.1 Alternating bit protocol.

3.4 Applications

In this section, we show the applications of CTC by verification of the alternating-bit protocol [19]. The alternating-bit protocol is a communication protocol and illustrated in Fig. 3.1.

The *Trans* and *Ack* lines may lose or duplicate message, and messages are sent tagged with the bits 0 or 1 alternately, and these bits also constitute the acknowledgments.

There are some variations of the classical alternating-bit protocol. We assume the message flows are bidirectional, the sender accepts messages from the outside world, and sends it to the replier via some line; and it also accepts messages from the replier, deliver it to the outside world and acknowledges it via some line. The role of the replier acts the same as the sender. But, for simplicities and without loss the generality, we only consider the dual one directional processes, we just suppose that the messages from the replier are always accompanied with the acknowledgments.

After accepting a message from the outside world, the sender sends it with bit b along the *Trans* line and sets a timer,

1. it may get a "time-out" from the timer, upon which it sends the message again with b;
2. it may get an acknowledgment b accompanied with the message from the replier via the *Ack* line, upon which it delivers the message to the outside world and is ready to accept another message tagged with bit $\hat{b} = 1 - b$;
3. it may get an acknowledgment $\hat{b}$ which it ignores.

After accepting a message from outside world, the replier also can send it to the sender, but we ignore this process and just assume the message is always accompanied with the acknowledgment to the sender, and just process the dual manner to the sender. After delivering a message to the outside world it acknowledges it with bit b along the Ack line and sets a timer,

1. it may get a "time-out" from the timer, upon which it acknowledges again with b;
2. it may get a new message with bit $\hat{b}$ from the $Trans$ line, upon which it is ready to deliver the new message and acknowledge with bit $\hat{b}$ accompanying with its messages from the outside world;
3. it may get a superfluous transmission of the previous message with bit b, which it ignores.

Now, we give the formal definitions as follows.

$$Send(b) \stackrel{\text{def}}{=} \overline{send_b}.\overline{time}.Sending(b)$$

$$Sending(b) \stackrel{\text{def}}{=} timeout.Send(b) + ack_b.timeout.Accept\,S(\hat{b}) + ack_{\hat{b}}.Deliver\,S(b)$$

$$Accept\,S(b) \stackrel{\text{def}}{=} accept\,S.Send(b)$$

$$Deliver\,S(b) \stackrel{\text{def}}{=} \overline{deliver\,S}.Sending(b)$$

$$Reply(b) \stackrel{\text{def}}{=} \overline{reply_b}.\overline{time}.Replying(b)$$

$$Replying(b) \stackrel{\text{def}}{=} timeout.Reply(b) + trans_{\hat{b}}.timeout.Deliver(\hat{b}) + trans_b.Accept\,R(b)$$

$$Deliver\,R(b) \stackrel{\text{def}}{=} \overline{deliver\,R}.Reply(b)$$

$$Accept\,R(b) \stackrel{\text{def}}{=} accept\,R.Replying(b)$$

$$Timer \stackrel{\text{def}}{=} \overline{time.timeout}.Timer$$

$$Ack(bs) \xrightarrow{\overline{ack_b}} Ack(s) \quad Trans(sb) \xrightarrow{\overline{trans_b}} Trans(s)$$

$$Ack(s) \xrightarrow{reply_b} Ack(sb) \quad Trans(s) \xrightarrow{send_b} Trans(bs)$$

$$Ack(sbt) \xrightarrow{\tau} Ack(st) \quad Trans(tbs) \xrightarrow{\tau} Trans(ts)$$

$$Ack(sbt) \xrightarrow{\tau} Ack(sbbt) \quad Trans(tbs) \xrightarrow{\tau} Trans(tbbs)$$

Then the complete system can be built by composing the components in parallel. That is, it can be expressed as follows, where ϵ is the empty sequence.

$$AB \stackrel{\text{def}}{=} Accept(\hat{b}) \parallel Trans(\epsilon) \parallel Ack(\epsilon) \parallel Reply(b) \parallel Timer$$

Now, we define the protocol specification to be a buffer as follows:

$$Buff \stackrel{\text{def}}{=} (accept\,S \parallel accept\,R).Buff'$$

$$Buff' \stackrel{\text{def}}{=} (\overline{deliver\,S} \parallel \overline{deliver\,R}).Buff$$

We need to prove that

$$AB \approx_s Buff$$

$$AB \approx_p Buff$$

$$AB \approx_{hp} Buff$$

$$AB \approx_{hhp} Buff$$

The deductive process is omitted, and we leave it as an excise to the readers.

3.5 Conclusions

We design a calculus for true concurrency (CTC). Indeed, we follow the way paved by Milner's well-known CCS [2,3] for interleaving bisimulation.

Fortunately, based on the concepts for true concurrency, CTC has good properties modulo several kinds of strongly truly concurrent bisimulations and weakly truly concurrent bisimulations. These properties include monoid laws, static laws, new expansion law for strongly truly concurrent bisimulations, τ laws for weakly truly concurrent bisimulations, and full congruences for strongly and weakly truly concurrent bisimulations, and also unique solution for recursion.

Note that, for true concurrency, an empty action playing the role of placeholder is needed, just like APTC with the shadow constant in subsection 4.6.1. We do not do this work and leave it as an exercise to the readers.

CTC is a peer in true concurrency to CCS in interleaving bisimulation semantics. It can be used widely in verification of computer systems with a truly concurrent flavor.

4

Algebraic laws for true concurrency

We find the algebraic laws for true concurrency following the way paved by ACP for bisimulation equivalence. And finally, we establish a whole axiomatization for true concurrency called $APTC$. The theory $APTC$ has four modules: $BATC$ (Basic Algebra for True Concurrency), $APTC$ (Algebra for Parallelism in True Concurrency), recursion and abstraction. With the help of placeholder in section 4.6.1, we get an intuitive result for true concurrency: $a \between b = a \cdot b + b \cdot a + a \parallel b + a \mid b$ modulo truly concurrent bisimilarities pomset bisimulation equivalence, step bisimulation equivalence and history-preserving bisimulation equivalence, with a, b are atomic actions (events), $\between$ is the whole true concurrency operator, $\cdot$ is the temporal causality operator, $+$ is a kind of structured conflict, $\parallel$ is the parallel operator and $\mid$ is the communication merge.

This chapter is organized as follows. We introduce $BATC$ in section 4.1, $APTC$ in section 4.2, recursion in section 4.3, and abstraction in section 4.4. In section 4.5, we show the applications of $APTC$ by an example called alternating bit protocol. We show the modularity and extension mechanism of $APTC$ in section 4.6. We establish an axiomatization for hhp-bisimilarity in section 4.7. Finally, in section 4.8, we conclude this chapter.

4.1 Basic algebra for true concurrency

In this section, we will discuss the algebraic laws for prime event structure $\mathcal{E}$, exactly for causality $\leq$ and conflict $\sharp$. We will follow the conventions of process algebra, using $\cdot$ instead of $\leq$ and $+$ instead of $\sharp$. The resulted algebra is called Basic Algebra for True Concurrency, abbreviated $BATC$.

4.1.1 Axiom system of $BATC$

In the following, let $e_1, e_2, e'_1, e'_2 \in \mathbb{E}$, and let variables x, y, z range over the set of terms for true concurrency, p, q, s range over the set of closed terms. The set of axioms of $BATC$ consists of the laws given in Table 4.1.

Intuitively, the axiom $A1$ says that the binary operator $+$ satisfies commutative law. The axiom $A2$ says that $+$ satisfies associativity. $A3$ says that $+$ satisfies idempotency. The axiom $A4$ is the right distributivity of the binary operator $\cdot$ to $+$. And $A5$ is the associativity of $\cdot$.

4.1.2 Properties of $BATC$

Definition 4.1 (Basic terms of $BATC$). The set of basic terms of $BATC$, $\mathcal{B}(BATC)$, is inductively defined as follows:

Table 4.1 Axioms of $BATC$.

No.	Axiom
$A1$	$x + y = y + x$
$A2$	$(x + y) + z = x + (y + z)$
$A3$	$x + x = x$
$A4$	$(x + y) \cdot z = x \cdot z + y \cdot z$
$A5$	$(x \cdot y) \cdot z = x \cdot (y \cdot z)$

Table 4.2 Term rewrite system of $BATC$.

No.	Rewriting rule
$RA3$	$x + x \rightarrow x$
$RA4$	$(x + y) \cdot z \rightarrow x \cdot z + y \cdot z$
$RA5$	$(x \cdot y) \cdot z \rightarrow x \cdot (y \cdot z)$

1. $\mathbb{E} \subset \mathcal{B}(BATC)$;
2. if $e \in \mathbb{E}, t \in \mathcal{B}(BATC)$ then $e \cdot t \in \mathcal{B}(BATC)$;
3. if $t, s \in \mathcal{B}(BATC)$ then $t + s \in \mathcal{B}(BATC)$.

Theorem 4.2 (Elimination theorem of $BATC$). *Let p be a closed $BATC$ term. Then there is a basic $BATC$ term q such that $BATC \vdash p = q$.*

Proof. (1) Firstly, suppose that the following ordering on the signature of $BATC$ is defined: $\cdot > +$ and the symbol $\cdot$ is given the lexicographical status for the first argument, then for each rewrite rule $p \rightarrow q$ in Table 4.2 relation $p >_{lpo} q$ can easily be proved. We obtain that the term rewrite system shown in Table 4.2 is strongly normalizing, for it has finitely many rewriting rules, and $>$ is a well-founded ordering on the signature of $BATC$, and if $s >_{lpo} t$, for each rewriting rule $s \rightarrow t$ is in Table 4.2 (see Theorem 2.12).

(2) Then we prove that the normal forms of closed $BATC$ terms are basic $BATC$ terms.

Suppose that p is a normal form of some closed $BATC$ term and suppose that p is not a basic term. Let p' denote the smallest sub-term of p which is not a basic term. It implies that each sub-term of p' is a basic term. Then we prove that p is not a term in normal form. It is sufficient to induct on the structure of p':

- Case $p' \equiv e, e \in \mathbb{E}$. p' is a basic term, which contradicts the assumption that p' is not a basic term, so this case should not occur.
- Case $p' \equiv p_1 \cdot p_2$. By induction on the structure of the basic term p_1:
  - Subcase $p_1 \in \mathbb{E}$. p' would be a basic term, which contradicts the assumption that p' is not a basic term;
  - Subcase $p_1 \equiv e \cdot p_1'$. $RA5$ rewriting rule can be applied. So p is not a normal form;
  - Subcase $p_1 \equiv p_1' + p_1''$. $RA4$ rewriting rule can be applied. So p is not a normal form.

Table 4.3 Single event transition rules of $BATC$.

$$\frac{}{e \xrightarrow{e} \surd}$$

$$\frac{x \xrightarrow{e} \surd}{x + y \xrightarrow{e} \surd} \quad \frac{x \xrightarrow{e} x'}{x + y \xrightarrow{e} x'} \quad \frac{y \xrightarrow{e} \surd}{x + y \xrightarrow{e} \surd} \quad \frac{y \xrightarrow{e} y'}{x + y \xrightarrow{e} y'}$$

$$\frac{x \xrightarrow{e} \surd}{x \cdot y \xrightarrow{e} y} \quad \frac{x \xrightarrow{e} x'}{x \cdot y \xrightarrow{e} x' \cdot y}$$

- Case $p' \equiv p_1 + p_2$. By induction on the structure of the basic terms both p_1 and p_2, all subcases will lead to that p' would be a basic term, which contradicts the assumption that p' is not a basic term. □

4.1.3 Structured operational semantics of $BATC$

In this subsection, we will define a term-deduction system which gives the operational semantics of $BATC$. We give the operational transition rules for operators $\cdot$ and $+$ as Table 4.3 shows. And the predicate $\xrightarrow{e} \surd$ represents successful termination after execution of the event e.

Theorem 4.3 (Congruence of $BATC$ with respect to bisimulation equivalence). *Bisimulation equivalence $\sim_{HM}$ is a congruence with respect to $BATC$.*

Proof. The axioms in Table 4.1 of $BATC$ are the same as the axioms of BPA (Basic Process Algebra) [1,2,4], so, bisimulation equivalence $\sim_{HM}$ is a congruence with respect to $BATC$. □

Theorem 4.4 (Soundness of $BATC$ modulo bisimulation equivalence). *Let x and y be $BATC$ terms. If $BATC \vdash x = y$, then $x \sim_{HM} y$.*

Proof. The axioms in Table 4.1 of $BATC$ are the same as the axioms of BPA (Basic Process Algebra) [1,2,4], so, $BATC$ is sound modulo bisimulation equivalence. □

Theorem 4.5 (Completeness of $BATC$ modulo bisimulation equivalence). *Let p and q be closed $BATC$ terms, if $p \sim_{HM} q$ then $p = q$.*

Proof. The axioms in Table 4.1 of $BATC$ are the same as the axioms of BPA (Basic Process Algebra) [1,2,4], so, $BATC$ is complete modulo bisimulation equivalence. □

The pomset transition rules are shown in Table 4.4, different to single event transition rules in Table 4.3, the pomset transition rules are labeled by pomsets, which are defined by causality $\cdot$ and conflict $+$.

Table 4.4   Pomset transition rules of $BATC$.

$$\frac{}{X \xrightarrow{X} \checkmark}$$

$$\frac{x \xrightarrow{X} \checkmark}{x + y \xrightarrow{X} \checkmark}(X \subseteq x) \quad \frac{x \xrightarrow{X} x'}{x + y \xrightarrow{X} x'}(X \subseteq x) \quad \frac{y \xrightarrow{Y} \checkmark}{x + y \xrightarrow{Y} \checkmark}(Y \subseteq y) \quad \frac{y \xrightarrow{Y} y'}{x + y \xrightarrow{Y} y'}(Y \subseteq y)$$

$$\frac{x \xrightarrow{X} \checkmark}{x \cdot y \xrightarrow{X} y}(X \subseteq x) \quad \frac{x \xrightarrow{X} x'}{x \cdot y \xrightarrow{X} x' \cdot y}(X \subseteq x)$$

Theorem 4.6 (Congruence of $BATC$ with respect to pomset bisimulation equivalence).
Pomset bisimulation equivalence $\sim_p$ is a congruence with respect to $BATC$.

Proof. It is easy to see that pomset bisimulation is an equivalent relation on $BATC$ terms, we only need to prove that $\sim_p$ is preserved by the operators $\cdot$ and $+$.

- Causality operator $\cdot$. Let x_1, x_2 and y_1, y_2 be $BATC$ processes, and $x_1 \sim_p y_1, x_2 \sim_p y_2$, it is sufficient to prove that $x_1 \cdot x_2 \sim_p y_1 \cdot y_2$.
  By the definition of pomset bisimulation $\sim_p$ (Definition 2.17), $x_1 \sim_p y_1$ means that

$$x_1 \xrightarrow{X_1} x_1' \quad y_1 \xrightarrow{Y_1} y_1'$$

  with $X_1 \subseteq x_1, Y_1 \subseteq y_1, X_1 \sim Y_1$ and $x_1' \sim_p y_1'$. The meaning of $x_2 \sim_p y_2$ is similar.
  By the pomset transition rules for causality operator $\cdot$ in Table 4.4, we can get

$$x_1 \cdot x_2 \xrightarrow{X_1} x_2 \quad y_1 \cdot y_2 \xrightarrow{Y_1} y_2$$

  with $X_1 \subseteq x_1, Y_1 \subseteq y_1, X_1 \sim Y_1$ and $x_2 \sim_p y_2$, so, we get $x_1 \cdot x_2 \sim_p y_1 \cdot y_2$, as desired.
  Or, we can get

$$x_1 \cdot x_2 \xrightarrow{X_1} x_1' \cdot x_2 \quad y_1 \cdot y_2 \xrightarrow{Y_1} y_1' \cdot y_2$$

  with $X_1 \subseteq x_1, Y_1 \subseteq y_1, X_1 \sim Y_1$ and $x_1' \sim_p y_1', x_2 \sim_p y_2$, so, we get $x_1 \cdot x_2 \sim_p y_1 \cdot y_2$, as desired.
- Conflict operator $+$. Let x_1, x_2 and y_1, y_2 be $BATC$ processes, and $x_1 \sim_p y_1, x_2 \sim_p y_2$, it is sufficient to prove that $x_1 + x_2 \sim_p y_1 + y_2$. The meanings of $x_1 \sim_p y_1$ and $x_2 \sim_p y_2$ are the same as the above case, according to the definition of pomset bisimulation $\sim_p$ in Definition 2.17.
  By the pomset transition rules for conflict operator $+$ in Table 4.4, we can get four cases:

$$x_1 + x_2 \xrightarrow{X_1} \checkmark \quad y_1 + y_2 \xrightarrow{Y_1} \checkmark$$

  with $X_1 \subseteq x_1, Y_1 \subseteq y_1, X_1 \sim Y_1$, so, we get $x_1 + x_2 \sim_p y_1 + y_2$, as desired.

Or, we can get

$$x_1 + x_2 \xrightarrow{X_1} x_1' \quad y_1 + y_2 \xrightarrow{Y_1} y_1'$$

with $X_1 \subseteq x_1$, $Y_1 \subseteq y_1$, $X_1 \sim Y_1$, and $x_1' \sim_p y_1'$, so, we get $x_1 + x_2 \sim_p y_1 + y_2$, as desired.
Or, we can get

$$x_1 + x_2 \xrightarrow{X_2} \surd \quad y_1 + y_2 \xrightarrow{Y_2} \surd$$

with $X_2 \subseteq x_2$, $Y_2 \subseteq y_2$, $X_2 \sim Y_2$, so, we get $x_1 + x_2 \sim_p y_1 + y_2$, as desired.
Or, we can get

$$x_1 + x_2 \xrightarrow{X_2} x_2' \quad y_1 + y_2 \xrightarrow{Y_2} y_2'$$

with $X_2 \subseteq x_2$, $Y_2 \subseteq y_2$, $X_2 \sim Y_2$, and $x_2' \sim_p y_2'$, so, we get $x_1 + x_2 \sim_p y_1 + y_2$, as desired. $\square$

Theorem 4.7 (Soundness of $BATC$ modulo pomset bisimulation equivalence). *Let x and y be $BATC$ terms. If $BATC \vdash x = y$, then $x \sim_p y$.*

Proof. Since pomset bisimulation $\sim_p$ is both an equivalent and a congruent relation, we only need to check if each axiom in Table 4.1 is sound modulo pomset bisimulation equivalence.

- **Axiom** $A1$. Let p, q be $BATC$ processes, and $p + q = q + p$, it is sufficient to prove that $p + q \sim_p q + p$. By the pomset transition rules for operator $+$ in Table 4.4, we get

$$\frac{p \xrightarrow{P} \surd}{p + q \xrightarrow{P} \surd}(P \subseteq p) \quad \frac{p \xrightarrow{P} \surd}{q + p \xrightarrow{P} \surd}(P \subseteq p)$$

$$\frac{p \xrightarrow{P} p'}{p + q \xrightarrow{P} p'}(P \subseteq p) \quad \frac{p \xrightarrow{P} p'}{q + p \xrightarrow{P} p'}(P \subseteq p)$$

$$\frac{q \xrightarrow{Q} \surd}{p + q \xrightarrow{Q} \surd}(Q \subseteq q) \quad \frac{q \xrightarrow{Q} \surd}{q + p \xrightarrow{Q} \surd}(Q \subseteq q)$$

$$\frac{q \xrightarrow{Q} q'}{p + q \xrightarrow{Q} q'}(Q \subseteq q) \quad \frac{q \xrightarrow{Q} q'}{q + p \xrightarrow{Q} q'}(Q \subseteq q)$$

So, $p + q \sim_p q + p$, as desired.
- **Axiom** $A2$. Let p, q, s be $BATC$ processes, and $(p + q) + s = p + (q + s)$, it is sufficient to prove that $(p + q) + s \sim_p p + (q + s)$. By the pomset transition rules for operator $+$ in Table 4.4, we get

$$\frac{p \xrightarrow{P} \surd}{(p + q) + s \xrightarrow{P} \surd}(P \subseteq p) \quad \frac{p \xrightarrow{P} \surd}{p + (q + s) \xrightarrow{P} \surd}(P \subseteq p)$$

$$\frac{p \xrightarrow{P} p'}{(p+q)+s \xrightarrow{P} p'}(P \subseteq p) \qquad \frac{p \xrightarrow{P} p'}{p+(q+s) \xrightarrow{P} p'}(P \subseteq p)$$

$$\frac{q \xrightarrow{Q} \surd}{(p+q)+s \xrightarrow{Q} \surd}(Q \subseteq q) \qquad \frac{q \xrightarrow{Q} \surd}{p+(q+s) \xrightarrow{Q} \surd}(Q \subseteq q)$$

$$\frac{q \xrightarrow{Q} q'}{(p+q)+s \xrightarrow{Q} q'}(Q \subseteq q) \qquad \frac{q \xrightarrow{Q} q'}{p+(q+s) \xrightarrow{Q} q'}(Q \subseteq q)$$

$$\frac{s \xrightarrow{S} \surd}{(p+q)+s \xrightarrow{S} \surd}(S \subseteq s) \qquad \frac{s \xrightarrow{S} \surd}{p+(q+s) \xrightarrow{S} \surd}(S \subseteq s)$$

$$\frac{s \xrightarrow{S} s'}{(p+q)+s \xrightarrow{S} s'}(S \subseteq s) \qquad \frac{s \xrightarrow{S} s'}{p+(q+s) \xrightarrow{S} s'}(S \subseteq s)$$

So, $(p+q)+s \sim_p p+(q+s)$, as desired.

- **Axiom** $A3$. Let p be a $BATC$ process, and $p + p = p$, it is sufficient to prove that $p + p \sim_p p$. By the pomset transition rules for operator $+$ in Table 4.4, we get

$$\frac{p \xrightarrow{P} \surd}{p+p \xrightarrow{P} \surd}(P \subseteq p) \qquad \frac{p \xrightarrow{P} \surd}{p \xrightarrow{P} \surd}(P \subseteq p)$$

$$\frac{p \xrightarrow{P} p'}{p+p \xrightarrow{P} p'}(P \subseteq p) \qquad \frac{p \xrightarrow{P} p'}{p \xrightarrow{P} p'}(P \subseteq p)$$

So, $p + p \sim_p p$, as desired.

- **Axiom** $A4$. Let p, q, s be $BATC$ processes, and $(p+q) \cdot s = p \cdot s + q \cdot s$, it is sufficient to prove that $(p+q) \cdot s \sim_p p \cdot s + q \cdot s$. By the pomset transition rules for operators $+$ and $\cdot$ in Table 4.4, we get

$$\frac{p \xrightarrow{P} \surd}{(p+q) \cdot s \xrightarrow{P} s}(P \subseteq p) \qquad \frac{p \xrightarrow{P} \surd}{p \cdot s + q \cdot s \xrightarrow{P} s}(P \subseteq p)$$

$$\frac{p \xrightarrow{P} p'}{(p+q) \cdot s \xrightarrow{P} p' \cdot s}(P \subseteq p) \qquad \frac{p \xrightarrow{P} p'}{p \cdot s + q \cdot s \xrightarrow{P} p' \cdot s}(P \subseteq p)$$

$$\frac{q \xrightarrow{Q} \surd}{(p+q) \cdot s \xrightarrow{Q} s}(Q \subseteq q) \qquad \frac{q \xrightarrow{Q} \surd}{p \cdot s + q \cdot s \xrightarrow{Q} s}(Q \subseteq q)$$

$$\frac{q \xrightarrow{Q} q'}{(p+q)\cdot s \xrightarrow{Q} q'\cdot s}(Q \subseteq q) \qquad \frac{q \xrightarrow{Q} q'}{p\cdot s+q\cdot s \xrightarrow{Q} q'\cdot s}(Q \subseteq q)$$

So, $(p+q)\cdot s \sim_p p\cdot s + q\cdot s$, as desired.

- **Axiom** $A5$. Let p,q,s be $BATC$ processes, and $(p\cdot q)\cdot s = p\cdot(q\cdot s)$, it is sufficient to prove that $(p\cdot q)\cdot s \sim_p p\cdot(q\cdot s)$. By the pomset transition rules for operator $\cdot$ in Table 4.4, we get

$$\frac{p \xrightarrow{P} \surd}{(p\cdot q)\cdot s \xrightarrow{P} q\cdot s}(P \subseteq p) \qquad \frac{p \xrightarrow{P} \surd}{p\cdot(q\cdot s) \xrightarrow{P} q\cdot s}(P \subseteq p)$$

$$\frac{p \xrightarrow{P} p'}{(p\cdot q)\cdot s \xrightarrow{P} (p'\cdot q)\cdot s}(P \subseteq p) \qquad \frac{p \xrightarrow{P} p'}{p\cdot(q\cdot s) \xrightarrow{P} p'\cdot(q\cdot s)}(P \subseteq p)$$

With an assumption $(p'\cdot q)\cdot s = p'\cdot(q\cdot s)$, so, $(p\cdot q)\cdot s \sim_p p\cdot(q\cdot s)$, as desired. □

Theorem 4.8 (Completeness of $BATC$ modulo pomset bisimulation equivalence). *Let p and q be closed $BATC$ terms, if $p \sim_p q$ then $p = q$.*

Proof. Firstly, by the elimination theorem of $BATC$, we know that for each closed $BATC$ term p, there exists a closed basic $BATC$ term p', such that $BATC \vdash p = p'$, so, we only need to consider closed basic $BATC$ terms.

The basic terms (see Definition 4.1) modulo associativity and commutativity (AC) of conflict $+$ (defined by axioms $A1$ and $A2$ in Table 4.1), and this equivalence is denoted by $=_{AC}$. Then, each equivalence class s modulo AC of $+$ has the following normal form

$$s_1 + \cdots + s_k$$

with each s_i either an atomic event or of the form $t_1 \cdot t_2$, and each s_i is called the summand of s.

Now, we prove that for normal forms n and n', if $n \sim_p n'$ then $n =_{AC} n'$. It is sufficient to induct on the sizes of n and n'.

- Consider a summand e of n. Then $n \xrightarrow{e} \surd$, so $n \sim_p n'$ implies $n' \xrightarrow{e} \surd$, meaning that n' also contains the summand e.
- Consider a summand $t_1 \cdot t_2$ of n. Then $n \xrightarrow{t_1} t_2$, so $n \sim_p n'$ implies $n' \xrightarrow{t_1} t_2'$ with $t_2 \sim_p t_2'$, meaning that n' contains a summand $t_1 \cdot t_2'$. Since t_2 and t_2' are normal forms and have sizes smaller than n and n', by the induction hypotheses $t_2 \sim_p t_2'$ implies $t_2 =_{AC} t_2'$.

So, we get $n =_{AC} n'$.

Finally, let s and t be basic terms, and $s \sim_p t$, there are normal forms n and n', such that $s = n$ and $t = n'$. The soundness theorem of $BATC$ modulo pomset bisimulation equivalence (see Theorem 4.7) yields $s \sim_p n$ and $t \sim_p n'$, so $n \sim_p s \sim_p t \sim_p n'$. Since if $n \sim_p n'$ then $n =_{AC} n'$, $s = n =_{AC} n' = t$, as desired. □

Table 4.5 Step transition rules of $BATC$.

$$\frac{}{X \xrightarrow{X} \surd}(\forall e_1, e_2 \in X \text{ are pairwise concurrent.})$$

$$\frac{x \xrightarrow{X} \surd}{x + y \xrightarrow{X} \surd}(X \subseteq x, \forall e_1, e_2 \in X \text{ are pairwise concurrent.})$$

$$\frac{x \xrightarrow{X} x'}{x + y \xrightarrow{X} x'}(X \subseteq x, \forall e_1, e_2 \in X \text{ are pairwise concurrent.})$$

$$\frac{y \xrightarrow{Y} \surd}{x + y \xrightarrow{Y} \surd}(Y \subseteq y, \forall e_1, e_2 \in Y \text{ are pairwise concurrent.})$$

$$\frac{y \xrightarrow{Y} y'}{x + y \xrightarrow{Y} y'}(Y \subseteq y, \forall e_1, e_2 \in Y \text{ are pairwise concurrent.})$$

$$\frac{x \xrightarrow{X} \surd}{x \cdot y \xrightarrow{X} y}(X \subseteq x, \forall e_1, e_2 \in X \text{ are pairwise concurrent.})$$

$$\frac{x \xrightarrow{X} x'}{x \cdot y \xrightarrow{X} x' \cdot y}(X \subseteq x, \forall e_1, e_2 \in X \text{ are pairwise concurrent.})$$

The step transition rules are defined in Table 4.5, different to pomset transition rules, the step transition rules are labeled by steps, in which every event is pairwise concurrent.

Theorem 4.9 (Congruence of $BATC$ with respect to step bisimulation equivalence). *Step bisimulation equivalence $\sim_s$ is a congruence with respect to $BATC$.*

Proof. It is easy to see that step bisimulation is an equivalent relation on $BATC$ terms, we only need to prove that $\sim_s$ is preserved by the operators $\cdot$ and $+$.

- Causality operator $\cdot$. Let x_1, x_2 and y_1, y_2 be $BATC$ processes, and $x_1 \sim_s y_1, x_2 \sim_s y_2$, it is sufficient to prove that $x_1 \cdot x_2 \sim_s y_1 \cdot y_2$.
  By the definition of step bisimulation $\sim_s$ (Definition 2.17), $x_1 \sim_s y_1$ means that

$$x_1 \xrightarrow{X_1} x_1' \quad y_1 \xrightarrow{Y_1} y_1'$$

with $X_1 \subseteq x_1, \forall e_1, e_2 \in X_1$ are pairwise concurrent, $Y_1 \subseteq y_1, \forall e_1, e_2 \in Y_1$ are pairwise concurrent, $X_1 \sim Y_1$ and $x_1' \sim_s y_1'$. The meaning of $x_2 \sim_s y_2$ is similar.

By the step transition rules for causality operator $\cdot$ in Table 4.5, we can get

$$x_1 \cdot x_2 \xrightarrow{X_1} x_2 \quad y_1 \cdot y_2 \xrightarrow{Y_1} y_2$$

with $X_1 \subseteq x_1$, $\forall e_1, e_2 \in X_1$ are pairwise concurrent, $Y_1 \subseteq y_1$, $\forall e_1, e_2 \in Y_1$ are pairwise concurrent, $X_1 \sim Y_1$ and $x_2 \sim_s y_2$, so, we get $x_1 \cdot x_2 \sim_s y_1 \cdot y_2$, as desired.
Or, we can get

$$x_1 \cdot x_2 \xrightarrow{X_1} x_1' \cdot x_2 \quad y_1 \cdot y_2 \xrightarrow{Y_1} y_1' \cdot y_2$$

with $X_1 \subseteq x_1$, $\forall e_1, e_2 \in X_1$ are pairwise concurrent, $Y_1 \subseteq y_1$, $\forall e_1, e_2 \in Y_1$ are pairwise concurrent, $X_1 \sim Y_1$ and $x_1' \sim_s y_1'$, $x_2 \sim_s y_2$, so, we get $x_1 \cdot x_2 \sim_s y_1 \cdot y_2$, as desired.

- Conflict operator $+$. Let x_1, x_2 and y_1, y_2 be $BATC$ processes, and $x_1 \sim_s y_1$, $x_2 \sim_s y_2$, it is sufficient to prove that $x_1 + x_2 \sim_s y_1 + y_2$. The meanings of $x_1 \sim_s y_1$ and $x_2 \sim_s y_2$ are the same as the above case, according to the definition of step bisimulation $\sim_s$ in Definition 2.17.

By the step transition rules for conflict operator $+$ in Table 4.5, we can get four cases:

$$x_1 + x_2 \xrightarrow{X_1} \surd \quad y_1 + y_2 \xrightarrow{Y_1} \surd$$

with $X_1 \subseteq x_1$, $\forall e_1, e_2 \in X_1$ are pairwise concurrent, $Y_1 \subseteq y_1$, $\forall e_1, e_2 \in Y_1$ are pairwise concurrent, $X_1 \sim Y_1$, so, we get $x_1 + x_2 \sim_s y_1 + y_2$, as desired.
Or, we can get

$$x_1 + x_2 \xrightarrow{X_1} x_1' \quad y_1 + y_2 \xrightarrow{Y_1} y_1'$$

with $X_1 \subseteq x_1$, $\forall e_1, e_2 \in X_1$ are pairwise concurrent, $Y_1 \subseteq y_1$, $\forall e_1, e_2 \in Y_1$ are pairwise concurrent, $X_1 \sim Y_1$, and $x_1' \sim_s y_1'$, so, we get $x_1 + x_2 \sim_s y_1 + y_2$, as desired.
Or, we can get

$$x_1 + x_2 \xrightarrow{X_2} \surd \quad y_1 + y_2 \xrightarrow{Y_2} \surd$$

with $X_2 \subseteq x_2$, $\forall e_1, e_2 \in X_2$ are pairwise concurrent, $Y_2 \subseteq y_2$, $\forall e_1, e_2 \in Y_2$ are pairwise concurrent, $X_2 \sim Y_2$, so, we get $x_1 + x_2 \sim_s y_1 + y_2$, as desired.
Or, we can get

$$x_1 + x_2 \xrightarrow{X_2} x_2' \quad y_1 + y_2 \xrightarrow{Y_2} y_2'$$

with $X_2 \subseteq x_2$, $\forall e_1, e_2 \in X_2$ are pairwise concurrent, $Y_2 \subseteq y_2$, $\forall e_1, e_2 \in Y_2$ are pairwise concurrent, $X_2 \sim Y_2$, and $x_2' \sim_s y_2'$, so, we get $x_1 + x_2 \sim_s y_1 + y_2$, as desired. $\qquad\square$

Theorem 4.10 (Soundness of $BATC$ modulo step bisimulation equivalence). *Let x and y be $BATC$ terms. If $BATC \vdash x = y$, then $x \sim_s y$.*

Proof. Since step bisimulation $\sim_s$ is both an equivalent and a congruent relation, we only need to check if each axiom in Table 4.1 is sound modulo step bisimulation equivalence.

- **Axiom** $A1$. Let p, q be $BATC$ processes, and $p + q = q + p$, it is sufficient to prove that $p + q \sim_s q + p$. By the step transition rules for operator $+$ in Table 4.5, we get

$$\frac{p \xrightarrow{P} \surd}{p + q \xrightarrow{P} \surd} (P \subseteq p, \forall e_1, e_2 \in P \text{ are pairwise concurrent.})$$

$$\frac{p \xrightarrow{P} \surd}{q + p \xrightarrow{P} \surd} (P \subseteq p, \forall e_1, e_2 \in P \text{ are pairwise concurrent.})$$

$$\frac{p \xrightarrow{P} p'}{p + q \xrightarrow{P} p'} (P \subseteq p, \forall e_1, e_2 \in P \text{ are pairwise concurrent.})$$

$$\frac{p \xrightarrow{P} p'}{q + p \xrightarrow{P} p'} (P \subseteq p, \forall e_1, e_2 \in P \text{ are pairwise concurrent.})$$

$$\frac{q \xrightarrow{Q} \surd}{p + q \xrightarrow{Q} \surd} (Q \subseteq q, \forall e_1, e_2 \in Q \text{ are pairwise concurrent.})$$

$$\frac{q \xrightarrow{Q} \surd}{q + p \xrightarrow{Q} \surd} (Q \subseteq q, \forall e_1, e_2 \in Q \text{ are pairwise concurrent.})$$

$$\frac{q \xrightarrow{Q} q'}{p + q \xrightarrow{Q} q'} (Q \subseteq q, \forall e_1, e_2 \in Q \text{ are pairwise concurrent.})$$

$$\frac{q \xrightarrow{Q} q'}{q + p \xrightarrow{Q} q'} (Q \subseteq q, \forall e_1, e_2 \in Q \text{ are pairwise concurrent.})$$

So, $p + q \sim_s q + p$, as desired.
- **Axiom** $A2$. Let p, q, s be $BATC$ processes, and $(p + q) + s = p + (q + s)$, it is sufficient to prove that $(p + q) + s \sim_s p + (q + s)$. By the step transition rules for operator $+$ in Table 4.5, we get

$$\frac{p \xrightarrow{P} \surd}{(p + q) + s \xrightarrow{P} \surd} (P \subseteq p, \forall e_1, e_2 \in P \text{ are pairwise concurrent.})$$

$$\frac{p \xrightarrow{P} \surd}{p + (q + s) \xrightarrow{P} \surd} (P \subseteq p, \forall e_1, e_2 \in P \text{ are pairwise concurrent.})$$

$$\frac{p \xrightarrow{P} p'}{(p+q)+s \xrightarrow{P} p'} (P \subseteq p, \forall e_1, e_2 \in P \text{ are pairwise concurrent.})$$

$$\frac{p \xrightarrow{P} p'}{p+(q+s) \xrightarrow{P} p'} (P \subseteq p, \forall e_1, e_2 \in P \text{ are pairwise concurrent.})$$

$$\frac{q \xrightarrow{Q} \surd}{(p+q)+s \xrightarrow{Q} \surd} (Q \subseteq q, \forall e_1, e_2 \in Q \text{ are pairwise concurrent.})$$

$$\frac{q \xrightarrow{Q} \surd}{p+(q+s) \xrightarrow{Q} \surd} (Q \subseteq q, \forall e_1, e_2 \in Q \text{ are pairwise concurrent.})$$

$$\frac{q \xrightarrow{Q} q'}{(p+q)+s \xrightarrow{Q} q'} (Q \subseteq q, \forall e_1, e_2 \in Q \text{ are pairwise concurrent.})$$

$$\frac{q \xrightarrow{Q} q'}{p+(q+s) \xrightarrow{Q} q'} (Q \subseteq q, \forall e_1, e_2 \in Q \text{ are pairwise concurrent.})$$

$$\frac{s \xrightarrow{S} \surd}{(p+q)+s \xrightarrow{S} \surd} (S \subseteq s, \forall e_1, e_2 \in S \text{ are pairwise concurrent.})$$

$$\frac{s \xrightarrow{S} \surd}{p+(q+s) \xrightarrow{S} \surd} (S \subseteq s, \forall e_1, e_2 \in S \text{ are pairwise concurrent.})$$

$$\frac{s \xrightarrow{S} s'}{(p+q)+s \xrightarrow{S} s'} (S \subseteq s, \forall e_1, e_2 \in S \text{ are pairwise concurrent.})$$

$$\frac{s \xrightarrow{S} s'}{p+(q+s) \xrightarrow{S} s'} (S \subseteq s, \forall e_1, e_2 \in S \text{ are pairwise concurrent.})$$

So, $(p+q)+s \sim_s p+(q+s)$, as desired.

- **Axiom** $A3$. Let p be a $BATC$ process, and $p+p = p$, it is sufficient to prove that $p+p \sim_s p$. By the step transition rules for operator $+$ in Table 4.5, we get

$$\frac{p \xrightarrow{P} \surd}{p+p \xrightarrow{P} \surd} (P \subseteq p, \forall e_1, e_2 \in P \text{ are pairwise concurrent.})$$

$$\frac{p \xrightarrow{P} \surd}{p \xrightarrow{P} \surd} (P \subseteq p, \forall e_1, e_2 \in P \text{ are pairwise concurrent.})$$

$$\frac{p \xrightarrow{P} p'}{p + p \xrightarrow{P} p'} (P \subseteq p, \forall e_1, e_2 \in P \text{ are pairwise concurrent.})$$

$$\frac{p \xrightarrow{P} p'}{p \xrightarrow{P} p'} (P \subseteq p, \forall e_1, e_2 \in P \text{ are pairwise concurrent.})$$

So, $p + p \sim_s p$, as desired.

- **Axiom** $A4$. Let p, q, s be $BATC$ processes, and $(p + q) \cdot s = p \cdot s + q \cdot s$, it is sufficient to prove that $(p + q) \cdot s \sim_s p \cdot s + q \cdot s$. By the step transition rules for operators $+$ and $\cdot$ in Table 4.5, we get

$$\frac{p \xrightarrow{P} \surd}{(p + q) \cdot s \xrightarrow{P} s} (P \subseteq p, \forall e_1, e_2 \in P \text{ are pairwise concurrent.})$$

$$\frac{p \xrightarrow{P} \surd}{p \cdot s + q \cdot s \xrightarrow{P} s} (P \subseteq p, \forall e_1, e_2 \in P \text{ are pairwise concurrent.})$$

$$\frac{p \xrightarrow{P} p'}{(p + q) \cdot s \xrightarrow{P} p' \cdot s} (P \subseteq p, \forall e_1, e_2 \in P \text{ are pairwise concurrent.})$$

$$\frac{p \xrightarrow{P} p'}{p \cdot s + q \cdot s \xrightarrow{P} p' \cdot s} (P \subseteq p, \forall e_1, e_2 \in P \text{ are pairwise concurrent.})$$

$$\frac{q \xrightarrow{Q} \surd}{(p + q) \cdot s \xrightarrow{Q} s} (Q \subseteq q, \forall e_1, e_2 \in Q \text{ are pairwise concurrent.})$$

$$\frac{q \xrightarrow{Q} \surd}{p \cdot s + q \cdot s \xrightarrow{Q} s} (Q \subseteq q, \forall e_1, e_2 \in Q \text{ are pairwise concurrent.})$$

$$\frac{q \xrightarrow{Q} q'}{(p + q) \cdot s \xrightarrow{Q} q' \cdot s} (Q \subseteq q, \forall e_1, e_2 \in Q \text{ are pairwise concurrent.})$$

$$\frac{q \xrightarrow{Q} q'}{p \cdot s + q \cdot s \xrightarrow{Q} q' \cdot s} (Q \subseteq q, \forall e_1, e_2 \in Q \text{ are pairwise concurrent.})$$

So, $(p + q) \cdot s \sim_s p \cdot s + q \cdot s$, as desired.

- **Axiom** $A5$. Let p, q, s be $BATC$ processes, and $(p \cdot q) \cdot s = p \cdot (q \cdot s)$, it is sufficient to prove that $(p \cdot q) \cdot s \sim_s p \cdot (q \cdot s)$. By the step transition rules for operator $\cdot$ in Table 4.5, we get

$$\frac{p \xrightarrow{P} \surd}{(p \cdot q) \cdot s \xrightarrow{P} q \cdot s} \quad (P \subseteq p, \forall e_1, e_2 \in P \text{ are pairwise concurrent.})$$

$$\frac{p \xrightarrow{P} \surd}{p \cdot (q \cdot s) \xrightarrow{P} q \cdot s} \quad (P \subseteq p, \forall e_1, e_2 \in P \text{ are pairwise concurrent.})$$

$$\frac{p \xrightarrow{P} p'}{(p \cdot q) \cdot s \xrightarrow{P} (p' \cdot q) \cdot s} \quad (P \subseteq p, \forall e_1, e_2 \in P \text{ are pairwise concurrent.})$$

$$\frac{p \xrightarrow{P} p'}{p \cdot (q \cdot s) \xrightarrow{P} p' \cdot (q \cdot s)} \quad (P \subseteq p, \forall e_1, e_2 \in P \text{ are pairwise concurrent.})$$

With an assumption $(p' \cdot q) \cdot s = p' \cdot (q \cdot s)$, so, $(p \cdot q) \cdot s \sim_s p \cdot (q \cdot s)$, as desired. $\qquad\square$

Theorem 4.11 (Completeness of $BATC$ modulo step bisimulation equivalence). *Let p and q be closed $BATC$ terms, if $p \sim_s q$ then $p = q$.*

Proof. Firstly, by the elimination theorem of $BATC$, we know that for each closed $BATC$ term p, there exists a closed basic $BATC$ term p', such that $BATC \vdash p = p'$, so, we only need to consider closed basic $BATC$ terms.

The basic terms (see Definition 4.1) modulo associativity and commutativity (AC) of conflict $+$ (defined by axioms $A1$ and $A2$ in Table 4.1), and this equivalence is denoted by $=_{AC}$. Then, each equivalence class s modulo AC of $+$ has the following normal form

$$s_1 + \cdots + s_k$$

with each s_i either an atomic event or of the form $t_1 \cdot t_2$, and each s_i is called the summand of s.

Now, we prove that for normal forms n and n', if $n \sim_s n'$ then $n =_{AC} n'$. It is sufficient to induct on the sizes of n and n'.

- Consider a summand e of n. Then $n \xrightarrow{e} \surd$, so $n \sim_s n'$ implies $n' \xrightarrow{e} \surd$, meaning that n' also contains the summand e.
- Consider a summand $t_1 \cdot t_2$ of n. Then $n \xrightarrow{t_1} t_2 (\forall e_1, e_2 \in t_1$ are pairwise concurrent), so $n \sim_s n'$ implies $n' \xrightarrow{t_1} t'_2 (\forall e_1, e_2 \in t_1$ are pairwise concurrent) with $t_2 \sim_s t'_2$, meaning that n' contains a summand $t_1 \cdot t'_2$. Since t_2 and t'_2 are normal forms and have sizes smaller than n and n', by the induction hypotheses if $t_2 \sim_s t'_2$ then $t_2 =_{AC} t'_2$.

Table 4.6   (Hereditary) hp-transition rules of $BATC$.

$$\frac{}{e \xrightarrow{e} \surd}$$

$$\frac{x \xrightarrow{e} \surd}{x + y \xrightarrow{e} \surd} \quad \frac{x \xrightarrow{e} x'}{x + y \xrightarrow{e} x'} \quad \frac{y \xrightarrow{e} \surd}{x + y \xrightarrow{e} \surd} \quad \frac{y \xrightarrow{e} y'}{x + y \xrightarrow{e} y'}$$

$$\frac{x \xrightarrow{e} \surd}{x \cdot y \xrightarrow{e} y} \quad \frac{x \xrightarrow{e} x'}{x \cdot y \xrightarrow{e} x' \cdot y}$$

So, we get $n =_{AC} n'$.

Finally, let s and t be basic terms, and $s \sim_s t$, there are normal forms n and n', such that $s = n$ and $t = n'$. The soundness theorem of $BATC$ modulo step bisimulation equivalence (see Theorem 4.10) yields $s \sim_s n$ and $t \sim_s n'$, so $n \sim_s s \sim_s t \sim_s n'$. Since if $n \sim_s n'$ then $n =_{AC} n'$, $s = n =_{AC} n' = t$, as desired. $\qquad\qquad\square$

The transition rules for (hereditary) hp-bisimulation of $BATC$ are defined in Table 4.6, they are the same as single event transition rules in Table 4.3.

Theorem 4.12 (Congruence of $BATC$ with respect to hp-bisimulation equivalence). *Hp-bisimulation equivalence $\sim_{hp}$ is a congruence with respect to $BATC$.*

Proof. It is easy to see that history-preserving bisimulation is an equivalent relation on $BATC$ terms, we only need to prove that $\sim_{hp}$ is preserved by the operators $\cdot$ and $+$.

- Causality operator $\cdot$. Let x_1, x_2 and y_1, y_2 be $BATC$ processes, and $x_1 \sim_{hp} y_1$, $x_2 \sim_{hp} y_2$, it is sufficient to prove that $x_1 \cdot x_2 \sim_{hp} y_1 \cdot y_2$.

  By the definition of hp-bisimulation $\sim_{hp}$ (Definition 2.21), $x_1 \sim_{hp} y_1$ means that there is a posetal relation $(C(x_1), f, C(y_1)) \in \sim_{hp}$, and

$$x_1 \xrightarrow{e_1} x_1' \quad y_1 \xrightarrow{e_2} y_1'$$

  with $(C(x_1'), f[e_1 \mapsto e_2], C(y_1')) \in \sim_{hp}$. The meaning of $x_2 \sim_{hp} y_2$ is similar.

  By the hp-transition rules for causality operator $\cdot$ in Table 4.6, we can get

$$x_1 \cdot x_2 \xrightarrow{e_1} x_2 \quad y_1 \cdot y_2 \xrightarrow{e_2} y_2$$

  with $x_2 \sim_{hp} y_2$, so, we get $x_1 \cdot x_2 \sim_{hp} y_1 \cdot y_2$, as desired.

  Or, we can get

$$x_1 \cdot x_2 \xrightarrow{e_1} x_1' \cdot x_2 \quad y_1 \cdot y_2 \xrightarrow{e_2} y_1' \cdot y_2$$

  with $x_1' \sim_{hp} y_1'$, $x_2 \sim_{hp} y_2$, so, we get $x_1 \cdot x_2 \sim_{hp} y_1 \cdot y_2$, as desired.

- Conflict operator $+$. Let x_1, x_2 and y_1, y_2 be $BATC$ processes, and $x_1 \sim_{hp} y_1$, $x_2 \sim_{hp} y_2$, it is sufficient to prove that $x_1 + x_2 \sim_{hp} y_1 + y_2$. The meanings of $x_1 \sim_{hp} y_1$ and $x_2 \sim_{hp} y_2$ are the same as the above case, according to the definition of hp-bisimulation $\sim_{hp}$ in Definition 2.21.

  By the hp-transition rules for conflict operator $+$ in Table 4.6, we can get four cases:

$$x_1 + x_2 \xrightarrow{e_1} \surd \quad y_1 + y_2 \xrightarrow{e_2} \surd$$

so, we get $x_1 + x_2 \sim_{hp} y_1 + y_2$, as desired.

Or, we can get

$$x_1 + x_2 \xrightarrow{e_1} x_1' \quad y_1 + y_2 \xrightarrow{e_2} y_1'$$

with $x_1' \sim_{hp} y_1'$, so, we get $x_1 + x_2 \sim_{hp} y_1 + y_2$, as desired.

Or, we can get

$$x_1 + x_2 \xrightarrow{e_1'} \surd \quad y_1 + y_2 \xrightarrow{e_2'} \surd$$

so, we get $x_1 + x_2 \sim_{hp} y_1 + y_2$, as desired.

Or, we can get

$$x_1 + x_2 \xrightarrow{e_1'} x_2' \quad y_1 + y_2 \xrightarrow{e_2'} y_2'$$

with $x_2' \sim_{hp} y_2'$, so, we get $x_1 + x_2 \sim_{hp} y_1 + y_2$, as desired. $\qquad \square$

Theorem 4.13 (Soundness of $BATC$ modulo hp-bisimulation equivalence). *Let x and y be $BATC$ terms. If $BATC \vdash x = y$, then $x \sim_{hp} y$.*

Proof. Since hp-bisimulation $\sim_{hp}$ is both an equivalent and a congruent relation, we only need to check if each axiom in Table 4.1 is sound modulo hp-bisimulation equivalence.

- **Axiom** $A1$. Let p, q be $BATC$ processes, and $p + q = q + p$, it is sufficient to prove that $p + q \sim_{hp} q + p$. By the hp-transition rules for operator $+$ in Table 4.6, we get

$$\frac{p \xrightarrow{e_1} \surd}{p + q \xrightarrow{e_1} \surd} \qquad \frac{p \xrightarrow{e_1} \surd}{q + p \xrightarrow{e_1} \surd}$$

$$\frac{p \xrightarrow{e_1} p'}{p + q \xrightarrow{e_1} p'} \qquad \frac{p \xrightarrow{e_1} p'}{q + p \xrightarrow{e_1} p'}$$

$$\frac{q \xrightarrow{e_2} \surd}{p + q \xrightarrow{e_2} \surd} \qquad \frac{q \xrightarrow{e_2} \surd}{q + p \xrightarrow{e_2} \surd}$$

$$\frac{q \xrightarrow{e_2} q'}{p + q \xrightarrow{e_2} q'} \qquad \frac{q \xrightarrow{e_2} q'}{q + p \xrightarrow{e_2} q'}$$

So, for $(C(p + q), f, C(q + p)) \in \sim_{hp}$, $(C((p + q)'), f[e_1 \mapsto e_1], C((q + p)')) \in \sim_{hp}$ and $(C((p+q)'), f[e_2 \mapsto e_2], C((q + p)')) \in \sim_{hp}$, that is, $p + q \sim_{hp} q + p$, as desired.

- **Axiom** $A2$. Let p, q, s be $BATC$ processes, and $(p + q) + s = p + (q + s)$, it is sufficient to prove that $(p + q) + s \sim_{hp} p + (q + s)$. By the hp-transition rules for operator $+$ in Table 4.6, we get

$$\frac{p \xrightarrow{e_1} \surd}{(p + q) + s \xrightarrow{e_1} \surd} \quad \frac{p \xrightarrow{e_1} \surd}{p + (q + s) \xrightarrow{e_1} \surd}$$

$$\frac{p \xrightarrow{e_1} p'}{(p + q) + s \xrightarrow{e_1} p'} \quad \frac{p \xrightarrow{e_1} p'}{p + (q + s) \xrightarrow{e_1} p'}$$

$$\frac{q \xrightarrow{e_2} \surd}{(p + q) + s \xrightarrow{e_2} \surd} \quad \frac{q \xrightarrow{e_2} \surd}{p + (q + s) \xrightarrow{e_2} \surd}$$

$$\frac{q \xrightarrow{e_2} q'}{(p + q) + s \xrightarrow{e_2} q'} \quad \frac{q \xrightarrow{e_2} q'}{p + (q + s) \xrightarrow{e_2} q'}$$

$$\frac{s \xrightarrow{e_3} \surd}{(p + q) + s \xrightarrow{e_3} \surd} \quad \frac{s \xrightarrow{e_3} \surd}{p + (q + s) \xrightarrow{e_3} \surd}$$

$$\frac{s \xrightarrow{e_3} s'}{(p + q) + s \xrightarrow{e_3} s'} \quad \frac{s \xrightarrow{e_3} s'}{p + (q + s) \xrightarrow{e_3} s'}$$

So, for $(C((p+q)+s), f, C(p+(q+s))) \in \sim_{hp}$, $(C(((p+q)+s)'), f[e_1 \mapsto e_1], C((p+(q+s))')) \in \sim_{hp}$ and $(C(((p+q)+s)'), f[e_2 \mapsto e_2], C((p+(q+s))')) \in \sim_{hp}$ and $(C(((p+q)+s)'), f[e_3 \mapsto e_3], C((p+(q+s))')) \in \sim_{hp}$, that is, $(p+q)+s \sim_{hp} p+(q+s)$, as desired.

- **Axiom** $A3$. Let p be a $BATC$ process, and $p + p = p$, it is sufficient to prove that $p + p \sim_{hp} p$. By the hp-transition rules for operator $+$ in Table 4.6, we get

$$\frac{p \xrightarrow{e_1} \surd}{p + p \xrightarrow{e_1} \surd} \quad \frac{p \xrightarrow{e_1} \surd}{p \xrightarrow{e_1} \surd}$$

$$\frac{p \xrightarrow{e_1} p'}{p + p \xrightarrow{e_1} p'} \quad \frac{p \xrightarrow{e_1} p'}{p \xrightarrow{e_1} p'}$$

So, for $(C(p + p), f, C(p)) \in \sim_{hp}$, $(C((p + p)'), f[e_1 \mapsto e_1], C((p)')) \in \sim_{hp}$, that is, $p + p \sim_{hp} p$, as desired.

- **Axiom** $A4$. Let p, q, s be $BATC$ processes, and $(p + q) \cdot s = p \cdot s + q \cdot s$, it is sufficient to prove that $(p + q) \cdot s \sim_{hp} p \cdot s + q \cdot s$. By the hp-transition rules for operators $+$ and $\cdot$ in

Table 4.6, we get

$$\frac{p \xrightarrow{e_1} \checkmark}{(p+q) \cdot s \xrightarrow{e_1} s} \qquad \frac{p \xrightarrow{e_1} \checkmark}{p \cdot s + q \cdot s \xrightarrow{e_1} s}$$

$$\frac{p \xrightarrow{e_1} p'}{(p+q) \cdot s \xrightarrow{e_1} p' \cdot s} \qquad \frac{p \xrightarrow{e_1} p'}{p \cdot s + q \cdot s \xrightarrow{e_1} p' \cdot s}$$

$$\frac{q \xrightarrow{e_2} \checkmark}{(p+q) \cdot s \xrightarrow{e_2} s} \qquad \frac{q \xrightarrow{e_2} \checkmark}{p \cdot s + q \cdot s \xrightarrow{e_2} s}$$

$$\frac{q \xrightarrow{e_2} q'}{(p+q) \cdot s \xrightarrow{e_2} q' \cdot s} \qquad \frac{q \xrightarrow{e_2} q'}{p \cdot s + q \cdot s \xrightarrow{Q} q' \cdot s}$$

So, for $(C((p+q) \cdot s), f, C(p \cdot s + q \cdot s)) \in \sim_{hp}$, $(C(((p+q) \cdot s)'), f[e_1 \mapsto e_1], C((p \cdot s + q \cdot s)')) \in \sim_{hp}$ and $(C(((p+q) \cdot s)'), f[e_2 \mapsto e_2], C((p \cdot s + q \cdot s)')) \in \sim_{hp}$, that is, $(p+q) \cdot s \sim_{hp} p \cdot s + q \cdot s$, as desired.

- **Axiom** $A5$. Let p, q, s be $BATC$ processes, and $(p \cdot q) \cdot s = p \cdot (q \cdot s)$, it is sufficient to prove that $(p \cdot q) \cdot s \sim_{hp} p \cdot (q \cdot s)$. By the hp-transition rules for operator $\cdot$ in Table 4.6, we get

$$\frac{p \xrightarrow{e_1} \checkmark}{(p \cdot q) \cdot s \xrightarrow{e_1} q \cdot s} \qquad \frac{p \xrightarrow{e_1} \checkmark}{p \cdot (q \cdot s) \xrightarrow{e_1} q \cdot s}$$

$$\frac{p \xrightarrow{e_1} p'}{(p \cdot q) \cdot s \xrightarrow{e_1} (p' \cdot q) \cdot s} \qquad \frac{p \xrightarrow{e_1} p'}{p \cdot (q \cdot s) \xrightarrow{e_1} p' \cdot (q \cdot s)}$$

With an assumption $(p' \cdot q) \cdot s = p' \cdot (q \cdot s)$, for $(C((p \cdot q) \cdot s), f, C(p \cdot (q \cdot s))) \in \sim_{hp}$, $(C(((p \cdot q) \cdot s)'), f[e_1 \mapsto e_1], C((p \cdot (q \cdot s))')) \in \sim_{hp}$, that is, so, $(p \cdot q) \cdot s \sim_{hp} p \cdot (q \cdot s)$, as desired.   □

Theorem 4.14 (Completeness of $BATC$ modulo hp-bisimulation equivalence). *Let p and q be closed $BATC$ terms, if $p \sim_{hp} q$ then $p = q$.*

Proof. Firstly, by the elimination theorem of $BATC$, we know that for each closed $BATC$ term p, there exists a closed basic $BATC$ term p', such that $BATC \vdash p = p'$, so, we only need to consider closed basic $BATC$ terms.

The basic terms (see Definition 4.1) modulo associativity and commutativity (AC) of conflict $+$ (defined by axioms $A1$ and $A2$ in Table 4.1), and this equivalence is denoted by $=_{AC}$. Then, each equivalence class s modulo AC of $+$ has the following normal form

$$s_1 + \cdots + s_k$$

with each s_i either an atomic event or of the form $t_1 \cdot t_2$, and each s_i is called the summand of s.

Now, we prove that for normal forms n and n', if $n \sim_{hp} n'$ then $n =_{AC} n'$. It is sufficient to induct on the sizes of n and n'.

- Consider a summand e of n. Then $n \xrightarrow{e} \surd$, so $n \sim_{hp} n'$ implies $n' \xrightarrow{e} \surd$, meaning that n' also contains the summand e.
- Consider a summand $e \cdot s$ of n. Then $n \xrightarrow{e} s$, so $n \sim_{hp} n'$ implies $n' \xrightarrow{e} t$ with $s \sim_{hp} t$, meaning that n' contains a summand $e \cdot t$. Since s and t are normal forms and have sizes smaller than n and n', by the induction hypotheses $s \sim_{hp} t$ implies $s =_{AC} t$.

So, we get $n =_{AC} n'$.

Finally, let s and t be basic terms, and $s \sim_{hp} t$, there are normal forms n and n', such that $s = n$ and $t = n'$. The soundness theorem of $BATC$ modulo hp-bisimulation equivalence (see Theorem 4.13) yields $s \sim_{hp} n$ and $t \sim_{hp} n'$, so $n \sim_{hp} s \sim_{hp} t \sim_{hp} n'$. Since if $n \sim_{hp} n'$ then $n =_{AC} n'$, $s = n =_{AC} n' = t$, as desired. $\qquad\square$

Theorem 4.15 (Congruence of $BATC$ with respect to hhp-bisimulation equivalence). *Hhp-bisimulation equivalence $\sim_{hhp}$ is a congruence with respect to $BATC$.*

Proof. It is easy to see that hhp-bisimulation is an equivalent relation on $BATC$ terms, we only need to prove that $\sim_{hhp}$ is preserved by the operators $\cdot$ and $+$.

- Causality operator $\cdot$. Let x_1, x_2 and y_1, y_2 be $BATC$ processes, and $x_1 \sim_{hhp} y_1, x_2 \sim_{hhp} y_2$, it is sufficient to prove that $x_1 \cdot x_2 \sim_{hhp} y_1 \cdot y_2$.
  By the definition of hhp-bisimulation $\sim_{hhp}$ (Definition 2.21), $x_1 \sim_{hhp} y_1$ means that there is a posetal relation $(C(x_1), f, C(y_1)) \in \sim_{hhp}$, and

  $$x_1 \xrightarrow{e_1} x_1' \quad y_1 \xrightarrow{e_2} y_1'$$

  with $(C(x_1'), f[e_1 \mapsto e_2], C(y_1')) \in \sim_{hhp}$. The meaning of $x_2 \sim_{hhp} y_2$ is similar.
  By the hhp-transition rules for causality operator $\cdot$ in Table 4.6, we can get

  $$x_1 \cdot x_2 \xrightarrow{e_1} x_2 \quad y_1 \cdot y_2 \xrightarrow{e_2} y_2$$

  with $x_2 \sim_{hhp} y_2$, so, we get $x_1 \cdot x_2 \sim_{hhp} y_1 \cdot y_2$, as desired.
  Or, we can get

  $$x_1 \cdot x_2 \xrightarrow{e_1} x_1' \cdot x_2 \quad y_1 \cdot y_2 \xrightarrow{e_2} y_1' \cdot y_2$$

  with $x_1' \sim_{hhp} y_1', x_2 \sim_{hhp} y_2$, so, we get $x_1 \cdot x_2 \sim_{hhp} y_1 \cdot y_2$, as desired.
- Conflict operator $+$. Let x_1, x_2 and y_1, y_2 be $BATC$ processes, and $x_1 \sim_{hhp} y_1, x_2 \sim_{hhp} y_2$, it is sufficient to prove that $x_1 + x_2 \sim_{hhp} y_1 + y_2$. The meanings of $x_1 \sim_{hhp} y_1$ and $x_2 \sim_{hhp} y_2$ are the same as the above case, according to the definition of hhp-bisimulation $\sim_{hhp}$ in Definition 2.21.
  By the hhp-transition rules for conflict operator $+$ in Table 4.6, we can get four cases:

  $$x_1 + x_2 \xrightarrow{e_1} \surd \quad y_1 + y_2 \xrightarrow{e_2} \surd$$

so, we get $x_1 + x_2 \sim_{hhp} y_1 + y_2$, as desired.
Or, we can get

$$x_1 + x_2 \xrightarrow{e_1} x_1' \quad y_1 + y_2 \xrightarrow{e_2} y_1'$$

with $x_1' \sim_{hhp} y_1'$, so, we get $x_1 + x_2 \sim_{hhp} y_1 + y_2$, as desired.
Or, we can get

$$x_1 + x_2 \xrightarrow{e_1'} \checkmark \quad y_1 + y_2 \xrightarrow{e_2'} \checkmark$$

so, we get $x_1 + x_2 \sim_{hhp} y_1 + y_2$, as desired.
Or, we can get

$$x_1 + x_2 \xrightarrow{e_1'} x_2' \quad y_1 + y_2 \xrightarrow{e_2'} y_2'$$

with $x_2' \sim_{hhp} y_2'$, so, we get $x_1 + x_2 \sim_{hhp} y_1 + y_2$, as desired. $\qquad\square$

Theorem 4.16 (Soundness of $BATC$ modulo hhp-bisimulation equivalence). *Let x and y be $BATC$ terms. If $BATC \vdash x = y$, then $x \sim_{hhp} y$.*

Proof. Since hhp-bisimulation $\sim_{hhp}$ is both an equivalent and a congruent relation, we only need to check if each axiom in Table 4.1 is sound modulo hhp-bisimulation equivalence.

- **Axiom** $A1$. Let p, q be $BATC$ processes, and $p + q = q + p$, it is sufficient to prove that $p + q \sim_{hhp} q + p$. By the hhp-transition rules for operator $+$ in Table 4.6, we get

$$\frac{p \xrightarrow{e_1} \checkmark}{p + q \xrightarrow{e_1} \checkmark} \quad \frac{p \xrightarrow{e_1} \checkmark}{q + p \xrightarrow{e_1} \checkmark}$$

$$\frac{p \xrightarrow{e_1} p'}{p + q \xrightarrow{e_1} p'} \quad \frac{p \xrightarrow{e_1} p'}{q + p \xrightarrow{e_1} p'}$$

$$\frac{q \xrightarrow{e_2} \checkmark}{p + q \xrightarrow{e_2} \checkmark} \quad \frac{q \xrightarrow{e_2} \checkmark}{q + p \xrightarrow{e_2} \checkmark}$$

$$\frac{q \xrightarrow{e_2} q'}{p + q \xrightarrow{e_2} q'} \quad \frac{q \xrightarrow{e_2} q'}{q + p \xrightarrow{e_2} q'}$$

So, for $(C(p + q), f, C(q + p)) \in \sim_{hhp}$, $(C((p + q)'), f[e_1 \mapsto e_1], C((q + p)')) \in \sim_{hhp}$ and $(C((p + q)'), f[e_2 \mapsto e_2], C((q + p)')) \in \sim_{hhp}$, that is, $p + q \sim_{hhp} q + p$, as desired.
- **Axiom** $A2$. Let p, q, s be $BATC$ processes, and $(p + q) + s = p + (q + s)$, it is sufficient to prove that $(p + q) + s \sim_{hhp} p + (q + s)$. By the hhp-transition rules for operator $+$ in

Table 4.6, we get

$$\frac{p \xrightarrow{e_1} \checkmark}{(p+q)+s \xrightarrow{e_1} \checkmark} \qquad \frac{p \xrightarrow{e_1} \checkmark}{p+(q+s) \xrightarrow{e_1} \checkmark}$$

$$\frac{p \xrightarrow{e_1} p'}{(p+q)+s \xrightarrow{e_1} p'} \qquad \frac{p \xrightarrow{e_1} p'}{p+(q+s) \xrightarrow{e_1} p'}$$

$$\frac{q \xrightarrow{e_2} \checkmark}{(p+q)+s \xrightarrow{e_2} \checkmark} \qquad \frac{q \xrightarrow{e_2} \checkmark}{p+(q+s) \xrightarrow{e_2} \checkmark}$$

$$\frac{q \xrightarrow{e_2} q'}{(p+q)+s \xrightarrow{e_2} q'} \qquad \frac{q \xrightarrow{e_2} q'}{p+(q+s) \xrightarrow{e_2} q'}$$

$$\frac{s \xrightarrow{e_3} \checkmark}{(p+q)+s \xrightarrow{e_3} \checkmark} \qquad \frac{s \xrightarrow{e_3} \checkmark}{p+(q+s) \xrightarrow{e_3} \checkmark}$$

$$\frac{s \xrightarrow{e_3} s'}{(p+q)+s \xrightarrow{e_3} s'} \qquad \frac{s \xrightarrow{e_3} s'}{p+(q+s) \xrightarrow{e_3} s'}$$

So, for $(C((p+q)+s), f, C(p+(q+s))) \in \sim_{hhp}$, $(C(((p+q)+s)'), f[e_1 \mapsto e_1], C((p+(q+s))'))\in\sim_{hhp}$ and $(C(((p+q)+s)'), f[e_2 \mapsto e_2], C((p+(q+s))'))\in\sim_{hhp}$ and $(C(((p+q)+s)'), f[e_3 \mapsto e_3], C((p+(q+s))'))\in\sim_{hhp}$, that is, $(p+q)+s \sim_{hhp} p+(q+s)$, as desired.

- **Axiom** $A3$. Let p be a $BATC$ process, and $p + p = p$, it is sufficient to prove that $p + p \sim_{hhp} p$. By the hhp-transition rules for operator $+$ in Table 4.6, we get

$$\frac{p \xrightarrow{e_1} \checkmark}{p+p \xrightarrow{e_1} \checkmark} \quad \frac{p \xrightarrow{e_1} \checkmark}{p \xrightarrow{e_1} \checkmark}$$

$$\frac{p \xrightarrow{e_1} p'}{p+p \xrightarrow{e_1} p'} \quad \frac{p \xrightarrow{e_1} p'}{p \xrightarrow{e_1} p'}$$

So, for $(C(p + p), f, C(p)) \in \sim_{hhp}$, $(C((p + p)'), f[e_1 \mapsto e_1], C((p)')) \in \sim_{hhp}$, that is, $p + p \sim_{hhp} p$, as desired.

- **Axiom** $A4$. Let p, q, s be $BATC$ processes, and $(p+q) \cdot s = p \cdot s + q \cdot s$, it is sufficient to prove that $(p+q) \cdot s \sim_{hhp} p \cdot s + q \cdot s$. By the hhp-transition rules for operators $+$ and $\cdot$ in Table 4.6, we get

$$\frac{p \xrightarrow{e_1} \checkmark}{(p+q) \cdot s \xrightarrow{e_1} s} \qquad \frac{p \xrightarrow{e_1} \checkmark}{p \cdot s + q \cdot s \xrightarrow{e_1} s}$$

$$\frac{p \xrightarrow{e_1} p'}{(p+q) \cdot s \xrightarrow{e_1} p' \cdot s} \qquad \frac{p \xrightarrow{e_1} p'}{p \cdot s + q \cdot s \xrightarrow{e_1} p' \cdot s}$$

$$\frac{q \xrightarrow{e_2} \surd}{(p+q) \cdot s \xrightarrow{e_2} s} \qquad \frac{q \xrightarrow{e_2} \surd}{p \cdot s + q \cdot s \xrightarrow{e_2} s}$$

$$\frac{q \xrightarrow{e_2} q'}{(p+q) \cdot s \xrightarrow{e_2} q' \cdot s} \qquad \frac{q \xrightarrow{e_2} q'}{p \cdot s + q \cdot s \xrightarrow{e_2} q' \cdot s}$$

So, for $(C((p+q) \cdot s), f, C(p \cdot s + q \cdot s)) \in \sim_{hhp}$, $(C(((p+q) \cdot s)'), f[e_1 \mapsto e_1], C((p \cdot s + q \cdot s)'))$ $\in \sim_{hhp}$ and $(C(((p+q) \cdot s)'), f[e_2 \mapsto e_2], C((p \cdot s + q \cdot s)')) \in \sim_{hhp}$, that is, $(p+q) \cdot s \sim_{hhp}$ $p \cdot s + q \cdot s$, as desired.

- **Axiom** $A5$. Let p, q, s be $BATC$ processes, and $(p \cdot q) \cdot s = p \cdot (q \cdot s)$, it is sufficient to prove that $(p \cdot q) \cdot s \sim_{hhp} p \cdot (q \cdot s)$. By the hhp-transition rules for operator $\cdot$ in Table 4.6, we get

$$\frac{p \xrightarrow{e_1} \surd}{(p \cdot q) \cdot s \xrightarrow{e_1} q \cdot s} \qquad \frac{p \xrightarrow{e_1} \surd}{p \cdot (q \cdot s) \xrightarrow{e_1} q \cdot s}$$

$$\frac{p \xrightarrow{e_1} p'}{(p \cdot q) \cdot s \xrightarrow{e_1} (p' \cdot q) \cdot s} \qquad \frac{p \xrightarrow{e_1} p'}{p \cdot (q \cdot s) \xrightarrow{e_1} p' \cdot (q \cdot s)}$$

With an assumption $(p' \cdot q) \cdot s = p' \cdot (q \cdot s)$, for $(C((p \cdot q) \cdot s), f, C(p \cdot (q \cdot s))) \in \sim_{hhp}$, $(C(((p \cdot q) \cdot s)'), f[e_1 \mapsto e_1], C((p \cdot (q \cdot s))')) \in \sim_{hhp}$, that is, so, $(p \cdot q) \cdot s \sim_{hhp} p \cdot (q \cdot s)$, as desired. $\square$

Theorem 4.17 (Completeness of $BATC$ modulo hhp-bisimulation equivalence). *Let p and q be closed $BATC$ terms, if $p \sim_{hhp} q$ then $p = q$.*

Proof. Firstly, by the elimination theorem of $BATC$, we know that for each closed $BATC$ term p, there exists a closed basic $BATC$ term p', such that $BATC \vdash p = p'$, so, we only need to consider closed basic $BATC$ terms.

The basic terms (see Definition 4.1) modulo associativity and commutativity (AC) of conflict $+$ (defined by axioms $A1$ and $A2$ in Table 4.1), and this equivalence is denoted by $=_{AC}$. Then, each equivalence class s modulo AC of $+$ has the following normal form

$$s_1 + \cdots + s_k$$

with each s_i either an atomic event or of the form $t_1 \cdot t_2$, and each s_i is called the summand of s.

Now, we prove that for normal forms n and n', if $n \sim_{hhp} n'$ then $n =_{AC} n'$. It is sufficient to induct on the sizes of n and n'.

- Consider a summand e of n. Then $n \xrightarrow{e} \surd$, so $n \sim_{hhp} n'$ implies $n' \xrightarrow{e} \surd$, meaning that n' also contains the summand e.

- Consider a summand $e \cdot s$ of n. Then $n \xrightarrow{e} s$, so $n \sim_{hhp} n'$ implies $n' \xrightarrow{e} t$ with $s \sim_{hhp} t$, meaning that n' contains a summand $e \cdot t$. Since s and t are normal forms and have sizes smaller than n and n', by the induction hypotheses $s \sim_{hhp} t$ implies $s =_{AC} t$.

So, we get $n =_{AC} n'$.

Finally, let s and t be basic terms, and $s \sim_{hhp} t$, there are normal forms n and n', such that $s = n$ and $t = n'$. The soundness theorem of $BATC$ modulo history-preserving bisimulation equivalence (see Theorem 4.16) yields $s \sim_{hhp} n$ and $t \sim_{hhp} n'$, so $n \sim_{hhp} s \sim_{hhp} t \sim_{hhp} n'$. Since if $n \sim_{hhp} n'$ then $n =_{AC} n'$, $s = n =_{AC} n' = t$, as desired.   □

4.2 Algebra for parallelism in true concurrency

In this section, we will discuss parallelism in true concurrency. We know that parallelism can be modeled by left merge and communication merge in ACP (Algebra of Communicating Process) [1,4] with an interleaving bisimulation semantics. Parallelism in true concurrency is quite different to that in interleaving bisimulation: it is a fundamental computational pattern (modeled by parallel operator ‖) and cannot be merged (replaced by other operators). The resulted algebra is called Algebra for Parallelism in True Concurrency, abbreviated $APTC$.

4.2.1 Parallelism as a fundamental computational pattern

Through several propositions, we show that parallelism is a fundamental computational pattern. Firstly, we give the transition rules for parallel operator ‖ as follows, it is suitable for all truly concurrent behavioral equivalences, including pomset bisimulation, step bisimulation, hp-bisimulation and hhp-bisimulation.

We will show that Milner's expansion law [1] does not hold modulo any truly concurrent behavioral equivalence, as the following proposition shows.

Proposition 4.18 (Milner's expansion law modulo truly concurrent behavioral equivalences). *Milner's expansion law does not hold modulo any truly concurrent behavioral equivalence, that is:*

1. *For atomic event e_1 and e_2,*
   a. $e_1 \parallel e_2 \nsim_p e_1 \cdot e_2 + e_2 \cdot e_1$;
   b. $e_1 \parallel e_2 \nsim_s e_1 \cdot e_2 + e_2 \cdot e_1$;
   c. $e_1 \parallel e_2 \nsim_{hp} e_1 \cdot e_2 + e_2 \cdot e_1$;
   d. $e_1 \parallel e_2 \nsim_{hhp} e_1 \cdot e_2 + e_2 \cdot e_1$;
2. *Specially, for auto-concurrency, let e be an atomic event,*
   a. $e \parallel e \nsim_p e \cdot e$;
   b. $e \parallel e \nsim_s e \cdot e$;
   c. $e \parallel e \nsim_{hp} e \cdot e$;
   d. $e \parallel e \nsim_{hhp} e \cdot e$.

Table 4.7   Transition rules of parallel operator ∥.

$$\frac{x \xrightarrow{e_1} \checkmark \quad y \xrightarrow{e_2} \checkmark}{x \parallel y \xrightarrow{\{e_1,e_2\}} \checkmark} \qquad \frac{x \xrightarrow{e_1} x' \quad y \xrightarrow{e_2} \checkmark}{x \parallel y \xrightarrow{\{e_1,e_2\}} x'}$$

$$\frac{x \xrightarrow{e_1} \checkmark \quad y \xrightarrow{e_2} y'}{x \parallel y \xrightarrow{\{e_1,e_2\}} y'} \qquad \frac{x \xrightarrow{e_1} x' \quad y \xrightarrow{e_2} y'}{x \parallel y \xrightarrow{\{e_1,e_2\}} x' \parallel y'}$$

Proof. In nature, it is caused by $e_1 \parallel e_2$ and $e_1 \cdot e_2 + e_2 \cdot e_1$ (specially $e \parallel e$ and $e \cdot e$) having different causality structure. They are based on the following obvious facts according to transition rules for parallel operator in Table 4.7:

1. $e_1 \parallel e_2 \xrightarrow{\{e_1,e_2\}} \checkmark$, while $e_1 \cdot e_2 + e_2 \cdot e_1 \nrightarrow^{\{e_1,e_2\}}$;
2. specially, $e \parallel e \xrightarrow{\{e,e\}} \checkmark$, while $e \cdot e \nrightarrow^{\{e,e\}}$. $\qquad\qquad\square$

In the following, we show that the elimination theorem does not hold for truly concurrent processes combined the operators $\cdot$, $+$ and $\parallel$. Firstly, we define the basic terms for $APTC$.

Definition 4.19 (Basic terms of $APTC$). The set of basic terms of $APTC$, $\mathcal{B}(APTC)$, is inductively defined as follows:

1. $\mathbb{E} \subset \mathcal{B}(APTC)$;
2. if $e \in \mathbb{E}, t \in \mathcal{B}(APTC)$ then $e \cdot t \in \mathcal{B}(APTC)$;
3. if $t, s \in \mathcal{B}(APTC)$ then $t + s \in \mathcal{B}(APTC)$;
4. if $t, s \in \mathcal{B}(APTC)$ then $t \parallel s \in \mathcal{B}(APTC)$.

Proposition 4.20 (About elimination theorem of $APTC$). **1.** *Let p be a closed $APTC$ term. Then there may not be a closed $BATC$ term q such that $APTC \vdash p = q$;*
2. *Let p be a closed $APTC$ term. Then there may not be a closed basic $APTC$ term q such that $APTC \vdash p = q$.*

Proof. **1.** By Proposition 4.18;
2. We show this property through two aspects:
  a. The left and right distributivity of $\cdot$ to $\parallel$, and $\parallel$ to $\cdot$, do not hold modulo any truly concurrent bisimulation equivalence.
  Left distributivity of $\cdot$ to $\parallel$: $(e_1 \cdot e_2) \parallel (e_1 \cdot e_3) \xrightarrow{\{e_1,e_1\}} e_2 \parallel e_3$, while $e_1 \cdot (e_2 \parallel e_3) \nrightarrow^{\{e_1,e_1\}}$.
  Right distributivity of $\cdot$ to $\parallel$: $(e_1 \cdot e_3) \parallel (e_2 \cdot e_3) \xrightarrow{\{e_1,e_2\}} e_3 \parallel e_3 \xrightarrow{\{e_3,e_3\}} \checkmark$, while $(e_1 \parallel e_2) \cdot e_3 \xrightarrow{\{e_1,e_2\}} e_3 \nrightarrow^{\{e_3,e_3\}}$.
  Left distributivity of $\parallel$ to $\cdot$: $(e_1 \parallel e_2) \cdot (e_1 \parallel e_3) \xrightarrow{\{e_1,e_2\}} e_1 \parallel e_3 \xrightarrow{\{e_1,e_3\}} \checkmark$, while $e_1 \parallel (e_2 \cdot e_3) \xrightarrow{\{e_1,e_2\}} e_3 \nrightarrow^{\{e_1,e_3\}}$.

FIGURE 4.1 Causality relation among parallel branches.

Right distributivity of $\parallel$ to $\cdot$: $(e_1 \parallel e_3) \cdot (e_2 \parallel e_3) \xrightarrow{\{e_1,e_3\}} e_2 \parallel e_3 \xrightarrow{\{e_2,e_3\}} \surd$, while $(e_1 \cdot e_2) \parallel$ $e_3 \xrightarrow{\{e_1,e_3\}} e_2 \not\rightarrow \{e_2,e_3\}$.

This means that there are not normal forms for the closed basic $APTC$ terms.

b. There are causality relations among different parallel branches can not be expressed by closed basic $APTC$ terms.

We consider the graph as Fig. 4.1 illustrates. There are four events labeled a, b, c, d, and there are three causality relations: c after a, d after b, and c after b. This graph can not be expressed by basic $APTC$ terms. a and b are in parallel, c after a, so c and a are in the same parallel branch; d after b, so d and b are in the same parallel branch; so c and d are in different parallel branches. But, c after b means that c and d are in the same parallel branch. This causes contradictions, it means that the graph in Fig. 4.1 can not be expressed by closed basic $APTC$ terms.   □

Until now, we see that parallelism acts as a fundamental computational pattern, and any elimination theorem does not hold any more. In nature, an event structure $\mathcal{E}$ (see Definition 2.13) is a graph defined by causality and conflict relations among events, while concurrency and consistency are implicitly defined by causality and conflict. The above conclusions say that an event structure $\mathcal{E}$ cannot be fully structured, the explicit parallel operator $\parallel$ in a fully structured event structure combined by $\cdot$, $+$ and $\parallel$ can not be replaced by $\cdot$ and $+$, and a fully structured event structure combined by $\cdot$, $+$ and $\parallel$ has no a normal form.

The above propositions mean that a perfectly sound and complete axiomatization of parallelism for truly concurrent bisimulation equivalence (like ACP [4] for bisimulation equivalence) *cannot* be established. Then, what can we do for $APTC$?

4.2.2 Axiom system of parallelism

Though a fully sound and complete axiomatization for $APTC$ seems impossible, we must and can do something, we believe. We also believe that the future is fully implied by the history, let us reconsider parallelism in interleaving bisimulation. In ACP [4], the full parallelism is captured by an auxiliary left merge and communication merge, left merge captures the interleaving bisimulation semantics, while communication merge expresses the communications among parallel branches. In true concurrency, if we try to define parallelism explicitly like $APTC$, the left merge captured Milner's expansion law does not hold

Table 4.8 Transition rules of communication operator |.

$$\frac{x \xrightarrow{e_1} \surd \quad y \xrightarrow{e_2} \surd}{x \mid y \xrightarrow{\gamma(e_1,e_2)} \surd} \qquad \frac{x \xrightarrow{e_1} x' \quad y \xrightarrow{e_2} \surd}{x \mid y \xrightarrow{\gamma(e_1,e_2)} x'}$$

$$\frac{x \xrightarrow{e_1} \surd \quad y \xrightarrow{e_2} y'}{x \mid y \xrightarrow{\gamma(e_1,e_2)} y'} \qquad \frac{x \xrightarrow{e_1} x' \quad y \xrightarrow{e_2} y'}{x \mid y \xrightarrow{\gamma(e_1,e_2)} x' \between y'}$$

any more, while communications among different parallel branches captured by communication merge still stand there. So, it is reasonable to assume that causality relations among different parallel branches are all communications among them. The communication between two parallel branches is defined as a communicating function between two communicating events $e_1, e_2 \in \mathbb{E}$, $\gamma(e_1, e_2) : \mathbb{E} \times \mathbb{E} \to \mathbb{E}$.

The communications among parallel branches are still defined by the communication operator |, which is expressed by four transition rules in Table 4.8. The whole parallelism semantics is modeled by the parallel operator ∥ and communication operator |, we denote the whole parallel operator as $\between$ (for the transition rules of $\between$, we omit them).

Note that the last transition rule for the parallel operator ∥ in Table 4.7 should be modified to the following one.

$$\frac{x \xrightarrow{e_1} x' \quad y \xrightarrow{e_2} y'}{x \parallel y \xrightarrow{\{e_1,e_2\}} x' \between y'}$$

By communication operator |, the causality relation among different parallel branches is structured (we will show the algebra laws on communication operator in the following). Now, let us consider conflicts in parallelism. The conflicts exist within the same parallel branches can be captured by + by a structured way, but, how to express conflicts among events in different parallel branches? The conflict relation is also a binary relation between two events $e_1, e_2 \in \mathbb{E}$, $\sharp(e_1, e_2) : \mathbb{E} \times \mathbb{E} \to \mathbb{E}$, and we know that $\sharp$ is irreflexive, symmetric and hereditary with respect to ·, that is, for all $e, e', e'' \in \mathbb{E}$, if $e \sharp e' \cdot e''$, then $e \sharp e''$ (see Definition 2.13).

These conflicts among different parallel branches must be eliminated to make the concurrent process structured. We are inspired by modeling of priority in ACP [4], the conflict elimination is also captured by two auxiliary operators, the unary conflict elimination operator Θ and the binary unless operator $\lhd$. The transition rules for Θ and $\lhd$ are expressed by ten transition rules in Table 4.9.

In four transition rules in Table 4.9, there is a new constant τ called silent step (see section 4.4), this makes the semantics of conflict elimination is really based on weakly true concurrency (see Definition 2.18 and Definition 2.22), and we should move it to section 4.4. But the movement would make $APTC$ incomplete (conflicts among different

Table 4.9   Transition rules of conflict elimination.

$$\frac{x \xrightarrow{e_1} \surd \quad (\sharp(e_1, e_2))}{\Theta(x) \xrightarrow{e_1} \surd} \qquad \frac{x \xrightarrow{e_2} \surd \quad (\sharp(e_1, e_2))}{\Theta(x) \xrightarrow{e_2} \surd}$$

$$\frac{x \xrightarrow{e_1} x' \quad (\sharp(e_1, e_2))}{\Theta(x) \xrightarrow{e_1} \Theta(x')} \qquad \frac{x \xrightarrow{e_2} x' \quad (\sharp(e_1, e_2))}{\Theta(x) \xrightarrow{e_2} \Theta(x')}$$

$$\frac{x \xrightarrow{e_1} \surd \quad y \nrightarrow^{e_2} \quad (\sharp(e_1, e_2))}{x \vartriangleleft y \xrightarrow{\tau} \surd} \qquad \frac{x \xrightarrow{e_1} x' \quad y \nrightarrow^{e_2} \quad (\sharp(e_1, e_2))}{x \vartriangleleft y \xrightarrow{\tau} x'}$$

$$\frac{x \xrightarrow{e_1} \surd \quad y \nrightarrow^{e_3} \quad (\sharp(e_1, e_2), e_2 \leq e_3)}{x \vartriangleleft y \xrightarrow{\tau} \surd} \qquad \frac{x \xrightarrow{e_1} x' \quad y \nrightarrow^{e_3} \quad (\sharp(e_1, e_2), e_2 \leq e_3)}{x \vartriangleleft y \xrightarrow{\tau} x'}$$

$$\frac{x \xrightarrow{e_3} \surd \quad y \nrightarrow^{e_2} \quad (\sharp(e_1, e_2), e_1 \leq e_3)}{x \vartriangleleft y \xrightarrow{\tau} \surd} \qquad \frac{x \xrightarrow{e_3} x' \quad y \nrightarrow^{e_2} \quad (\sharp(e_1, e_2), e_1 \leq e_3)}{x \vartriangleleft y \xrightarrow{\tau} x'}$$

parallel branches cannot be expressed), let us forget this regret and just remember that τ can be eliminated, without anything on weakly true concurrency.

Ok, causality relations and conflict relations among events in different parallel branches are structured. In the following, we prove the congruence theorem.

Theorem 4.21 (Congruence theorem of $APTC$). *Truly concurrent bisimulation equivalences $\sim_p$, $\sim_s$, $\sim_{hp}$ and $\sim_{hhp}$ are all congruences with respect to $APTC$.*

Proof. (1) Case pomset bisimulation equivalence $\sim_p$.

- Case parallel operator $\parallel$. Let x_1, x_2 and y_1, y_2 be $APTC$ processes, and $x_1 \sim_p y_1$, $x_2 \sim_p y_2$, it is sufficient to prove that $x_1 \parallel x_2 \sim_p y_1 \parallel y_2$.

  By the definition of pomset bisimulation $\sim_p$ (Definition 2.17), $x_1 \sim_p y_1$ means that

  $$x_1 \xrightarrow{X_1} x_1' \quad y_1 \xrightarrow{Y_1} y_1'$$

  with $X_1 \subseteq x_1$, $Y_1 \subseteq y_1$, $X_1 \sim Y_1$ and $x_1' \sim_p y_1'$. The meaning of $x_2 \sim_p y_2$ is similar.

  By the pomset transition rules for parallel operator $\parallel$ in Table 4.7, we can get

  $$x_1 \parallel x_2 \xrightarrow{\{X_1, X_2\}} \surd \quad y_1 \parallel y_2 \xrightarrow{\{Y_1, Y_2\}} \surd$$

  with $X_1 \subseteq x_1$, $Y_1 \subseteq y_1$, $X_2 \subseteq x_2$, $Y_2 \subseteq y_2$, $X_1 \sim Y_1$ and $X_2 \sim Y_2$, so, we get $x_1 \parallel x_2 \sim_p y_1 \parallel y_2$, as desired.

  Or, we can get

  $$x_1 \parallel x_2 \xrightarrow{\{X_1, X_2\}} x_1' \quad y_1 \parallel y_2 \xrightarrow{\{Y_1, Y_2\}} y_1'$$

with $X_1 \subseteq x_1$, $Y_1 \subseteq y_1$, $X_2 \subseteq x_2$, $Y_2 \subseteq y_2$, $X_1 \sim Y_1$, $X_2 \sim Y_2$, and $x_1' \sim_p y_1'$, so, we get $x_1 \parallel x_2 \sim_p y_1 \parallel y_2$, as desired.

Or, we can get

$$x_1 \parallel x_2 \xrightarrow{\{X_1, X_2\}} x_2' \quad y_1 \parallel y_2 \xrightarrow{\{Y_1, Y_2\}} y_2'$$

with $X_1 \subseteq x_1$, $Y_1 \subseteq y_1$, $X_2 \subseteq x_2$, $Y_2 \subseteq y_2$, $X_1 \sim Y_1$, $X_2 \sim Y_2$, and $x_2' \sim_p y_2'$, so, we get $x_1 \parallel x_2 \sim_p y_1 \parallel y_2$, as desired.

Or, we can get

$$x_1 \parallel x_2 \xrightarrow{\{X_1, X_2\}} x_1' \between x_2' \quad y_1 \parallel y_2 \xrightarrow{\{Y_1, Y_2\}} y_1' \between y_2'$$

with $X_1 \subseteq x_1$, $Y_1 \subseteq y_1$, $X_2 \subseteq x_2$, $Y_2 \subseteq y_2$, $X_1 \sim Y_1$, $X_2 \sim Y_2$, $x_1' \sim_p y_1'$ and $x_2' \sim_p y_2'$, and also the assumption $x_1' \between x_2' \sim_p y_1' \between y_2'$, so, we get $x_1 \parallel x_2 \sim_p y_1 \parallel y_2$, as desired.

- Case communication operator $|$. It can be proved similarly to the case of parallel operator $\parallel$, we omit it. Note that, a communication is defined between two single communicating events.
- Case conflict elimination operator Θ. It can be proved similarly to the above cases, we omit it. Note that the conflict elimination operator Θ is a unary operator.
- Case unless operator $\triangleleft$. It can be proved similarly to the case of parallel operator $\parallel$, we omit it. Note that, a conflict relation is defined between two single events.

(2) The cases of step bisimulation $\sim_s$, hp-bisimulation $\sim_{hp}$ and hhp-bisimulation $\sim_{hhp}$ can be proven similarly, we omit them. □

So, we design the axioms of parallelism in Table 4.10, including algebraic laws for parallel operator $\parallel$, communication operator $|$, conflict elimination operator Θ and unless operator $\triangleleft$, and also the whole parallel operator $\between$. Since the communication between two communicating events in different parallel branches may cause deadlock (a state of inactivity), which is caused by mismatch of two communicating events or the imperfectness of the communication channel. We introduce a new constant δ to denote the deadlock, and let the atomic event $e \in \mathbb{E} \cup \{\delta\}$.

We explain the intuitions of the axioms of parallelism in Table 4.10 in the following. The axiom $A6$ says that the deadlock δ is redundant in the process term $t + \delta$. $A7$ says that the deadlock blocks all behaviors of the process term $\delta \cdot t$.

The axiom $P1$ is the definition of the whole parallelism $\between$, which says that $s \between t$ either is the form of $s \parallel t$ or $s \mid t$. $P2$ says that $\parallel$ satisfies commutative law, while $P3$ says that $\parallel$ satisfies associativity. $P4$, $P5$ and $P6$ are the defining axioms of $\parallel$, say the $s \parallel t$ executes s and t concurrently. $P7$ and $P8$ are the right and left distributivity of $\parallel$ to $+$. $P9$ and $P10$ say that both $\delta \parallel t$ and $t \parallel \delta$ all block any event.

$C11$, $C12$, $C13$, and $C14$ are the defining axioms of the communication operator $|$ which say that $s \mid t$ makes a communication between s and t. $C15$ and $C16$ are the right and left distributivity of $|$ to $+$. $C17$ and $C18$ say that both $\delta \mid t$ and $t \mid \delta$ all block any event.

$CE19$ and $CE20$ say that the conflict elimination operator Θ leaves atomic events and the deadlock unchanged. $CE21 - CE24$ are the functions of Θ acting on the operators $+$, $\cdot$,

Table 4.10 Axioms of parallelism.

No.	Axiom
A6	$x + \delta = x$
A7	$\delta \cdot x = \delta$
P1	$x \between y = x \parallel y + x \mid y$
P2	$x \parallel y = y \parallel x$
P3	$(x \parallel y) \parallel z = x \parallel (y \parallel z)$
P4	$e_1 \parallel (e_2 \cdot y) = (e_1 \parallel e_2) \cdot y$
P5	$(e_1 \cdot x) \parallel e_2 = (e_1 \parallel e_2) \cdot x$
P6	$(e_1 \cdot x) \parallel (e_2 \cdot y) = (e_1 \parallel e_2) \cdot (x \between y)$
P7	$(x + y) \parallel z = (x \parallel z) + (y \parallel z)$
P8	$x \parallel (y + z) = (x \parallel y) + (x \parallel z)$
P9	$\delta \parallel x = \delta$
P10	$x \parallel \delta = \delta$
C11	$e_1 \mid e_2 = \gamma(e_1, e_2)$
C12	$e_1 \mid (e_2 \cdot y) = \gamma(e_1, e_2) \cdot y$
C13	$(e_1 \cdot x) \mid e_2 = \gamma(e_1, e_2) \cdot x$
C14	$(e_1 \cdot x) \mid (e_2 \cdot y) = \gamma(e_1, e_2) \cdot (x \between y)$
C15	$(x + y) \mid z = (x \mid z) + (y \mid z)$
C16	$x \mid (y + z) = (x \mid y) + (x \mid z)$
C17	$\delta \mid x = \delta$
C18	$x \mid \delta = \delta$
CE19	$\Theta(e) = e$
CE20	$\Theta(\delta) = \delta$
CE21	$\Theta(x + y) = \Theta(x) + \Theta(y)$
CE22	$\Theta(x \cdot y) = \Theta(x) \cdot \Theta(y)$
CE23	$\Theta(x \parallel y) = ((\Theta(x) \triangleleft y) \parallel y) + ((\Theta(y) \triangleleft x) \parallel x)$
CE24	$\Theta(x \mid y) = ((\Theta(x) \triangleleft y) \mid y) + ((\Theta(y) \triangleleft x) \mid x)$
U25	$(\sharp(e_1, e_2)) \quad e_1 \triangleleft e_2 = \tau$
U26	$(\sharp(e_1, e_2), e_2 \leq e_3) \quad e_1 \triangleleft e_3 = \tau$
U27	$(\sharp(e_1, e_2), e_2 \leq e_3) \quad e_3 \triangleleft e_1 = \tau$
U28	$e \triangleleft \delta = e$
U29	$\delta \triangleleft e = \delta$
U30	$(x + y) \triangleleft z = (x \triangleleft z) + (y \triangleleft z)$
U31	$(x \cdot y) \triangleleft z = (x \triangleleft z) \cdot (y \triangleleft z)$
U32	$(x \parallel y) \triangleleft z = (x \triangleleft z) \parallel (y \triangleleft z)$
U33	$(x \mid y) \triangleleft z = (x \triangleleft z) \mid (y \triangleleft z)$
U34	$x \triangleleft (y + z) = (x \triangleleft y) \triangleleft z$
U35	$x \triangleleft (y \cdot z) = (x \triangleleft y) \triangleleft z$
U36	$x \triangleleft (y \parallel z) = (x \triangleleft y) \triangleleft z$
U37	$x \triangleleft (y \mid z) = (x \triangleleft y) \triangleleft z$

$\parallel$ and $\mid$. $U25$, $U26$, and $U27$ are the defining laws of the unless operator $\vartriangleleft$, in $U25$, $U26$, and $U27$, there is a new constant τ, the silent step, we will discuss τ in details in section 4.4, in these two axioms, we just need to remember that τ really keeps silent. $U28$ says that the deadlock δ cannot block any event in the process term $e \vartriangleleft \delta$, while $U29$ says that $\delta \vartriangleleft e$ does not exhibit any behavior. $U30 - U37$ are the disguised right and left distributivity of $\vartriangleleft$ to the operators $+$, $\cdot$, $\parallel$, and $\mid$.

4.2.3 Properties of parallelism

Based on the definition of basic terms for $APTC$ (see Definition 4.19) and axioms of parallelism (see Table 4.10), we can prove the elimination theorem of parallelism.

Theorem 4.22 (Elimination theorem of parallelism). *Let p be a closed $APTC$ term. Then there is a basic $APTC$ term q such that $APTC \vdash p = q$.*

Proof. (1) Firstly, suppose that the following ordering on the signature of $APTC$ is defined: $\parallel > \cdot > +$ and the symbol $\cdot$ is given the lexicographical status for the first argument, then for each rewrite rule $p \to q$ in Table 4.11 relation $p >_{lpo} q$ can easily be proved. We obtain that the term rewrite system shown in Table 4.11 is strongly normalizing, for it has finitely many rewriting rules, and $>$ is a well-founded ordering on the signature of $APTC$, and if $s >_{lpo} t$, for each rewriting rule $s \to t$ is in Table 4.11 (see Theorem 2.12).

(2) Then we prove that the normal forms of closed $APTC$ terms are basic $APTC$ terms.

Suppose that p is a normal form of some closed $APTC$ term and suppose that p is not a basic $APTC$ term. Let p' denote the smallest sub-term of p which is not a basic $APTC$ term. It implies that each sub-term of p' is a basic $APTC$ term. Then we prove that p is not a term in normal form. It is sufficient to induct on the structure of p':

- Case $p' \equiv e, e \in \mathbb{E}$. p' is a basic $APTC$ term, which contradicts the assumption that p' is not a basic $APTC$ term, so this case should not occur.
- Case $p' \equiv p_1 \cdot p_2$. By induction on the structure of the basic $APTC$ term p_1:
  - Subcase $p_1 \in \mathbb{E}$. p' would be a basic $APTC$ term, which contradicts the assumption that p' is not a basic $APTC$ term;
  - Subcase $p_1 \equiv e \cdot p_1'$. $RA5$ rewriting rule in Table 4.2 can be applied. So p is not a normal form;
  - Subcase $p_1 \equiv p_1' + p_1''$. $RA4$ rewriting rule in Table 4.2 can be applied. So p is not a normal form;
  - Subcase $p_1 \equiv p_1' \parallel p_1''$. p' would be a basic $APTC$ term, which contradicts the assumption that p' is not a basic $APTC$ term;
  - Subcase $p_1 \equiv p_1' \mid p_1''$. $RC11$ rewrite rule in Table 4.11 can be applied. So p is not a normal form;
  - Subcase $p_1 \equiv \Theta(p_1')$. $RCE19$ and $RCE20$ rewrite rules in Table 4.11 can be applied. So p is not a normal form.
- Case $p' \equiv p_1 + p_2$. By induction on the structure of the basic $APTC$ terms both p_1 and p_2, all subcases will lead to that p' would be a basic $APTC$ term, which contradicts the assumption that p' is not a basic $APTC$ term.

Table 4.11   Term rewrite system of $APTC$.

No.	Rewriting rule
$RA6$	$x + \delta \to x$
$RA7$	$\delta \cdot x \to \delta$
$RP1$	$x \between y \to x \parallel y + x \mid y$
$RP2$	$x \parallel y \to y \parallel x$
$RP3$	$(x \parallel y) \parallel z \to x \parallel (y \parallel z)$
$RP4$	$e_1 \parallel (e_2 \cdot y) \to (e_1 \parallel e_2) \cdot y$
$RP5$	$(e_1 \cdot x) \parallel e_2 \to (e_1 \parallel e_2) \cdot x$
$RP6$	$(e_1 \cdot x) \parallel (e_2 \cdot y) \to (e_1 \parallel e_2) \cdot (x \between y)$
$RP7$	$(x + y) \parallel z \to (x \parallel z) + (y \parallel z)$
$RP8$	$x \parallel (y + z) \to (x \parallel y) + (x \parallel z)$
$RP9$	$\delta \parallel x \to \delta$
$RP10$	$x \parallel \delta \to \delta$
$RC11$	$e_1 \mid e_2 \to \gamma(e_1, e_2)$
$RC12$	$e_1 \mid (e_2 \cdot y) \to \gamma(e_1, e_2) \cdot y$
$RC13$	$(e_1 \cdot x) \mid e_2 \to \gamma(e_1, e_2) \cdot x$
$RC14$	$(e_1 \cdot x) \mid (e_2 \cdot y) \to \gamma(e_1, e_2) \cdot (x \between y)$
$RC15$	$(x + y) \mid z \to (x \mid z) + (y \mid z)$
$RC16$	$x \mid (y + z) \to (x \mid y) + (x \mid z)$
$RC17$	$\delta \mid x \to \delta$
$RC18$	$x \mid \delta \to \delta$
$RCE19$	$\Theta(e) \to e$
$RCE20$	$\Theta(\delta) \to \delta$
$RCE21$	$\Theta(x + y) \to \Theta(x) + \Theta(y)$
$RCE22$	$\Theta(x \cdot y) \to \Theta(x) \cdot \Theta(y)$
$RCE23$	$\Theta(x \parallel y) \to ((\Theta(x) \triangleleft y) \parallel y) + ((\Theta(y) \triangleleft x) \parallel x)$
$RCE24$	$\Theta(x \mid y) \to ((\Theta(x) \triangleleft y) \mid y) + ((\Theta(y) \triangleleft x) \mid x)$
$RU25$	$(\sharp(e_1, e_2))\ \ e_1 \triangleleft e_2 \to \tau$
$RU26$	$(\sharp(e_1, e_2), e_2 \cdot e_3)\ \ e_1 \triangleleft e_3 \to \tau$
$RU27$	$(\sharp(e_1, e_2), e_2 \cdot e_3)\ \ e3 \triangleleft e_1 \to \tau$
$RU28$	$e \triangleleft \delta \to e$
$RU29$	$\delta \triangleleft e \to \delta$
$RU30$	$(x + y) \triangleleft z \to (x \triangleleft z) + (y \triangleleft z)$
$RU31$	$(x \cdot y) \triangleleft z \to (x \triangleleft z) \cdot (y \triangleleft z)$
$RU32$	$(x \parallel y) \triangleleft z \to (x \triangleleft z) \parallel (y \triangleleft z)$
$RU33$	$(x \mid y) \triangleleft z \to (x \triangleleft z) \mid (y \triangleleft z)$
$RU34$	$x \triangleleft (y + z) \to (x \triangleleft y) \triangleleft z$
$RU35$	$x \triangleleft (y \cdot z) \to (x \triangleleft y) \triangleleft z$
$RU36$	$x \triangleleft (y \parallel z) \to (x \triangleleft y) \triangleleft z$
$RU37$	$x \triangleleft (y \mid z) \to (x \triangleleft y) \triangleleft z$

- Case $p' \equiv p_1 \parallel p_2$. By induction on the structure of the basic $APTC$ terms both p_1 and p_2, all subcases will lead to that p' would be a basic $APTC$ term, which contradicts the assumption that p' is not a basic $APTC$ term.
- Case $p' \equiv p_1 \mid p_2$. By induction on the structure of the basic $APTC$ terms both p_1 and p_2, all subcases will lead to that p' would be a basic $APTC$ term, which contradicts the assumption that p' is not a basic $APTC$ term.
- Case $p' \equiv \Theta(p_1)$. By induction on the structure of the basic $APTC$ term p_1, $RCE19 - RCE24$ rewrite rules in Table 4.11 can be applied. So p is not a normal form.
- Case $p' \equiv p_1 \triangleleft p_2$. By induction on the structure of the basic $APTC$ terms both p_1 and p_2, all subcases will lead to that p' would be a basic $APTC$ term, which contradicts the assumption that p' is not a basic $APTC$ term.   □

4.2.4 Structured operational semantics of parallelism

It is quite a challenge to prove the algebraic laws in Table 4.10 is sound/complete or un-sound/incomplete modulo truly concurrent behavioral equivalences (pomset bisimulation equivalence, step bisimulation equivalence, hp-bisimulation equivalence and hhp-bisimulation equivalence), in this subsection, we try to do these.

Theorem 4.23 (Generalization of the algebra for parallelism with respect to $BATC$). *The algebra for parallelism is a generalization of $BATC$.*

Proof. It follows from the following three facts.

1. The transition rules of $BATC$ in section 4.1 are all source-dependent;
2. The sources of the transition rules for the algebra for parallelism contain an occurrence of $\between$, or $\parallel$, or $\mid$, or Θ, or $\triangleleft$;
3. The transition rules of $APTC$ are all source-dependent.

So, the algebra for parallelism is a generalization of $BATC$, that is, $BATC$ is an embedding of the algebra for parallelism, as desired.   □

Theorem 4.24 (Soundness of parallelism modulo step bisimulation equivalence). *Let x and y be $APTC$ terms. If $APTC \vdash x = y$, then $x \sim_s y$.*

Proof. Since step bisimulation $\sim_s$ is both an equivalent and a congruent relation with respect to the operators $\between$, $\parallel$, $\mid$, Θ and $\triangleleft$, we only need to check if each axiom in Table 4.10 is sound modulo step bisimulation equivalence.

Though transition rules in Table 4.7, 4.8, and 4.9 are defined in the flavor of single event, they can be modified into a step (a set of events within which each event is pairwise concurrent), we omit them. If we treat a single event as a step containing just one event, the proof of this soundness theorem does not exist any problem, so we use this way and still use the transition rules in Table 4.7, 4.8, and 4.9.

We omit the defining axioms, including axioms $P1$, $C11$, $CE19$, $CE20$, $U25 - U27$ (the soundness of $U25$ and $U27$ is remained to section 4.4); we also omit the trivial axioms related to δ, including axioms $A6$, $A7$, $P9$, $P10$, $C17$, $C18$, $U28$, and $U29$; in the following, we only prove the soundness of the non-trivial axioms, including axioms $P2 - P8$, $C12 - C16$, $CE21 - CE24$ and $U30 - U37$.

- **Axiom** $P2$. Let p, q be $APTC$ processes, and $p \parallel q = q \parallel p$, it is sufficient to prove that $p \parallel q \sim_s q \parallel p$. By the transition rules for operator $\parallel$ in Table 4.7, we get

$$\frac{p \xrightarrow{e_1} \surd \quad q \xrightarrow{e_2} \surd}{p \parallel q \xrightarrow{\{e_1, e_2\}} \surd} \qquad \frac{p \xrightarrow{e_1} \surd \quad q \xrightarrow{e_2} \surd}{q \parallel p \xrightarrow{\{e_1, e_2\}} \surd}$$

$$\frac{p \xrightarrow{e_1} p' \quad q \xrightarrow{e_2} \surd}{p \parallel q \xrightarrow{\{e_1, e_2\}} p'} \qquad \frac{p \xrightarrow{e_1} p' \quad q \xrightarrow{e_2} \surd}{q \parallel p \xrightarrow{\{e_1, e_2\}} p'}$$

$$\frac{p \xrightarrow{e_1} \surd \quad q \xrightarrow{e_2} q'}{p \parallel q \xrightarrow{\{e_1, e_2\}} q'} \qquad \frac{p \xrightarrow{e_1} \surd \quad q \xrightarrow{e_2} q'}{q \parallel p \xrightarrow{\{e_1, e_2\}} q'}$$

$$\frac{p \xrightarrow{e_1} p' \quad q \xrightarrow{e_2} q'}{p \parallel q \xrightarrow{\{e_1, e_2\}} p' \between q'} \qquad \frac{p \xrightarrow{e_1} p' \quad q \xrightarrow{e_2} q'}{q \parallel p \xrightarrow{\{e_1, e_2\}} q' \between p'}$$

So, with the assumption $p' \between q' = q' \between p'$, $p \parallel q \sim_s q \parallel p$, as desired.

- **Axiom** $P3$. Let p, q, r be $APTC$ processes, and $(p \parallel q) \parallel r = p \parallel (q \parallel r)$, it is sufficient to prove that $(p \parallel q) \parallel r \sim_s p \parallel (q \parallel r)$. By the transition rules for operator $\parallel$ in Table 4.7, we get

$$\frac{p \xrightarrow{e_1} \surd \quad q \xrightarrow{e_2} \surd \quad r \xrightarrow{e_3} \surd}{(p \parallel q) \parallel r \xrightarrow{\{e_1, e_2, e_3\}} \surd} \qquad \frac{p \xrightarrow{e_1} \surd \quad q \xrightarrow{e_2} \surd \quad r \xrightarrow{e_3} \surd}{p \parallel (q \parallel r) \xrightarrow{\{e_1, e_2, e_3\}} \surd}$$

$$\frac{p \xrightarrow{e_1} p' \quad q \xrightarrow{e_2} \surd \quad r \xrightarrow{e_3} \surd}{(p \parallel q) \parallel r \xrightarrow{\{e_1, e_2, e_3\}} p'} \qquad \frac{p \xrightarrow{e_1} p' \quad q \xrightarrow{e_2} \surd \quad r \xrightarrow{e_3} \surd}{p \parallel (q \parallel r) \xrightarrow{\{e_1, e_2, e_3\}} p'}$$

There are also two cases that two process terms successfully terminate, we omit them.

$$\frac{p \xrightarrow{e_1} p' \quad q \xrightarrow{e_2} q' \quad r \xrightarrow{e_3} \surd}{(p \parallel q) \parallel r \xrightarrow{\{e_1, e_2, e_3\}} p' \between q'} \qquad \frac{p \xrightarrow{e_1} p' \quad q \xrightarrow{e_2} q' \quad r \xrightarrow{e_3} \surd}{p \parallel (q \parallel r) \xrightarrow{\{e_1, e_2, e_3\}} p' \between q'}$$

There are also other cases that just one process term successfully terminate, we also omit them.

$$\frac{p \xrightarrow{e_1} p' \quad q \xrightarrow{e_2} q' \quad r \xrightarrow{e_3} r'}{(p \parallel q) \parallel r' \xrightarrow{\{e_1, e_2, e_3\}} (p' \between q') \between r'} \qquad \frac{p \xrightarrow{e_1} p' \quad q \xrightarrow{e_2} q' \quad r \xrightarrow{e_3} r'}{p \parallel (q \parallel r) \xrightarrow{\{e_1, e_2, e_3\}} p' \between (q' \between r')}$$

So, with the assumption $(p' \between q') \between r' = p' \between (q' \between r')$, $(p \parallel q) \parallel r \sim_s p \parallel (q \parallel r)$, as desired.

- **Axiom** $P4$. Let q be an $APTC$ process, and $e_1 \parallel (e_2 \cdot q) = (e_1 \parallel e_2) \cdot q$, it is sufficient to prove that $e_1 \parallel (e_2 \cdot q) \sim_s (e_1 \parallel e_2) \cdot q$. By the transition rules for operator $\parallel$ in Table 4.7, we get

$$\frac{e_1 \xrightarrow{e_1} \surd \quad e_2 \cdot q \xrightarrow{e_2} q}{e_1 \parallel (e_2 \cdot q) \xrightarrow{\{e_1,e_2\}} q}$$

$$\frac{e_1 \xrightarrow{e_1} \surd \quad e_2 \xrightarrow{e_2} \surd}{(e_1 \parallel e_2) \cdot q \xrightarrow{\{e_1,e_2\}} q}$$

So, $e_1 \parallel (e_2 \cdot q) \sim_s (e_1 \parallel e_2) \cdot q$, as desired.

- **Axiom** $P5$. Let p be an $APTC$ process, and $(e_1 \cdot p) \parallel e_2 = (e_1 \parallel e_2) \cdot p$, it is sufficient to prove that $(e_1 \cdot p) \parallel e_2 \sim_s (e_1 \parallel e_2) \cdot p$. By the transition rules for operator $\parallel$ in Table 4.7, we get

$$\frac{e_1 \cdot p \xrightarrow{e_1} p \quad e_2 \xrightarrow{e_2} \surd}{(e_1 \cdot p) \parallel e_2 \xrightarrow{\{e_1,e_2\}} p}$$

$$\frac{e_1 \xrightarrow{e_1} \surd \quad e_2 \xrightarrow{e_2} \surd}{(e_1 \parallel e_2) \cdot p \xrightarrow{\{e_1,e_2\}} p}$$

So, $(e_1 \cdot p) \parallel e_2 \sim_s (e_1 \parallel e_2) \cdot p$, as desired.

- **Axiom** $P6$. Let p, q be $APTC$ processes, and $(e_1 \cdot p) \parallel (e_2 \cdot q) = (e_1 \parallel e_2) \cdot (p \between q)$, it is sufficient to prove that $(e_1 \cdot p) \parallel (e_2 \cdot q) \sim_s (e_1 \parallel e_2) \cdot (p \between q)$. By the transition rules for operator $\parallel$ in Table 4.7, we get

$$\frac{e_1 \cdot p \xrightarrow{e_1} p \quad e_2 \cdot q \xrightarrow{e_2} q}{(e_1 \cdot p) \parallel (e_2 \cdot q) \xrightarrow{\{e_1,e_2\}} p \between q}$$

$$\frac{e_1 \xrightarrow{e_1} \surd \quad e_2 \xrightarrow{e_2} \surd}{(e_1 \parallel e_2) \cdot (p \between q) \xrightarrow{\{e_1,e_2\}} p \between q}$$

So, $(e_1 \cdot p) \parallel (e_2 \cdot q) \sim_s (e_1 \parallel e_2) \cdot (p \between q)$, as desired.

- **Axiom** $P7$. Let p, q, r be $APTC$ processes, and $(p+q) \parallel r = (p \parallel r) + (q \parallel r)$, it is sufficient to prove that $(p + q) \parallel r \sim_s (p \parallel r) + (q \parallel r)$. By the transition rules for operators $+$ and $\parallel$ in Table 4.5 and 4.7, we get

$$\frac{p \xrightarrow{e_1} \surd \quad r \xrightarrow{e_2} \surd}{(p+q) \parallel r \xrightarrow{\{e_1,e_2\}} \surd} \qquad \frac{p \xrightarrow{e_1} \surd \quad r \xrightarrow{e_2} \surd}{(p \parallel r) + (q \parallel r) \xrightarrow{\{e_1,e_2\}} \surd}$$

$$\frac{q \xrightarrow{e_1} \checkmark \quad r \xrightarrow{e_2} \checkmark}{(p+q) \parallel r \xrightarrow{\{e_1,e_2\}} \checkmark} \qquad \frac{q \xrightarrow{e_1} \checkmark \quad r \xrightarrow{e_2} \checkmark}{(p \parallel r)+(q \parallel r) \xrightarrow{\{e_1,e_2\}} \checkmark}$$

$$\frac{p \xrightarrow{e_1} p' \quad r \xrightarrow{e_2} \checkmark}{(p+q) \parallel r \xrightarrow{\{e_1,e_2\}} p'} \qquad \frac{p \xrightarrow{e_1} p' \quad r \xrightarrow{e_2} \checkmark}{(p \parallel r)+(q \parallel r) \xrightarrow{\{e_1,e_2\}} p'}$$

$$\frac{q \xrightarrow{e_1} q' \quad r \xrightarrow{e_2} \checkmark}{(p+q) \parallel r \xrightarrow{\{e_1,e_2\}} q'} \qquad \frac{q \xrightarrow{e_1} q' \quad r \xrightarrow{e_2} \checkmark}{(p \parallel r)+(q \parallel r) \xrightarrow{\{e_1,e_2\}} q'}$$

$$\frac{p \xrightarrow{e_1} \checkmark \quad r \xrightarrow{e_2} r'}{(p+q) \parallel r \xrightarrow{\{e_1,e_2\}} r'} \qquad \frac{p \xrightarrow{e_1} \checkmark \quad r \xrightarrow{e_2} r'}{(p \parallel r)+(q \parallel r) \xrightarrow{\{e_1,e_2\}} r'}$$

$$\frac{q \xrightarrow{e_1} \checkmark \quad r \xrightarrow{e_2} r'}{(p+q) \parallel r \xrightarrow{\{e_1,e_2\}} r'} \qquad \frac{q \xrightarrow{e_1} \checkmark \quad r \xrightarrow{e_2} r'}{(p \parallel r)+(q \parallel r) \xrightarrow{\{e_1,e_2\}} r'}$$

$$\frac{p \xrightarrow{e_1} p' \quad r \xrightarrow{e_2} r'}{(p+q) \parallel r \xrightarrow{\{e_1,e_2\}} p' \between r'} \qquad \frac{p \xrightarrow{e_1} p' \quad r \xrightarrow{e_2} r'}{(p \parallel r)+(q \parallel r) \xrightarrow{\{e_1,e_2\}} p' \between r'}$$

$$\frac{q \xrightarrow{e_1} q' \quad r \xrightarrow{e_2} r'}{(p+q) \parallel r \xrightarrow{\{e_1,e_2\}} q' \between r'} \qquad \frac{q \xrightarrow{e_1} q' \quad r \xrightarrow{e_2} r'}{(p \parallel r)+(q \parallel r) \xrightarrow{\{e_1,e_2\}} q' \between r'}$$

So, $(p+q) \parallel r \sim_s (p \parallel r)+(q \parallel r)$, as desired.

- **Axiom** $P8$. Let p,q,r be $APTC$ processes, and $p \parallel (q+r) = (p \parallel q)+(p \parallel r)$, it is sufficient to prove that $p \parallel (q+r) \sim_s (p \parallel q)+(p \parallel r)$. By the transition rules for operators $+$ and $\parallel$ in Table 4.5 and 4.7, we get

$$\frac{p \xrightarrow{e_1} \checkmark \quad q \xrightarrow{e_2} \checkmark}{p \parallel (q+r) \xrightarrow{\{e_1,e_2\}} \checkmark} \qquad \frac{p \xrightarrow{e_1} \checkmark \quad q \xrightarrow{e_2} \checkmark}{(p \parallel q)+(p \parallel r) \xrightarrow{\{e_1,e_2\}} \checkmark}$$

$$\frac{p \xrightarrow{e_1} \checkmark \quad r \xrightarrow{e_2} \checkmark}{p \parallel (q+r) \xrightarrow{\{e_1,e_2\}} \checkmark} \qquad \frac{p \xrightarrow{e_1} \checkmark \quad r \xrightarrow{e_2} \checkmark}{(p \parallel q)+(p \parallel r) \xrightarrow{\{e_1,e_2\}} \checkmark}$$

$$\frac{p \xrightarrow{e_1} p' \quad q \xrightarrow{e_2} \checkmark}{p \parallel (q+r) \xrightarrow{\{e_1,e_2\}} p'} \qquad \frac{p \xrightarrow{e_1} p' \quad q \xrightarrow{e_2} \checkmark}{(p \parallel q)+(p \parallel r) \xrightarrow{\{e_1,e_2\}} p'}$$

$$\frac{p \xrightarrow{e_1} p' \quad r \xrightarrow{e_2} \checkmark}{p \parallel (q+r) \xrightarrow{\{e_1,e_2\}} p'} \qquad \frac{p \xrightarrow{e_1} p' \quad r \xrightarrow{e_2} \checkmark}{(p \parallel q)+(p \parallel r) \xrightarrow{\{e_1,e_2\}} p'}$$

$$\frac{p \xrightarrow{e_1} \surd \quad q \xrightarrow{e_2} q'}{p \parallel (q + r) \xrightarrow{\{e_1, e_2\}} q'} \qquad \frac{p \xrightarrow{e_1} \surd \quad q \xrightarrow{e_2} q'}{(p \parallel q) + (p \parallel r) \xrightarrow{\{e_1, e_2\}} q'}$$

$$\frac{p \xrightarrow{e_1} \surd \quad r \xrightarrow{e_2} r'}{p \parallel (q + r) \xrightarrow{\{e_1, e_2\}} r'} \qquad \frac{p \xrightarrow{e_1} \surd \quad r \xrightarrow{e_2} r'}{(p \parallel q) + (p \parallel r) \xrightarrow{\{e_1, e_2\}} r'}$$

$$\frac{p \xrightarrow{e_1} p' \quad q \xrightarrow{e_2} q'}{p \parallel (q + r) \xrightarrow{\{e_1, e_2\}} p' \between q'} \qquad \frac{p \xrightarrow{e_1} p' \quad q \xrightarrow{e_2} q'}{(p \parallel q) + (p \parallel r) \xrightarrow{\{e_1, e_2\}} p' \between q'}$$

$$\frac{p \xrightarrow{e_1} p' \quad r \xrightarrow{e_2} r'}{p \parallel (q + r) \xrightarrow{\{e_1, e_2\}} p' \between r'} \qquad \frac{p \xrightarrow{e_1} p' \quad r \xrightarrow{e_2} r'}{(p \parallel q) + (p \parallel r) \xrightarrow{\{e_1, e_2\}} p' \between r'}$$

So, $p \parallel (q + r) \sim_s (p \parallel q) + (p \parallel r)$, as desired.

- **Axiom** $C12$. Let q be an $APTC$ process, and $e_1 \mid (e_2 \cdot q) = \gamma(e_1, e_2) \cdot q$, it is sufficient to prove that $e_1 \mid (e_2 \cdot q) \sim_s \gamma(e_1, e_2) \cdot q$. By the transition rules for operator $\mid$ in Table 4.8, we get

$$\frac{e_1 \xrightarrow{e_1} \surd \quad e_2 \cdot q \xrightarrow{e_2} q}{e_1 \mid (e_2 \cdot q) \xrightarrow{\gamma(e_1, e_2)} q}$$

$$\frac{e_1 \xrightarrow{e_1} \surd \quad e_2 \xrightarrow{e_2} \surd}{\gamma(e_1, e_2) \cdot q \xrightarrow{\gamma(e_1, e_2)} q}$$

So, $e_1 \mid (e_2 \cdot q) \sim_s \gamma(e_1, e_2) \cdot q$, as desired.

- **Axiom** $C13$. Let p be an $APTC$ process, and $(e_1 \cdot p) \mid e_2 = \gamma(e_1, e_2) \cdot p$, it is sufficient to prove that $(e_1 \cdot p) \mid e_2 \sim_s \gamma(e_1, e_2) \cdot p$. By the transition rules for operator $\mid$ in Table 4.8, we get

$$\frac{e_1 \cdot p \xrightarrow{e_1} p \quad e_2 \xrightarrow{e_2} \surd}{(e_1 \cdot p) \mid e_2 \xrightarrow{\gamma(e_1, e_2)} p}$$

$$\frac{e_1 \xrightarrow{e_1} \surd \quad e_2 \xrightarrow{e_2} \surd}{\gamma(e_1, e_2) \cdot p \xrightarrow{\gamma(e_1, e_2)} p}$$

So, $(e_1 \cdot p) \mid e_2 \sim_s \gamma(e_1, e_2) \cdot p$, as desired.

- **Axiom** $C14$. Let p, q be $APTC$ processes, and $(e_1 \cdot p) \mid (e_2 \cdot q) = \gamma(e_1, e_2) \cdot (p \between q)$, it is sufficient to prove that $(e_1 \cdot p) \mid (e_2 \cdot q) \sim_s \gamma(e_1, e_2) \cdot (p \between q)$. By the transition rules for operator $\mid$ in Table 4.8, we get

$$\frac{e_1 \cdot p \xrightarrow{e_1} p \quad e_2 \cdot q \xrightarrow{e_2} q}{(e_1 \cdot p) \mid (e_2 \cdot q) \xrightarrow{\gamma(e_1, e_2)} p \between q}$$

$$\frac{e_1 \xrightarrow{e_1} \surd \quad e_2 \xrightarrow{e_2} \surd}{\gamma(e_1,e_2) \cdot (p \between q) \xrightarrow{\gamma(e_1,e_2)} p \between q}$$

So, $(e_1 \cdot p) \mid (e_2 \cdot q) \sim_s \gamma(e_1,e_2) \cdot (p \between q)$, as desired.

- **Axiom** $C15$. Let p, q, r be $APTC$ processes, and $(p+q) \mid r = (p \mid r) + (q \mid r)$, it is sufficient to prove that $(p+q) \mid r \sim_s (p \mid r) + (q \mid r)$. By the transition rules for operators $+$ and $\mid$ in Table 4.5 and 4.8, we get

$$\frac{p \xrightarrow{e_1} \surd \quad r \xrightarrow{e_2} \surd}{(p+q) \mid r \xrightarrow{\gamma(e_1,e_2)} \surd} \qquad \frac{p \xrightarrow{e_1} \surd \quad r \xrightarrow{e_2} \surd}{(p \mid r) + (q \mid r) \xrightarrow{\gamma(e_1,e_2)} \surd}$$

$$\frac{q \xrightarrow{e_1} \surd \quad r \xrightarrow{e_2} \surd}{(p+q) \mid r \xrightarrow{\gamma(e_1,e_2)} \surd} \qquad \frac{q \xrightarrow{e_1} \surd \quad r \xrightarrow{e_2} \surd}{(p \mid r) + (q \mid r) \xrightarrow{\gamma(e_1,e_2)} \surd}$$

$$\frac{p \xrightarrow{e_1} p' \quad r \xrightarrow{e_2} \surd}{(p+q) \mid r \xrightarrow{\gamma(e_1,e_2)} p'} \qquad \frac{p \xrightarrow{e_1} p' \quad r \xrightarrow{e_2} \surd}{(p \mid r) + (q \mid r) \xrightarrow{\gamma(e_1,e_2)} p'}$$

$$\frac{q \xrightarrow{e_1} q' \quad r \xrightarrow{e_2} \surd}{(p+q) \mid r \xrightarrow{\gamma(e_1,e_2)} q'} \qquad \frac{q \xrightarrow{e_1} q' \quad r \xrightarrow{e_2} \surd}{(p \mid r) + (q \mid r) \xrightarrow{\gamma(e_1,e_2)} q'}$$

$$\frac{p \xrightarrow{e_1} \surd \quad r \xrightarrow{e_2} r'}{(p+q) \mid r \xrightarrow{\gamma(e_1,e_2)} r'} \qquad \frac{p \xrightarrow{e_1} \surd \quad r \xrightarrow{e_2} r'}{(p \mid r) + (q \mid r) \xrightarrow{\gamma(e_1,e_2)} r'}$$

$$\frac{q \xrightarrow{e_1} \surd \quad r \xrightarrow{e_2} r'}{(p+q) \mid r \xrightarrow{\gamma(e_1,e_2)} r'} \qquad \frac{q \xrightarrow{e_1} \surd \quad r \xrightarrow{e_2} r'}{(p \mid r) + (q \mid r) \xrightarrow{\gamma(e_1,e_2)} r'}$$

$$\frac{p \xrightarrow{e_1} p' \quad r \xrightarrow{e_2} r'}{(p+q) \mid r \xrightarrow{\gamma(e_1,e_2)} p' \between r'} \qquad \frac{p \xrightarrow{e_1} p' \quad r \xrightarrow{e_2} r'}{(p \mid r) + (q \mid r) \xrightarrow{\gamma(e_1,e_2)} p' \between r'}$$

$$\frac{q \xrightarrow{e_1} q' \quad r \xrightarrow{e_2} r'}{(p+q) \mid r \xrightarrow{\gamma(e_1,e_2)} q' \between r'} \qquad \frac{q \xrightarrow{e_1} q' \quad r \xrightarrow{e_2} r'}{(p \mid r) + (q \mid r) \xrightarrow{\gamma(e_1,e_2)} q' \between r'}$$

So, $(p+q) \mid r \sim_s (p \mid r) + (q \mid r)$, as desired.

- **Axiom** $C16$. Let p, q, r be $APTC$ processes, and $p \mid (q+r) = (p \mid q) + (p \mid r)$, it is sufficient to prove that $p \mid (q+r) \sim_s (p \mid q) + (p \mid r)$. By the transition rules for operators $+$ and $\mid$ in Table 4.5 and 4.8, we get

$$\frac{p \xrightarrow{e_1} \surd \quad q \xrightarrow{e_2} \surd}{p \mid (q+r) \xrightarrow{\gamma(e_1,e_2)} \surd} \qquad \frac{p \xrightarrow{e_1} \surd \quad q \xrightarrow{e_2} \surd}{(p \mid q) + (p \mid r) \xrightarrow{\gamma(e_1,e_2)} \surd}$$

$$\frac{p \xrightarrow{e_1} \surd \quad r \xrightarrow{e_2} \surd}{p \mid (q+r) \xrightarrow{\gamma(e_1,e_2)} \surd} \qquad \frac{p \xrightarrow{e_1} \surd \quad r \xrightarrow{e_2} \surd}{(p \mid q)+(p \mid r) \xrightarrow{\gamma(e_1,e_2)} \surd}$$

$$\frac{p \xrightarrow{e_1} p' \quad q \xrightarrow{e_2} \surd}{p \mid (q+r) \xrightarrow{\gamma(e_1,e_2)} p'} \qquad \frac{p \xrightarrow{e_1} p' \quad q \xrightarrow{e_2} \surd}{(p \mid q)+(p \mid r) \xrightarrow{\gamma(e_1,e_2)} p'}$$

$$\frac{p \xrightarrow{e_1} p' \quad r \xrightarrow{e_2} \surd}{p \mid (q+r) \xrightarrow{\gamma(e_1,e_2)} p'} \qquad \frac{p \xrightarrow{e_1} p' \quad r \xrightarrow{e_2} \surd}{(p \mid q)+(p \mid r) \xrightarrow{\gamma(e_1,e_2)} p'}$$

$$\frac{p \xrightarrow{e_1} \surd \quad q \xrightarrow{e_2} q'}{p \mid (q+r) \xrightarrow{\gamma(e_1,e_2)} q'} \qquad \frac{p \xrightarrow{e_1} \surd \quad q \xrightarrow{e_2} q'}{(p \mid q)+(p \mid r) \xrightarrow{\gamma(e_1,e_2)} q'}$$

$$\frac{p \xrightarrow{e_1} \surd \quad r \xrightarrow{e_2} r'}{p \mid (q+r) \xrightarrow{\gamma(e_1,e_2)} r'} \qquad \frac{p \xrightarrow{e_1} \surd \quad r \xrightarrow{e_2} r'}{(p \mid q)+(p \mid r) \xrightarrow{\gamma(e_1,e_2)} r'}$$

$$\frac{p \xrightarrow{e_1} p' \quad q \xrightarrow{e_2} q'}{p \mid (q+r) \xrightarrow{\gamma(e_1,e_2)} p' \between q'} \qquad \frac{p \xrightarrow{e_1} p' \quad q \xrightarrow{e_2} q'}{(p \mid q)+(p \mid r) \xrightarrow{\gamma(e_1,e_2)} p' \between q'}$$

$$\frac{p \xrightarrow{e_1} p' \quad r \xrightarrow{e_2} r'}{p \mid (q+r) \xrightarrow{\gamma(e_1,e_2)} p' \between r'} \qquad \frac{p \xrightarrow{e_1} p' \quad r \xrightarrow{e_2} r'}{(p \mid q)+(p \mid r) \xrightarrow{\gamma(e_1,e_2)} p' \between r'}$$

So, $p \mid (q+r) \sim_s (p \mid q)+(p \mid r)$, as desired.

- **Axiom** $CE21$. Let p, q be $APTC$ processes, and $\Theta(p+q) = \Theta(p)+\Theta(q)$, it is sufficient to prove that $\Theta(p+q) \sim_s \Theta(p)+\Theta(q)$. By the transition rules for operators $+$ in Table 4.5, and Θ and $\triangleleft$ in Table 4.9, we get

$$\frac{p \xrightarrow{e_1} \surd(\sharp(e_1,e_2))}{\Theta(p+q) \xrightarrow{e_1} \surd} \qquad \frac{p \xrightarrow{e_1} \surd(\sharp(e_1,e_2))}{\Theta(p)+\Theta(q) \xrightarrow{e_1} \surd}$$

$$\frac{q \xrightarrow{e_2} \surd(\sharp(e_1,e_2))}{\Theta(p+q) \xrightarrow{e_2} \surd} \qquad \frac{q \xrightarrow{e_2} \surd(\sharp(e_1,e_2))}{\Theta(p)+\Theta(q) \xrightarrow{e_2} \surd}$$

$$\frac{p \xrightarrow{e_1} p'(\sharp(e_1,e_2))}{\Theta(p+q) \xrightarrow{e_1} \Theta(p')} \qquad \frac{p \xrightarrow{e_1} p'(\sharp(e_1,e_2))}{\Theta(p)+\Theta(q) \xrightarrow{e_1} \Theta(p')}$$

$$\frac{q \xrightarrow{e_2} q'(\sharp(e_1,e_2))}{\Theta(p+q) \xrightarrow{e_2} \Theta(q')} \qquad \frac{q \xrightarrow{e_2} q'(\sharp(e_1,e_2))}{\Theta(p)+\Theta(q) \xrightarrow{e_2} \Theta(q')}$$

So, $\Theta(p+q) \sim_s \Theta(p) + \Theta(q)$, as desired.

- **Axiom** $CE22$. Let p, q be $APTC$ processes, and $\Theta(p \cdot q) = \Theta(p) \cdot \Theta(q)$, it is sufficient to prove that $\Theta(p \cdot q) \sim_s \Theta(p) \cdot \Theta(q)$. By the transition rules for operators $\cdot$ in Table 4.5, and Θ in Table 4.9, we get

$$\frac{p \xrightarrow{e_1} \surd}{\Theta(p \cdot q) \xrightarrow{e_1} \Theta(q)} \qquad \frac{p \xrightarrow{e_1} \surd}{\Theta(p) \cdot \Theta(q) \xrightarrow{e_1} \Theta(q)}$$

$$\frac{p \xrightarrow{e_1} p'}{\Theta(p \cdot q) \xrightarrow{e_1} \Theta(p' \cdot q)} \qquad \frac{p \xrightarrow{e_1} p'}{\Theta(p) \cdot \Theta(q) \xrightarrow{e_1} \Theta(p') \cdot \Theta(q)}$$

So, with the assumption $\Theta(p' \cdot q) = \Theta(p') \cdot \Theta(q)$, $\Theta(p \cdot q) \sim_s \Theta(p) \cdot \Theta(q)$, as desired.

- **Axiom** $CE23$. Let p, q be $APTC$ processes, and $\Theta(p \parallel q) = ((\Theta(p) \triangleleft q) \parallel q) + ((\Theta(q) \triangleleft p) \parallel p)$, it is sufficient to prove that $\Theta(p \parallel q) \sim_s ((\Theta(p) \triangleleft q) \parallel q) + ((\Theta(q) \triangleleft p) \parallel p)$. By the transition rules for operators $+$ in Table 4.5, and Θ and $\triangleleft$ in Table 4.9, and $\parallel$ in Table 4.7 we get

$$\frac{p \xrightarrow{e_1} \surd \quad q \xrightarrow{e_2} \surd}{\Theta(p \parallel q) \xrightarrow{\{e_1,e_2\}} \surd}$$

$$\frac{p \xrightarrow{e_1} \surd \quad q \xrightarrow{e_2} \surd}{((\Theta(p) \triangleleft q) \parallel q) + ((\Theta(q) \triangleleft p) \parallel p) \xrightarrow{\{e_1,e_2\}} \surd}$$

$$\frac{p \xrightarrow{e_1} p' \quad q \xrightarrow{e_2} \surd}{\Theta(p \parallel q) \xrightarrow{\{e_1,e_2\}} \Theta(p')}$$

$$\frac{p \xrightarrow{e_1} p' \quad q \xrightarrow{e_2} \surd}{((\Theta(p) \triangleleft q) \parallel q) + ((\Theta(q) \triangleleft p) \parallel p) \xrightarrow{\{e_1,e_2\}} \Theta(p')}$$

$$\frac{p \xrightarrow{e_1} \surd \quad q \xrightarrow{e_2} q'}{\Theta(p \parallel q) \xrightarrow{\{e_1,e_2\}} \Theta(q')}$$

$$\frac{p \xrightarrow{e_1} \surd \quad q \xrightarrow{e_2} q'}{((\Theta(p) \triangleleft q) \parallel q) + ((\Theta(q) \triangleleft p) \parallel p) \xrightarrow{\{e_1,e_2\}} \Theta(q')}$$

$$\frac{p \xrightarrow{e_1} p' \quad q \xrightarrow{e_2} q'}{\Theta(p \parallel q) \xrightarrow{\{e_1,e_2\}} \Theta(p' \between q')}$$

$$\frac{p \xrightarrow{e_1} p' \quad q \xrightarrow{e_2} q'}{((\Theta(p) \triangleleft q) \parallel q) + ((\Theta(q) \triangleleft p) \parallel p) \xrightarrow{\{e_1, e_2\}} ((\Theta(p') \triangleleft q') \between q') + ((\Theta(q') \triangleleft p') \between p')}$$

So, with the assumption $\Theta(p' \between q') = ((\Theta(p') \triangleleft q') \between q') + ((\Theta(q') \triangleleft p') \between p')$, $\Theta(p \parallel q) \sim_s ((\Theta(p) \triangleleft q) \parallel q) + ((\Theta(q) \triangleleft p) \parallel p)$, as desired.

- **Axiom** $CE24$. Let p, q be $APTC$ processes, and $\Theta(p \mid q) = ((\Theta(p) \triangleleft q) \mid q) + ((\Theta(q) \triangleleft p) \mid p)$, it is sufficient to prove that $\Theta(p \mid q) \sim_s ((\Theta(p) \triangleleft q) \mid q) + ((\Theta(q) \triangleleft p) \mid p)$. By the transition rules for operators $+$ in Table 4.5, and Θ and $\triangleleft$ in Table 4.9, and $\mid$ in Table 4.8 we get

$$\frac{p \xrightarrow{e_1} \checkmark \quad q \xrightarrow{e_2} \checkmark}{\Theta(p \mid q) \xrightarrow{\gamma(e_1, e_2)} \checkmark}$$

$$\frac{p \xrightarrow{e_1} \checkmark \quad q \xrightarrow{e_2} \checkmark}{((\Theta(p) \triangleleft q) \mid q) + ((\Theta(q) \triangleleft p) \mid p) \xrightarrow{\gamma(e_1, e_2)} \checkmark}$$

$$\frac{p \xrightarrow{e_1} p' \quad q \xrightarrow{e_2} \checkmark}{\Theta(p \mid q) \xrightarrow{\gamma(e_1, e_2)} \Theta(p')}$$

$$\frac{p \xrightarrow{e_1} p' \quad q \xrightarrow{e_2} \checkmark}{((\Theta(p) \triangleleft q) \mid q) + ((\Theta(q) \triangleleft p) \mid p) \xrightarrow{\gamma(e_1, e_2)} \Theta(p')}$$

$$\frac{p \xrightarrow{e_1} \checkmark \quad q \xrightarrow{e_2} q'}{\Theta(p \mid q) \xrightarrow{\gamma(e_1, e_2)} \Theta(q')}$$

$$\frac{p \xrightarrow{e_1} \checkmark \quad q \xrightarrow{e_2} q'}{((\Theta(p) \triangleleft q) \mid q) + ((\Theta(q) \triangleleft p) \mid p) \xrightarrow{\gamma(e_1, e_2)} \Theta(q')}$$

$$\frac{p \xrightarrow{e_1} p' \quad q \xrightarrow{e_2} q'}{\Theta(p \mid q) \xrightarrow{\gamma(e_1, e_2)} \Theta(p' \between q')}$$

$$\frac{p \xrightarrow{e_1} p' \quad q \xrightarrow{e_2} q'}{((\Theta(p) \triangleleft q) \mid q) + ((\Theta(q) \triangleleft p) \mid p) \xrightarrow{\gamma(e_1, e_2)} ((\Theta(p') \triangleleft q') \between q') + ((\Theta(q') \triangleleft p') \between p')}$$

So, with the assumption $\Theta(p' \between q') = ((\Theta(p') \triangleleft q') \between q') + ((\Theta(q') \triangleleft p') \between p')$, $\Theta(p \mid q) \sim_s ((\Theta(p) \triangleleft q) \mid q) + ((\Theta(q) \triangleleft p) \mid p)$, as desired.

- **Axiom** $U30$. Let p, q, r be $APTC$ processes, and $(p + q) \triangleleft r = (p \triangleleft r) + (q \triangleleft r)$, it is sufficient to prove that $(p + q) \triangleleft r \sim_s (p \triangleleft r) + (q \triangleleft r)$. By the transition rules for operators $+$ and $\triangleleft$

in Table 4.5 and 4.9, we get

$$\frac{p \xrightarrow{e_1} \surd}{(p+q) \triangleleft r \xrightarrow{e_1} \surd} \qquad \frac{p \xrightarrow{e_1} \surd}{(p \triangleleft r) + (q \triangleleft r) \xrightarrow{e_1} \surd}$$

$$\frac{q \xrightarrow{e_2} \surd}{(p+q) \triangleleft r \xrightarrow{e_2} \surd} \qquad \frac{q \xrightarrow{e_2} \surd}{(p \triangleleft r) + (q \triangleleft r) \xrightarrow{e_2} \surd}$$

$$\frac{p \xrightarrow{e_1} p'}{(p+q) \triangleleft r \xrightarrow{e_1} p' \triangleleft r} \qquad \frac{p \xrightarrow{e_1} p'}{(p \triangleleft r) + (q \triangleleft r) \xrightarrow{e_1} p' \triangleleft r}$$

$$\frac{q \xrightarrow{e_2} q'}{(p+q) \triangleleft r \xrightarrow{e_2} q' \triangleleft r} \qquad \frac{q \xrightarrow{e_2} q'}{(p \triangleleft r) + (q \triangleleft r) \xrightarrow{e_2} q' \triangleleft r}$$

Let us forget anything about τ. So, $(p+q) \triangleleft r \sim_s (p \triangleleft r) + (q \triangleleft r)$, as desired.

- **Axiom** $U31$. Let p, q, r be $APTC$ processes, and $(p \cdot q) \triangleleft r = (p \triangleleft r) \cdot (q \triangleleft r)$, it is sufficient to prove that $(p \cdot q) \triangleleft r \sim_s (p \triangleleft r) \cdot (q \triangleleft r)$. By the transition rules for operators $\cdot$ and $\triangleleft$ in Table 4.5 and 4.9, we get

$$\frac{p \xrightarrow{e_1} \surd}{(p \cdot q) \triangleleft r \xrightarrow{e_1} q \triangleleft r} \qquad \frac{p \xrightarrow{e_1} \surd}{(p \triangleleft r) \cdot (q \triangleleft r) \xrightarrow{e_1} q \triangleleft r}$$

$$\frac{p \xrightarrow{e_1} p'}{(p \cdot q) \triangleleft r \xrightarrow{e_1} (p' \cdot q) \triangleleft r} \qquad \frac{p \xrightarrow{e_1} p'}{(p \triangleleft r) \cdot (q \triangleleft r) \xrightarrow{e_1} (p' \triangleleft r) \cdot (q \triangleleft r)}$$

Let us forget anything about τ. With the assumption $(p' \cdot q) \triangleleft r = (p' \triangleleft r) \cdot (q \triangleleft r)$, so, $(p \cdot q) \triangleleft r \sim_s (p \triangleleft r) \cdot (q \triangleleft r)$, as desired.

- **Axiom** $U32$. Let p, q, r be $APTC$ processes, and $(p \parallel q) \triangleleft r = (p \triangleleft r) \parallel (q \triangleleft r)$, it is sufficient to prove that $(p \parallel q) \triangleleft r \sim_s (p \triangleleft r) \parallel (q \triangleleft r)$. By the transition rules for operators $\parallel$ and $\triangleleft$ in Table 4.7 and 4.9, we get

$$\frac{p \xrightarrow{e_1} \surd \quad q \xrightarrow{e_2} \surd}{(p \parallel q) \triangleleft r \xrightarrow{\{e_1, e_2\}} \surd} \qquad \frac{p \xrightarrow{e_1} \surd \quad q \xrightarrow{e_2} \surd}{(p \triangleleft r) \parallel (q \triangleleft r) \xrightarrow{\{e_1, e_2\}} \surd}$$

$$\frac{p \xrightarrow{e_1} p' \quad q \xrightarrow{e_2} \surd}{(p \parallel q) \triangleleft r \xrightarrow{\{e_1, e_2\}} p' \triangleleft r} \qquad \frac{p \xrightarrow{e_1} p' \quad q \xrightarrow{e_2} \surd}{(p \triangleleft r) \parallel (q \triangleleft r) \xrightarrow{\{e_1, e_2\}} p' \triangleleft r}$$

$$\frac{p \xrightarrow{e_1} \surd \quad q \xrightarrow{e_2} q'}{(p \parallel q) \triangleleft r \xrightarrow{\{e_1, e_2\}} q' \triangleleft r} \qquad \frac{p \xrightarrow{e_1} \surd \quad q \xrightarrow{e_2} q'}{(p \triangleleft r) \parallel (q \triangleleft r) \xrightarrow{\{e_1, e_2\}} q' \triangleleft r}$$

$$\frac{p \xrightarrow{e_1} p' \quad q \xrightarrow{e_2} q'}{(p \parallel q) \triangleleft r \xrightarrow{\{e_1,e_2\}} (p' \between q') \triangleleft r} \qquad \frac{p \xrightarrow{e_1} p' \quad q \xrightarrow{e_2} q'}{(p \triangleleft r) \parallel (q \triangleleft r) \xrightarrow{\{e_1,e_2\}} (p' \triangleleft r) \between (q' \triangleleft r)}$$

Let us forget anything about τ. With the assumption $(p' \between q') \triangleleft r = (p' \triangleleft r) \between (q' \triangleleft r)$, so, $(p \parallel q) \triangleleft r \sim_s (p \triangleleft r) \parallel (q \triangleleft r)$, as desired.

- **Axiom** $U33$. Let p, q, r be $APTC$ processes, and $(p \mid q) \triangleleft r = (p \triangleleft r) \mid (q \triangleleft r)$, it is sufficient to prove that $(p \mid q) \triangleleft r \sim_s (p \triangleleft r) \mid (q \triangleleft r)$. By the transition rules for operators $\mid$ and $\triangleleft$ in Table 4.8 and 4.9, we get

$$\frac{p \xrightarrow{e_1} \surd \quad q \xrightarrow{e_2} \surd}{(p \mid q) \triangleleft r \xrightarrow{\gamma(e_1,e_2)} \surd} \qquad \frac{p \xrightarrow{e_1} \surd \quad q \xrightarrow{e_2} \surd}{(p \triangleleft r) \mid (q \triangleleft r) \xrightarrow{\gamma(e_1,e_2)} \surd}$$

$$\frac{p \xrightarrow{e_1} p' \quad q \xrightarrow{e_2} \surd}{(p \mid q) \triangleleft r \xrightarrow{\gamma(e_1,e_2)} p' \triangleleft r} \qquad \frac{p \xrightarrow{e_1} p' \quad q \xrightarrow{e_2} \surd}{(p \triangleleft r) \mid (q \triangleleft r) \xrightarrow{\gamma(e_1,e_2)} p' \triangleleft r}$$

$$\frac{p \xrightarrow{e_1} \surd \quad q \xrightarrow{e_2} q'}{(p \mid q) \triangleleft r \xrightarrow{\gamma(e_1,e_2)} q' \triangleleft r} \qquad \frac{p \xrightarrow{e_1} \surd \quad q \xrightarrow{e_2} q'}{(p \triangleleft r) \mid (q \triangleleft r) \xrightarrow{\gamma(e_1,e_2)} q' \triangleleft r}$$

$$\frac{p \xrightarrow{e_1} p' \quad q \xrightarrow{e_2} q'}{(p \mid q) \triangleleft r \xrightarrow{\gamma(e_1,e_2)} (p' \between q') \triangleleft r} \qquad \frac{p \xrightarrow{e_1} p' \quad q \xrightarrow{e_2} q'}{(p \triangleleft r) \mid (q \triangleleft r) \xrightarrow{\gamma(e_1,e_2)} (p' \triangleleft r) \between (q' \triangleleft r)}$$

Let us forget anything about τ. With the assumption $(p' \between q') \triangleleft r = (p' \triangleleft r) \between (q' \triangleleft r)$, so, $(p \mid q) \triangleleft r \sim_s (p \triangleleft r) \mid (q \triangleleft r)$, as desired.

- **Axiom** $U34$. Let p, q, r be $APTC$ processes, and $p \triangleleft (q + r) = (p \triangleleft q) \triangleleft r$, it is sufficient to prove that $p \triangleleft (q + r) \sim_s (p \triangleleft q) \triangleleft r$. By the transition rules for operators $+$ and $\triangleleft$ in Table 4.5 and 4.9, we get

$$\frac{p \xrightarrow{e_1} \surd}{p \triangleleft (q + r) \xrightarrow{e_1} \surd} \qquad \frac{p \xrightarrow{e_1} \surd}{(p \triangleleft q) \triangleleft r \xrightarrow{e_1} \surd}$$

$$\frac{p \xrightarrow{e_1} p'}{p \triangleleft (q + r) \xrightarrow{e_1} p' \triangleleft (q + r)} \qquad \frac{p \xrightarrow{e_1} p'}{(p \triangleleft q) \triangleleft r \xrightarrow{e_1} (p' \triangleleft q) \triangleleft r}$$

Let us forget anything about τ. With the assumption $p' \triangleleft (q + r) = (p' \triangleleft q) \triangleleft r$, so, $p \triangleleft (q + r) \sim_s (p \triangleleft q) \triangleleft r$, as desired.

- **Axiom** $U35$. Let p, q, r be $APTC$ processes, and $p \triangleleft (q \cdot r) = (p \triangleleft q) \triangleleft r$, it is sufficient to prove that $p \triangleleft (q \cdot r) \sim_s (p \triangleleft q) \triangleleft r$. By the transition rules for operators $\cdot$ and $\triangleleft$ in Table 4.5 and 4.9, we get

$$\frac{p \xrightarrow{e_1} \surd}{p \triangleleft (q \cdot r) \xrightarrow{e_1} \surd} \qquad \frac{p \xrightarrow{e_1} \surd}{(p \triangleleft q) \triangleleft r \xrightarrow{e_1} \surd}$$

$$\frac{p \xrightarrow{e_1} p'}{p \triangleleft (q \cdot r) \xrightarrow{e_1} p' \triangleleft (q \cdot r)} \qquad \frac{p \xrightarrow{e_1} p'}{(p \triangleleft q) \triangleleft r \xrightarrow{e_1} (p' \triangleleft q) \triangleleft r}$$

Let us forget anything about τ. With the assumption $p' \triangleleft (q \cdot r) = (p' \triangleleft q) \triangleleft r$, so, $p \triangleleft (q \cdot r) \sim_s (p \triangleleft q) \triangleleft r$, as desired.

- **Axiom** $U36$. Let p, q, r be $APTC$ processes, and $p \triangleleft (q \parallel r) = (p \triangleleft q) \triangleleft r$, it is sufficient to prove that $p \triangleleft (q \parallel r) \sim_s (p \triangleleft q) \triangleleft r$. By the transition rules for operators $\parallel$ and $\triangleleft$ in Table 4.7 and 4.9, we get

$$\frac{p \xrightarrow{e_1} \surd}{p \triangleleft (q \parallel r) \xrightarrow{e_1} \surd} \qquad \frac{p \xrightarrow{e_1} \surd}{(p \triangleleft q) \triangleleft r \xrightarrow{e_1} \surd}$$

$$\frac{p \xrightarrow{e_1} p'}{p \triangleleft (q \parallel r) \xrightarrow{e_1} p' \triangleleft (q \parallel r)} \qquad \frac{p \xrightarrow{e_1} p'}{(p \triangleleft q) \triangleleft r \xrightarrow{e_1} (p' \triangleleft q) \triangleleft r}$$

Let us forget anything about τ. With the assumption $p' \triangleleft (q \parallel r) = (p' \triangleleft q) \triangleleft r$, so, $p \triangleleft (q \parallel r) \sim_s (p \triangleleft q) \triangleleft r$, as desired.

- **Axiom** $U37$. Let p, q, r be $APTC$ processes, and $p \triangleleft (q \mid r) = (p \triangleleft q) \triangleleft r$, it is sufficient to prove that $p \triangleleft (q \mid r) \sim_s (p \triangleleft q) \triangleleft r$. By the transition rules for operators $\mid$ and $\triangleleft$ in Table 4.8 and 4.9, we get

$$\frac{p \xrightarrow{e_1} \surd}{p \triangleleft (q \mid r) \xrightarrow{e_1} \surd} \qquad \frac{p \xrightarrow{e_1} \surd}{(p \triangleleft q) \triangleleft r \xrightarrow{e_1} \surd}$$

$$\frac{p \xrightarrow{e_1} p'}{p \triangleleft (q \mid r) \xrightarrow{e_1} p' \triangleleft (q \mid r)} \qquad \frac{p \xrightarrow{e_1} p'}{(p \triangleleft q) \triangleleft r \xrightarrow{e_1} (p' \triangleleft q) \triangleleft r}$$

Let us forget anything about τ. With the assumption $p' \triangleleft (q \mid r) = (p' \triangleleft q) \triangleleft r$, so, $p \triangleleft (q \mid r) \sim_s (p \triangleleft q) \triangleleft r$, as desired.   □

Theorem 4.25 (Completeness of parallelism modulo step bisimulation equivalence). *Let p and q be closed $APTC$ terms, if $p \sim_s q$ then $p = q$.*

Proof. Firstly, by the elimination theorem of $APTC$ (see Theorem 4.22), we know that for each closed $APTC$ term p, there exists a closed basic $APTC$ term p', such that $APTC \vdash p = p'$, so, we only need to consider closed basic $APTC$ terms.

The basic terms (see Definition 4.19) modulo associativity and commutativity (AC) of conflict $+$ (defined by axioms $A1$ and $A2$ in Table 4.1) and associativity and commutativity (AC) of parallel $\parallel$ (defined by axioms $P2$ and $P3$ in Table 4.10), and these equivalences are denoted by $=_{AC}$. Then, each equivalence class s modulo AC of $+$ and $\parallel$ has the following normal form

$$s_1 + \cdots + s_k$$

with each s_i either an atomic event or of the form

$$t_1 \cdots \cdots t_m$$

with each t_j either an atomic event or of the form

$$u_1 \parallel \cdots \parallel u_l$$

with each u_l an atomic event, and each s_i is called the summand of s.

Now, we prove that for normal forms n and n', if $n \sim_s n'$ then $n =_{AC} n'$. It is sufficient to induct on the sizes of n and n'.

- Consider a summand e of n. Then $n \xrightarrow{e} \surd$, so $n \sim_s n'$ implies $n' \xrightarrow{e} \surd$, meaning that n' also contains the summand e.
- Consider a summand $t_1 \cdot t_2$ of n,
  - if $t_1 \equiv e'$, then $n \xrightarrow{e'} t_2$, so $n \sim_s n'$ implies $n' \xrightarrow{e'} t_2'$ with $t_2 \sim_s t_2'$, meaning that n' contains a summand $e' \cdot t_2'$. Since t_2 and t_2' are normal forms and have sizes smaller than n and n', by the induction hypotheses if $t_2 \sim_s t_2'$ then $t_2 =_{AC} t_2'$;
  - if $t_1 \equiv e_1 \parallel \cdots \parallel e_l$, then $n \xrightarrow{\{e_1, \cdots, e_l\}} t_2$, so $n \sim_s n'$ implies $n' \xrightarrow{\{e_1, \cdots, e_l\}} t_2'$ with $t_2 \sim_s t_2'$, meaning that n' contains a summand $(e_1 \parallel \cdots \parallel e_l) \cdot t_2'$. Since t_2 and t_2' are normal forms and have sizes smaller than n and n', by the induction hypotheses if $t_2 \sim_s t_2'$ then $t_2 =_{AC} t_2'$.

So, we get $n =_{AC} n'$.

Finally, let s and t be basic $APTC$ terms, and $s \sim_s t$, there are normal forms n and n', such that $s = n$ and $t = n'$. The soundness theorem of parallelism modulo step bisimulation equivalence (see Theorem 4.24) yields $s \sim_s n$ and $t \sim_s n'$, so $n \sim_s s \sim_s t \sim_s n'$. Since if $n \sim_s n'$ then $n =_{AC} n'$, $s = n =_{AC} n' = t$, as desired. $\qquad \square$

Theorem 4.26 (Soundness of parallelism modulo pomset bisimulation equivalence). *Let x and y be $APTC$ terms. If $APTC \vdash x = y$, then $x \sim_p y$.*

Proof. Since pomset bisimulation $\sim_p$ is both an equivalent and a congruent relation with respect to the operators $\between$, $\parallel$, $\mid$, Θ and $\triangleleft$, we only need to check if each axiom in Table 4.10 is sound modulo pomset bisimulation equivalence.

From the definition of pomset bisimulation (see Definition 2.17), we know that pomset bisimulation is defined by pomset transitions, which are labeled by pomsets. In a pomset transition, the events in the pomset are either within causality relations (defined by $\cdot$) or in concurrency (implicitly defined by $\cdot$ and $+$, and explicitly defined by $\between$), of course, they are pairwise consistent (without conflicts). In Theorem 4.24, we have already proven the case that all events are pairwise concurrent, so, we only need to prove the case of events in causality. Without loss of generality, we take a pomset of $P = \{e_1, e_2 : e_1 \cdot e_2\}$. Then the pomset transition labeled by the above P is just composed of one single event transition labeled by e_1 succeeded by another single event transition labeled by e_2, that is, $\xrightarrow{P} = \xrightarrow{e_1} \xrightarrow{e_2}$.

Similarly to the proof of soundness of parallelism modulo step bisimulation equivalence (see Theorem 4.24), we can prove that each axiom in Table 4.10 is sound modulo pomset bisimulation equivalence, we omit them.   □

Theorem 4.27 (Completeness of parallelism modulo pomset bisimulation equivalence). *Let p and q be closed $APTC$ terms, if $p \sim_p q$ then $p = q$.*

Proof. Firstly, by the elimination theorem of $APTC$ (see Theorem 4.22), we know that for each closed $APTC$ term p, there exists a closed basic $APTC$ term p', such that $APTC \vdash p = p'$, so, we only need to consider closed basic $APTC$ terms.

The basic terms (see Definition 4.19) modulo associativity and commutativity (AC) of conflict $+$ (defined by axioms $A1$ and $A2$ in Table 4.1) and associativity and commutativity (AC) of parallel $\parallel$ (defined by axioms $P2$ and $P3$ in Table 4.10), and these equivalences are denoted by $=_{AC}$. Then, each equivalence class s modulo AC of $+$ and $\parallel$ has the following normal form

$$s_1 + \cdots + s_k$$

with each s_i either an atomic event or of the form

$$t_1 \cdots\cdots t_m$$

with each t_j either an atomic event or of the form

$$u_1 \parallel \cdots \parallel u_l$$

with each u_l an atomic event, and each s_i is called the summand of s.

Now, we prove that for normal forms n and n', if $n \sim_p n'$ then $n =_{AC} n'$. It is sufficient to induct on the sizes of n and n'.

- Consider a summand e of n. Then $n \xrightarrow{e} \surd$, so $n \sim_p n'$ implies $n' \xrightarrow{e} \surd$, meaning that n' also contains the summand e.
- Consider a summand $t_1 \cdot t_2$ of n,
  - if $t_1 \equiv e'$, then $n \xrightarrow{e'} t_2$, so $n \sim_p n'$ implies $n' \xrightarrow{e'} t_2'$ with $t_2 \sim_p t_2'$, meaning that n' contains a summand $e' \cdot t_2'$. Since t_2 and t_2' are normal forms and have sizes smaller than n and n', by the induction hypotheses if $t_2 \sim_p t_2'$ then $t_2 =_{AC} t_2'$;
  - if $t_1 \equiv e_1 \parallel \cdots \parallel e_l$, then $n \xrightarrow{\{e_1, \cdots, e_l\}} t_2$, so $n \sim_p n'$ implies $n' \xrightarrow{\{e_1, \cdots, e_l\}} t_2'$ with $t_2 \sim_p t_2'$, meaning that n' contains a summand $(e_1 \parallel \cdots \parallel e_l) \cdot t_2'$. Since t_2 and t_2' are normal forms and have sizes smaller than n and n', by the induction hypotheses if $t_2 \sim_p t_2'$ then $t_2 =_{AC} t_2'$.

So, we get $n =_{AC} n'$.

Finally, let s and t be basic $APTC$ terms, and $s \sim_p t$, there are normal forms n and n', such that $s = n$ and $t = n'$. The soundness theorem of parallelism modulo pomset bisimulation equivalence (see Theorem 4.26) yields $s \sim_p n$ and $t \sim_p n'$, so $n \sim_p s \sim_p t \sim_p n'$. Since if $n \sim_p n'$ then $n =_{AC} n'$, $s = n =_{AC} n' = t$, as desired.   □

Theorem 4.28 (Soundness of parallelism modulo hp-bisimulation equivalence). *Let x and y be $APTC$ terms. If $APTC \vdash x = y$, then $x \sim_{hp} y$.*

Proof. Since hp-bisimulation $\sim_{hp}$ is both an equivalent and a congruent relation with respect to the operators $(\!|, \|, |, \Theta$ and $\lhd$, we only need to check if each axiom in Table 4.10 is sound modulo hp-bisimulation equivalence.

From the definition of hp-bisimulation (see Definition 2.21), we know that hp-bisimulation is defined on the posetal product (C_1, f, C_2), $f : C_1 \rightarrow C_2$ isomorphism. Two process terms s related to C_1 and t related to C_2, and $f : C_1 \rightarrow C_2$ isomorphism. Initially, $(C_1, f, C_2) = (\emptyset, \emptyset, \emptyset)$, and $(\emptyset, \emptyset, \emptyset) \in \sim_{hp}$. When $s \xrightarrow{e} s'$ ($C_1 \xrightarrow{e} C_1'$), there will be $t \xrightarrow{e} t'$ ($C_2 \xrightarrow{e} C_2'$), and we define $f' = f[e \mapsto e]$. Then, if $(C_1, f, C_2) \in \sim_{hp}$, then $(C_1', f', C_2') \in \sim_{hp}$.

Similarly to the proof of soundness of parallelism modulo pomset bisimulation equivalence (see Theorem 4.26), we can prove that each axiom in Table 4.10 is sound modulo hp-bisimulation equivalence, we just need additionally to check the above conditions on hp-bisimulation, we omit them. $\qquad\square$

Theorem 4.29 (Completeness of parallelism modulo hp-bisimulation equivalence). *Let p and q be closed $APTC$ terms, if $p \sim_{hp} q$ then $p = q$.*

Proof. Firstly, by the elimination theorem of $APTC$ (see Theorem 4.22), we know that for each closed $APTC$ term p, there exists a closed basic $APTC$ term p', such that $APTC \vdash p = p'$, so, we only need to consider closed basic $APTC$ terms.

The basic terms (see Definition 4.19) modulo associativity and commutativity (AC) of conflict $+$ (defined by axioms $A1$ and $A2$ in Table 4.1) and associativity and commutativity (AC) of parallel $\|$ (defined by axioms $P2$ and $P3$ in Table 4.10), and these equivalences are denoted by $=_{AC}$. Then, each equivalence class s modulo AC of $+$ and $\|$ has the following normal form

$$s_1 + \cdots + s_k$$

with each s_i either an atomic event or of the form

$$t_1 \cdots \cdots t_m$$

with each t_j either an atomic event or of the form

$$u_1 \| \cdots \| u_l$$

with each u_l an atomic event, and each s_i is called the summand of s.

Now, we prove that for normal forms n and n', if $n \sim_{hp} n'$ then $n =_{AC} n'$. It is sufficient to induct on the sizes of n and n'.

- Consider a summand e of n. Then $n \xrightarrow{e} \sqrt{}$, so $n \sim_{hp} n'$ implies $n' \xrightarrow{e} \sqrt{}$, meaning that n' also contains the summand e.
- Consider a summand $t_1 \cdot t_2$ of n,

- if $t_1 \equiv e'$, then $n \xrightarrow{e'} t_2$, so $n \sim_{hp} n'$ implies $n' \xrightarrow{e'} t_2'$ with $t_2 \sim_{hp} t_2'$, meaning that n' contains a summand $e' \cdot t_2'$. Since t_2 and t_2' are normal forms and have sizes smaller than n and n', by the induction hypotheses if $t_2 \sim_{hp} t_2'$ then $t_2 =_{AC} t_2'$;

- if $t_1 \equiv e_1 \parallel \cdots \parallel e_l$, then $n \xrightarrow{\{e_1, \cdots, e_l\}} t_2$, so $n \sim_{hp} n'$ implies $n' \xrightarrow{\{e_1, \cdots, e_l\}} t_2'$ with $t_2 \sim_{hp} t_2'$, meaning that n' contains a summand $(e_1 \parallel \cdots \parallel e_l) \cdot t_2'$. Since t_2 and t_2' are normal forms and have sizes smaller than n and n', by the induction hypotheses if $t_2 \sim_{hp} t_2'$ then $t_2 =_{AC} t_2'$.

So, we get $n =_{AC} n'$.

Finally, let s and t be basic $APTC$ terms, and $s \sim_{hp} t$, there are normal forms n and n', such that $s = n$ and $t = n'$. The soundness theorem of parallelism modulo hp-bisimulation equivalence (see Theorem 4.26) yields $s \sim_{hp} n$ and $t \sim_{hp} n'$, so $n \sim_{hp} s \sim_{hp} t \sim_{hp} n'$. Since if $n \sim_{hp} n'$ then $n =_{AC} n'$, $s = n =_{AC} n' = t$, as desired. $\square$

Proposition 4.30 (About Soundness and Completeness of parallelism modulo hhp-bisimulation equivalence). *1. Let x and y be $APTC$ terms. If $APTC \vdash x = y \nRightarrow x \sim_{hhp} y$;*
2. If p and q are closed $APTC$ terms, then $p \sim_{hhp} q \nRightarrow p = q$.

Proof. Imperfectly, the algebraic laws in Table 4.10 are not sound and complete modulo hhp-bisimulation equivalence, we just need enumerate several key axioms in Table 4.10 are not sound modulo hhp-bisimulation equivalence.

From the definition of hhp-bisimulation (see Definition 2.21), we know that an hhp-bisimulation is a downward closed hp-bisimulation. That is, for any posetal products (C_1, f, C_2) and (C_1', f, C_2'), if $(C_1, f, C_2) \subseteq (C_1', f', C_2')$ pointwise and $(C_1', f', C_2') \in \sim_{hhp}$, then $(C_1, f, C_2) \in \sim_{hhp}$.

Now, let us consider the axioms $P7$ and $P8$ (the right and left distributivity of $\parallel$ to $+$). Let $s_1 = (a + b) \parallel c$, $t_1 = (a \parallel c) + (b \parallel c)$, and $s_2 = a \parallel (b + c)$, $t_2 = (a \parallel b) + (a \parallel c)$. We know that $s_1 \sim_{hp} t_1$ and $s_2 \sim_{hp} t_2$ (by Theorem 4.28), we prove that $s_1 \nsim_{hhp} t_1$ and $s_2 \nsim_{hhp} t_2$. Let $(C(s_1), f_1, C(t_1))$ and $(C(s_2), f_2, C(t_2))$ be the corresponding posetal products.

- **Axiom $P7$.** $s_1 \xrightarrow{\{a,c\}} \sqrt{}(s_1') (C(s_1) \xrightarrow{\{a,c\}} C(s_1'))$, then $t_1 \xrightarrow{\{a,c\}} \sqrt{}(t_1') (C(t_1) \xrightarrow{\{a,c\}} C(t_1'))$, we define $f_1' = f_1[a \mapsto a, c \mapsto c]$, obviously, $(C(s_1), f_1, C(t_1)) \in \sim_{hp}$ and $(C(s_1'), f_1', C(t_1')) \in \sim_{hp}$. But, $(C(s_1), f_1, C(t_1)) \in \sim_{hhp}$ and $(C(s_1'), f_1', C(t_1')) \in \nsim_{hhp}$, just because they are not downward closed. Let $(C(s_1''), f_1'', C(t_1''))$, and $f_1'' = f_1[c \mapsto c]$, $s_1 \xrightarrow{c} s_1'' (C(s_1) \xrightarrow{c} C(s_1''))$, $t_1 \xrightarrow{c} t_1''$ $(C(t_1) \xrightarrow{c} C(t_1''))$, it is easy to see that $(C(s_1''), f_1'', C(t_1'')) \subseteq (C(s_1'), f_1', C(t_1'))$ pointwise, while $(C(s_1''), f_1'', C(t_1'')) \notin \sim_{hp}$, because s_1'' and $C(s_1'')$ exist, but t_1'' and $C(t_1'')$ do not exist.

- **Axiom $P8$.** $s_2 \xrightarrow{\{a,c\}} \sqrt{}(s_2') (C(s_2) \xrightarrow{\{a,c\}} C(s_2'))$, then $t_2 \xrightarrow{\{a,c\}} \sqrt{}(t_2') (C(t_2) \xrightarrow{\{a,c\}} C(t_2'))$, we define $f_2' = f_2[a \mapsto a, c \mapsto c]$, obviously, $(C(s_2), f_2, C(t_2)) \in \sim_{hp}$ and $(C(s_2'), f_2', C(t_2')) \in \sim_{hp}$. But, $(C(s_2), f_2, C(t_2)) \in \sim_{hhp}$ and $(C(s_2'), f_2', C(t_2')) \in \nsim_{hhp}$, just because they are not downward closed. Let $(C(s_2''), f_2'', C(t_2''))$, and $f_2'' = f_2[a \mapsto a]$, $s_2 \xrightarrow{a} s_2'' (C(s_2) \xrightarrow{a} C(s_2''))$, $t_2 \xrightarrow{a} t_2''$ $(C(t_2) \xrightarrow{a} C(t_2''))$, it is easy to see that $(C(s_2''), f_2'', C(t_2'')) \subseteq (C(s_2'), f_2', C(t_2'))$ pointwise, while $(C(s_2''), f_2'', C(t_2'')) \notin \sim_{hp}$, because s_2'' and $C(s_2'')$ exist, but t_2'' and $C(t_2'')$ do not exist.

The unsoundness of parallelism modulo hhp-bisimulation equivalence makes the completeness of parallelism modulo hhp-bisimulation equivalence meaningless. Further more, unsoundness of $P7$ and $P8$ lead to the elimination theorem of $APTC$ (see Theorem 4.22) failing, so, the non-existence of normal form also makes the completeness impossible.   □

Actually, a finite sound and complete axiomatization for parallel composition ‖ modulo hhp-bisimulation equivalence does not exist, about the axiomatization for hhp-bisimilarity, please refer to section 4.7 for details.

In following sections, we will discuss nothing about hhp-bisimulation, because the following encapsulation, recursion and abstraction are based on the algebraic laws in this section.

Finally, let us explain the so-called absorption law [18] in a straightforward way. Process term $P = a \parallel (b+c) + a \parallel b + b \parallel (a+c)$, and process term $Q = a \parallel (b+c) + b \parallel (a+c)$, equated by the absorption law.

Modulo $\sim_s$, $\sim_p$, and $\sim_{hp}$, by use of the axioms of $BATC$ and $APTC$, we have the following deductions:

$$P = a \parallel (b+c) + a \parallel b + b \parallel (a+c)$$
$$\stackrel{P8}{=} a \parallel b + a \parallel c + a \parallel b + b \parallel a + b \parallel c$$
$$\stackrel{P2}{=} a \parallel b + a \parallel c + a \parallel b + a \parallel b + b \parallel c$$
$$\stackrel{A3}{=} a \parallel b + a \parallel c + b \parallel c$$

$$Q = a \parallel (b+c) + b \parallel (a+c)$$
$$\stackrel{P8}{=} a \parallel b + a \parallel c + b \parallel a + b \parallel c$$
$$\stackrel{P2}{=} a \parallel b + a \parallel c + a \parallel b + b \parallel c$$
$$\stackrel{A3}{=} a \parallel b + a \parallel c + b \parallel c$$

It means that $P = Q$ modulo $\sim_s$, $\sim_p$, and $\sim_{hp}$, that is, $P \sim_s Q$, $P \sim_p Q$ and $P \sim_{hp} Q$. But, $P \neq Q$ modulo $\sim_{hhp}$, which means that $P \nsim_{hhp} Q$.

4.2.5 Encapsulation

The mismatch of two communicating events in different parallel branches can cause deadlock, so the deadlocks in the concurrent processes should be eliminated. Like ACP [4], we also introduce the unary encapsulation operator ∂_H for set H of atomic events, which renames all atomic events in H into δ. The whole algebra including parallelism for true concurrency in the above subsections, deadlock δ and encapsulation operator ∂_H, is called Algebra for Parallelism in True Concurrency, abbreviated $APTC$.

The transition rules of encapsulation operator ∂_H are shown in Table 4.12.

Table 4.12   Transition rules of encapsulation operator ∂_H.

$$\frac{x \xrightarrow{e} \surd}{\partial_H(x) \xrightarrow{e} \surd} \quad (e \notin H) \qquad \frac{x \xrightarrow{e} x'}{\partial_H(x) \xrightarrow{e} \partial_H(x')} \quad (e \notin H)$$

Table 4.13   Axioms of encapsulation operator.

No.	Axiom
$D1$	$e \notin H \quad \partial_H(e) = e$
$D2$	$e \in H \quad \partial_H(e) = \delta$
$D3$	$\partial_H(\delta) = \delta$
$D4$	$\partial_H(x + y) = \partial_H(x) + \partial_H(y)$
$D5$	$\partial_H(x \cdot y) = \partial_H(x) \cdot \partial_H(y)$
$D6$	$\partial_H(x \parallel y) = \partial_H(x) \parallel \partial_H(y)$

Based on the transition rules for encapsulation operator ∂_H in Table 4.12, we design the axioms as Table 4.13 shows.

The axioms $D1 - D3$ are the defining laws for the encapsulation operator ∂_H, $D1$ leaves atomic events outside H unchanged, $D2$ renames atomic events in H into δ, and $D3$ says that it leaves δ unchanged. $D4 - D6$ say that in term $\partial_H(t)$, all transitions of t labeled with atomic events in H are blocked.

Theorem 4.31 (Conservativity of $APTC$ with respect to the algebra for parallelism). *$APTC$ is a conservative extension of the algebra for parallelism.*

Proof. It follows from the following two facts (see Theorem 2.8).

1. The transition rules of the algebra for parallelism in the above subsections are all source-dependent;
2. The sources of the transition rules for the encapsulation operator contain an occurrence of ∂_H.

So, $APTC$ is a conservative extension of the algebra for parallelism, as desired. $\square$

Theorem 4.32 (Congruence theorem of encapsulation operator ∂_H). *Truly concurrent bisimulation equivalences $\sim_p$, $\sim_s$, $\sim_{hp}$, and $\sim_{hhp}$ are all congruences with respect to encapsulation operator ∂_H.*

Proof. (1) Case pomset bisimulation equivalence $\sim_p$.

Let x and y be $APTC$ processes, and $x \sim_p y$, it is sufficient to prove that $\partial_H(x) \sim_p \partial_H(y)$.

Table 4.14 Term rewrite system of encapsulation operator ∂_H.

No.	Rewriting rule
$RD1$	$e \notin H \quad \partial_H(e) \to e$
$RD2$	$e \in H \quad \partial_H(e) \to \delta$
$RD3$	$\partial_H(\delta) \to \delta$
$RD4$	$\partial_H(x + y) \to \partial_H(x) + \partial_H(y)$
$RD5$	$\partial_H(x \cdot y) \to \partial_H(x) \cdot \partial_H(y)$
$RD6$	$\partial_H(x \parallel y) \to \partial_H(x) \parallel \partial_H(y)$

By the definition of pomset bisimulation $\sim_p$ (Definition 2.17), $x \sim_p y$ means that

$$x \xrightarrow{X} x' \quad y \xrightarrow{Y} y'$$

with $X \subseteq x$, $Y \subseteq y$, $X \sim Y$ and $x' \sim_p y'$.

By the pomset transition rules for encapsulation operator ∂_H in Table 4.12, we can get

$$\partial_H(x) \xrightarrow{X} \checkmark(X \nsubseteq H) \quad \partial_H(y) \xrightarrow{Y} \checkmark(Y \nsubseteq H)$$

with $X \subseteq x$, $Y \subseteq y$, and $X \sim Y$, so, we get $\partial_H(x) \sim_p \partial_H(y)$, as desired.

Or, we can get

$$\partial_H(x) \xrightarrow{X} \partial_H(x')(X \nsubseteq H) \quad \partial_H(y) \xrightarrow{Y} \partial_H(y')(Y \nsubseteq H)$$

with $X \subseteq x$, $Y \subseteq y$, $X \sim Y$, $x' \sim_p y'$ and the assumption $\partial_H(x') \sim_p \partial_H(y')$, so, we get $\partial_H(x) \sim_p \partial_H(y)$, as desired.

(2) The cases of step bisimulation $\sim_s$, hp-bisimulation $\sim_{hp}$ and hhp-bisimulation $\sim_{hhp}$ can be proven similarly, we omit them. $\square$

Theorem 4.33 (Elimination theorem of $APTC$). *Let p be a closed $APTC$ term including the encapsulation operator ∂_H. Then there is a basic $APTC$ term q such that $APTC \vdash p = q$.*

Proof. (1) Firstly, suppose that the following ordering on the signature of $APTC$ is defined: $\parallel > \cdot > +$ and the symbol $\cdot$ is given the lexicographical status for the first argument, then for each rewrite rule $p \to q$ in Table 4.14 relation $p >_{lpo} q$ can easily be proved. We obtain that the term rewrite system shown in Table 4.14 is strongly normalizing, for it has finitely many rewriting rules, and $>$ is a well-founded ordering on the signature of $APTC$, and if $s >_{lpo} t$, for each rewriting rule $s \to t$ is in Table 4.14 (see Theorem 2.12).

(2) Then we prove that the normal forms of closed $APTC$ terms including encapsulation operator ∂_H are basic $APTC$ terms.

Suppose that p is a normal form of some closed $APTC$ term and suppose that p is not a basic $APTC$ term. Let p' denote the smallest sub-term of p which is not a basic $APTC$

term. It implies that each sub-term of p' is a basic $APTC$ term. Then we prove that p is not a term in normal form. It is sufficient to induct on the structure of p', following from Theorem 4.20, we only prove the new case $p' \equiv \partial_H(p_1)$:

- Case $p_1 \equiv e$. The transition rules $RD1$ or $RD2$ can be applied, so p is not a normal form;
- Case $p_1 \equiv \delta$. The transition rules $RD3$ can be applied, so p is not a normal form;
- Case $p_1 \equiv p_1' + p_1''$. The transition rules $RD4$ can be applied, so p is not a normal form;
- Case $p_1 \equiv p_1' \cdot p_1''$. The transition rules $RD5$ can be applied, so p is not a normal form;
- Case $p_1 \equiv p_1' \parallel p_1''$. The transition rules $RD6$ can be applied, so p is not a normal form.

$\square$

Theorem 4.34 (Soundness of $APTC$ modulo step bisimulation equivalence). *Let x and y be $APTC$ terms including encapsulation operator ∂_H. If $APTC \vdash x = y$, then $x \sim_s y$.*

Proof. Since step bisimulation $\sim_s$ is both an equivalent and a congruent relation with respect to the operator ∂_H, we only need to check if each axiom in Table 4.13 is sound modulo step bisimulation equivalence.

Though transition rules in Table 4.12 are defined in the flavor of single event, they can be modified into a step (a set of events within which each event is pairwise concurrent), we omit them. If we treat a single event as a step containing just one event, the proof of this soundness theorem does not exist any problem, so we use this way and still use the transition rules in Table 4.12.

We omit the defining axioms, including axioms $D1 - D3$, and we only prove the soundness of the non-trivial axioms, including axioms $D4 - D6$.

- **Axiom** $D4$. Let p, q be $APTC$ processes, and $\partial_H(p + q) = \partial_H(p) + \partial_H(q)$, it is sufficient to prove that $\partial_H(p + q) \sim_s \partial_H(p) + \partial_H(q)$. By the transition rules for operator $+$ in Table 4.5 and ∂_H in Table 4.12, we get

$$\frac{p \xrightarrow{e_1} \checkmark \quad (e_1 \notin H)}{\partial_H(p + q) \xrightarrow{e_1} \checkmark} \qquad \frac{p \xrightarrow{e_1} \checkmark \quad (e_1 \notin H)}{\partial_H(p) + \partial_H(q) \xrightarrow{e_1} \checkmark}$$

$$\frac{q \xrightarrow{e_2} \checkmark \quad (e_2 \notin H)}{\partial_H(p + q) \xrightarrow{e_2} \checkmark} \qquad \frac{q \xrightarrow{e_2} \checkmark \quad (e_2 \notin H)}{\partial_H(p) + \partial_H(q) \xrightarrow{e_2} \checkmark}$$

$$\frac{p \xrightarrow{e_1} p' \quad (e_1 \notin H)}{\partial_H(p + q) \xrightarrow{e_1} \partial_H(p')} \qquad \frac{p \xrightarrow{e_1} p' \quad (e_1 \notin H)}{\partial_H(p) + \partial_H(q) \xrightarrow{e_1} \partial_H(p')}$$

$$\frac{q \xrightarrow{e_2} q' \quad (e_2 \notin H)}{\partial_H(p + q) \xrightarrow{e_2} \partial_H(q')} \qquad \frac{q \xrightarrow{e_2} q' \quad (e_2 \notin H)}{\partial_H(p) + \partial_H(q) \xrightarrow{e_2} \partial_H(q')}$$

So, $\partial_H(p + q) \sim_s \partial_H(p) + \partial_H(q)$, as desired.

- **Axiom** $D5$. Let p, q be $APTC$ processes, and $\partial_H(p \cdot q) = \partial_H(p) \cdot \partial_H(q)$, it is sufficient to prove that $\partial_H(p \cdot q) \sim_s \partial_H(p) \cdot \partial_H(q)$. By the transition rules for operator $\cdot$ in Table 4.5 and ∂_H in Table 4.12, we get

$$\frac{p \xrightarrow{e_1} \surd \quad (e_1 \notin H)}{\partial_H(p \cdot q) \xrightarrow{e_1} \partial_H(q)} \qquad \frac{p \xrightarrow{e_1} \surd \quad (e_1 \notin H)}{\partial_H(p) \cdot \partial_H(q) \xrightarrow{e_1} \partial_H(q)}$$

$$\frac{p \xrightarrow{e_1} p' \quad (e_1 \notin H)}{\partial_H(p \cdot q) \xrightarrow{e_1} \partial_H(p' \cdot q)} \qquad \frac{p \xrightarrow{e_1} p' \quad (e_1 \notin H)}{\partial_H(p) \cdot \partial_H(q) \xrightarrow{e_1} \partial_H(p') \cdot \partial_H(q)}$$

So, with the assumption $\partial_H(p' \cdot q) = \partial_H(p') \cdot \partial_H(q)$, $\partial_H(p \cdot q) \sim_s \partial_H(p) \cdot \partial_H(q)$, as desired.

- **Axiom** $D6$. Let p, q be $APTC$ processes, and $\partial_H(p \parallel q) = \partial_H(p) \parallel \partial_H(q)$, it is sufficient to prove that $\partial_H(p \parallel q) \sim_s \partial_H(p) \parallel \partial_H(q)$. By the transition rules for operator $\parallel$ in Table 4.7 and ∂_H in Table 4.12, we get

$$\frac{p \xrightarrow{e_1} \surd \quad q \xrightarrow{e_2} \surd \quad (e_1, e_2 \notin H)}{\partial_H(p \parallel q) \xrightarrow{\{e_1,e_2\}} \surd} \qquad \frac{p \xrightarrow{e_1} \surd \quad q \xrightarrow{e_2} \surd \quad (e_1, e_2 \notin H)}{\partial_H(p) \parallel \partial_H(q) \xrightarrow{\{e_1,e_2\}} \surd}$$

$$\frac{p \xrightarrow{e_1} p' \quad q \xrightarrow{e_2} \surd \quad (e_1, e_2 \notin H)}{\partial_H(p \parallel q) \xrightarrow{\{e_1,e_2\}} \partial_H(p')} \qquad \frac{p \xrightarrow{e_1} p' \quad q \xrightarrow{e_2} \surd \quad (e_1, e_2 \notin H)}{\partial_H(p) \parallel \partial_H(q) \xrightarrow{\{e_1,e_2\}} \partial_H(p')}$$

$$\frac{p \xrightarrow{e_1} \surd \quad q \xrightarrow{e_2} q' \quad (e_1, e_2 \notin H)}{\partial_H(p \parallel q) \xrightarrow{\{e_1,e_2\}} \partial_H(q')} \qquad \frac{p \xrightarrow{e_1} \surd \quad q \xrightarrow{e_2} q' \quad (e_1, e_2 \notin H)}{\partial_H(p) \parallel \partial_H(q) \xrightarrow{\{e_1,e_2\}} \partial_H(q')}$$

$$\frac{p \xrightarrow{e_1} p' \quad q \xrightarrow{e_2} q' \quad (e_1, e_2 \notin H)}{\partial_H(p \parallel q) \xrightarrow{\{e_1,e_2\}} \partial_H(p' \between q')} \qquad \frac{p \xrightarrow{e_1} p' \quad q \xrightarrow{e_2} q' \quad (e_1, e_2 \notin H)}{\partial_H(p) \parallel \partial_H(q) \xrightarrow{\{e_1,e_2\}} \partial_H(p') \between \partial_H(q')}$$

So, with the assumption $\partial_H(p' \between q') = \partial_H(p') \between \partial_H(q')$, $\partial_H(p \parallel q) \sim_s \partial_H(p) \parallel \partial_H(q)$, as desired. $\qquad \square$

Theorem 4.35 (Completeness of $APTC$ modulo step bisimulation equivalence). *Let p and q be closed $APTC$ terms including encapsulation operator ∂_H, if $p \sim_s q$ then $p = q$.*

Proof. Firstly, by the elimination theorem of $APTC$ (see Theorem 4.33), we know that the normal form of $APTC$ does not contain ∂_H, and for each closed $APTC$ term p, there exists a closed basic $APTC$ term p', such that $APTC \vdash p = p'$, so, we only need to consider closed basic $APTC$ terms.

Similarly to Theorem 4.25, we can prove that for normal forms n and n', if $n \sim_s n'$ then $n =_{AC} n'$.

Finally, let s and t be basic $APTC$ terms, and $s \sim_s t$, there are normal forms n and n', such that $s = n$ and $t = n'$. The soundness theorem of $APTC$ modulo step bisimulation equivalence (see Theorem 4.34) yields $s \sim_s n$ and $t \sim_s n'$, so $n \sim_s s \sim_s t \sim_s n'$. Since if $n \sim_s n'$ then $n =_{AC} n'$, $s = n =_{AC} n' = t$, as desired.   □

Theorem 4.36 (Soundness of $APTC$ modulo pomset bisimulation equivalence). *Let x and y be $APTC$ terms including encapsulation operator ∂_H. If $APTC \vdash x = y$, then $x \sim_p y$.*

Proof. Since pomset bisimulation $\sim_p$ is both an equivalent and a congruent relation with respect to the operator ∂_H, we only need to check if each axiom in Table 4.13 is sound modulo pomset bisimulation equivalence.

From the definition of pomset bisimulation (see Definition 2.17), we know that pomset bisimulation is defined by pomset transitions, which are labeled by pomsets. In a pomset transition, the events in the pomset are either within causality relations (defined by ·) or in concurrency (implicitly defined by · and +, and explicitly defined by ⟲), of course, they are pairwise consistent (without conflicts). In Theorem 4.34, we have already proven the case that all events are pairwise concurrent, so, we only need to prove the case of events in causality. Without loss of generality, we take a pomset of $P = \{e_1, e_2 : e_1 \cdot e_2\}$. Then the pomset transition labeled by the above P is just composed of one single event transition labeled by e_1 succeeded by another single event transition labeled by e_2, that is, $\xrightarrow{P} = \xrightarrow{e_1} \xrightarrow{e_2}$.

Similarly to the proof of soundness of $APTC$ modulo step bisimulation equivalence (see Theorem 4.34), we can prove that each axiom in Table 4.13 is sound modulo pomset bisimulation equivalence, we omit them.   □

Theorem 4.37 (Completeness of $APTC$ modulo pomset bisimulation equivalence). *Let p and q be closed $APTC$ terms including encapsulation operator ∂_H, if $p \sim_p q$ then $p = q$.*

Proof. Firstly, by the elimination theorem of $APTC$ (see Theorem 4.33), we know that the normal form of $APTC$ does not contain ∂_H, and for each closed $APTC$ term p, there exists a closed basic $APTC$ term p', such that $APTC \vdash p = p'$, so, we only need to consider closed basic $APTC$ terms.

Similarly to Theorem 4.35, we can prove that for normal forms n and n', if $n \sim_p n'$ then $n =_{AC} n'$.

Finally, let s and t be basic $APTC$ terms, and $s \sim_p t$, there are normal forms n and n', such that $s = n$ and $t = n'$. The soundness theorem of $APTC$ modulo pomset bisimulation equivalence (see Theorem 4.36) yields $s \sim_p n$ and $t \sim_p n'$, so $n \sim_p s \sim_p t \sim_p n'$. Since if $n \sim_p n'$ then $n =_{AC} n'$, $s = n =_{AC} n' = t$, as desired.   □

Theorem 4.38 (Soundness of $APTC$ modulo hp-bisimulation equivalence). *Let x and y be $APTC$ terms including encapsulation operator ∂_H. If $APTC \vdash x = y$, then $x \sim_{hp} y$.*

Proof. Since hp-bisimulation $\sim_{hp}$ is both an equivalent and a congruent relation with respect to the operator ∂_H, we only need to check if each axiom in Table 4.13 is sound modulo hp-bisimulation equivalence.

From the definition of hp-bisimulation (see Definition 2.21), we know that hp-bisimulation is defined on the posetal product (C_1, f, C_2), $f : C_1 \to C_2$ isomorphism. Two process terms s related to C_1 and t related to C_2, and $f : C_1 \to C_2$ isomorphism. Initially,

$(C_1, f, C_2) = (\emptyset, \emptyset, \emptyset)$, and $(\emptyset, \emptyset, \emptyset) \in \sim_{hp}$. When $s \xrightarrow{e} s'$ ($C_1 \xrightarrow{e} C_1'$), there will be $t \xrightarrow{e} t'$ ($C_2 \xrightarrow{e} C_2'$), and we define $f' = f[e \mapsto e]$. Then, if $(C_1, f, C_2) \in \sim_{hp}$, then $(C_1', f', C_2') \in \sim_{hp}$.

Similarly to the proof of soundness of $APTC$ modulo pomset bisimulation equivalence (see Theorem 4.36), we can prove that each axiom in Table 4.13 is sound modulo hp-bisimulation equivalence, we just need additionally to check the above conditions on hp-bisimulation, we omit them.   □

Theorem 4.39 (Completeness of $APTC$ modulo hp-bisimulation equivalence). *Let p and q be closed $APTC$ terms including encapsulation operator ∂_H, if $p \sim_{hp} q$ then $p = q$.*

Proof. Firstly, by the elimination theorem of $APTC$ (see Theorem 4.33), we know that the normal form of $APTC$ does not contain ∂_H, and for each closed $APTC$ term p, there exists a closed basic $APTC$ term p', such that $APTC \vdash p = p'$, so, we only need to consider closed basic $APTC$ terms.

Similarly to Theorem 4.37, we can prove that for normal forms n and n', if $n \sim_{hp} n'$ then $n =_{AC} n'$.

Finally, let s and t be basic $APTC$ terms, and $s \sim_{hp} t$, there are normal forms n and n', such that $s = n$ and $t = n'$. The soundness theorem of $APTC$ modulo hp-bisimulation equivalence (see Theorem 4.38) yields $s \sim_{hp} n$ and $t \sim_{hp} n'$, so $n \sim_{hp} s \sim_{hp} t \sim_{hp} n'$. Since if $n \sim_{hp} n'$ then $n =_{AC} n'$, $s = n =_{AC} n' = t$, as desired.   □

4.3 Recursion

In this section, we introduce recursion to capture infinite processes based on $APTC$. Since in $APTC$, there are three basic operators $\cdot$, $+$ and $\parallel$, the recursion must be adapted this situation to include $\parallel$.

In the following, E, F, G are recursion specifications, X, Y, Z are recursive variables.

4.3.1 Guarded recursive specifications

Definition 4.40 (Recursive specification). A recursive specification is a finite set of recursive equations

$$X_1 = t_1(X_1, \cdots, X_n)$$

$$\cdots$$

$$X_n = t_n(X_1, \cdots, X_n)$$

where the left-hand sides of X_i are called recursion variables, and the right-hand sides $t_i(X_1, \cdots, X_n)$ are process terms in $APTC$ with possible occurrences of the recursion variables $X_1, \cdots, X_n$.

Definition 4.41 (Solution). Processes $p_1, \cdots, p_n$ are solutions for a recursive specification $\{X_i = t_i(X_1, \cdots, X_n) | i \in \{1, \cdots, n\}\}$ (with respect to truly concurrent bisimulation equivalences $\sim_s (\sim_p, \sim_{hp})$) if $p_i \sim_s (\sim_p, \sim_{hp}) t_i(p_1, \cdots, p_n)$ for $i \in \{1, \cdots, n\}$.

Definition 4.42 (Guarded recursive specification). A recursive specification

$$X_1 = t_1(X_1, \cdots, X_n)$$

$$\cdots$$

$$X_n = t_n(X_1, \cdots, X_n)$$

is guarded if the right-hand sides of its recursive equations can be adapted to the form by applications of the axioms in $APTC$ and replacing recursion variables by the right-hand sides of their recursive equations,

$$(a_{11} \parallel \cdots \parallel a_{1i_1}) \cdot s_1(X_1, \cdots, X_n) + \cdots + (a_{k1} \parallel \cdots \parallel a_{ki_k}) \cdot s_k(X_1, \cdots, X_n) + (b_{11} \parallel \cdots \parallel b_{1j_1}) + \cdots$$
$$+ (b_{1j_1} \parallel \cdots \parallel b_{lj_l})$$

where $a_{11}, \cdots, a_{1i_1}, a_{k1}, \cdots, a_{ki_k}, b_{11}, \cdots, b_{1j_1}, b_{1j_1}, \cdots, b_{lj_l} \in \mathbb{E}$, and the sum above is allowed to be empty, in which case it represents the deadlock δ.

Definition 4.43 (Linear recursive specification). A recursive specification is linear if its recursive equations are of the form

$$(a_{11} \parallel \cdots \parallel a_{1i_1})X_1 + \cdots + (a_{k1} \parallel \cdots \parallel a_{ki_k})X_k + (b_{11} \parallel \cdots \parallel b_{1j_1}) + \cdots + (b_{1j_1} \parallel \cdots \parallel b_{lj_l})$$

where $a_{11}, \cdots, a_{1i_1}, a_{k1}, \cdots, a_{ki_k}, b_{11}, \cdots, b_{1j_1}, b_{1j_1}, \cdots, b_{lj_l} \in \mathbb{E}$, and the sum above is allowed to be empty, in which case it represents the deadlock δ.

For a guarded recursive specifications E with the form

$$X_1 = t_1(X_1, \cdots, X_n)$$

$$\cdots$$

$$X_n = t_n(X_1, \cdots, X_n)$$

the behavior of the solution $\langle X_i | E \rangle$ for the recursion variable X_i in E, where $i \in \{1, \cdots, n\}$, is exactly the behavior of their right-hand sides $t_i(X_1, \cdots, X_n)$, which is captured by the two transition rules in Table 4.15.

Theorem 4.44 (Conservativity of $APTC$ with guarded recursion). *$APTC$ with guarded recursion is a conservative extension of $APTC$.*

Proof. Since the transition rules of $APTC$ are source-dependent, and the transition rules for guarded recursion in Table 4.15 contain only a fresh constant in their source, so the transition rules of $APTC$ with guarded recursion are conservative extensions of those of $APTC$. □

Theorem 4.45 (Congruence theorem of $APTC$ with guarded recursion). *Truly concurrent bisimulation equivalences $\sim_p$, $\sim_s$, and $\sim_{hp}$ are all congruences with respect to $APTC$ with guarded recursion.*

Table 4.15 Transition rules of guarded recursion.

$$\frac{t_i(\langle X_1|E\rangle,\cdots,\langle X_n|E\rangle)\xrightarrow{\{e_1,\cdots,e_k\}}\surd}{\langle X_i|E\rangle\xrightarrow{\{e_1,\cdots,e_k\}}\surd}$$

$$\frac{t_i(\langle X_1|E\rangle,\cdots,\langle X_n|E\rangle)\xrightarrow{\{e_1,\cdots,e_k\}}y}{\langle X_i|E\rangle\xrightarrow{\{e_1,\cdots,e_k\}}y}$$

Table 4.16 Recursive definition and specification principle.

No.	Axiom
RDP	$\langle X_i|E\rangle = t_i(\langle X_1|E\rangle,\cdots,\langle X_n|E\rangle)$ $(i\in\{1,\cdots,n\})$
RSP	if $y_i = t_i(y_1,\cdots,y_n)$ for $i\in\{1,\cdots,n\}$, then $y_i = \langle X_i|E\rangle$ $(i\in\{1,\cdots,n\})$

Proof. It follows the following two facts:

1. in a guarded recursive specification, right-hand sides of its recursive equations can be adapted to the form by applications of the axioms in $APTC$ and replacing recursion variables by the right-hand sides of their recursive equations;
2. truly concurrent bisimulation equivalences $\sim_p$, $\sim_s$, and $\sim_{hp}$ are all congruences with respect to all operators of $APTC$.   □

4.3.2 Recursive definition and specification principles

The RDP (Recursive Definition Principle) and the RSP (Recursive Specification Principle) are shown in Table 4.16.

RDP follows immediately from the two transition rules for guarded recursion, which express that $\langle X_i|E\rangle$ and $t_i(\langle X_1|E\rangle,\cdots,\langle X_n|E\rangle)$ have the same initial transitions for $i\in\{1,\cdots,n\}$. RSP follows from the fact that guarded recursive specifications have only one solution.

Theorem 4.46 (Elimination theorem of $APTC$ with linear recursion). *Each process term in $APTC$ with linear recursion is equal to a process term $\langle X_1|E\rangle$ with E a linear recursive specification.*

Proof. By applying structural induction with respect to term size, each process term t_1 in $APTC$ with linear recursion generates a process can be expressed in the form of equations

$$t_i = (a_{i11} \parallel \cdots \parallel a_{i1i_1})t_{i1} + \cdots + (a_{ik_i1} \parallel \cdots \parallel a_{ik_ii_k})t_{ik_i} + (b_{i11} \parallel \cdots \parallel b_{i1i_1}) + \cdots + (b_{il_i1} \parallel \cdots \parallel b_{il_ii_l})$$

for $i \in \{1, \cdots, n\}$. Let the linear recursive specification E consist of the recursive equations

$$X_i = (a_{i11} \parallel \cdots \parallel a_{i1l_i})X_{i1} + \cdots + (a_{ik_i1} \parallel \cdots \parallel a_{ik_il_k})X_{ik_i} + (b_{i11} \parallel \cdots \parallel b_{i1l_1}) + \cdots$$
$$+ (b_{il_i1} \parallel \cdots \parallel b_{il_il_i})$$

for $i \in \{1, \cdots, n\}$. Replacing X_i by t_i for $i \in \{1, \cdots, n\}$ is a solution for E, RSP yields $t_1 = \langle X_1|E \rangle$. $\qquad\qquad\qquad\qquad\qquad\qquad\qquad\qquad\qquad\qquad\qquad\qquad\qquad\qquad\qquad\qquad\qquad\qquad$ $\square$

Theorem 4.47 (Soundness of $APTC$ with guarded recursion). *Let x and y be $APTC$ with guarded recursion terms. If $APTC$ with guarded recursion $\vdash x = y$, then*

1. $x \sim_s y$;
2. $x \sim_p y$;
3. $x \sim_{hp} y$.

Proof. (1) Soundness of $APTC$ with guarded recursion with respect to step bisimulation $\sim_s$.

Since step bisimulation $\sim_s$ is both an equivalent and a congruent relation with respect to $APTC$ with guarded recursion, we only need to check if each axiom in Table 4.16 is sound modulo step bisimulation equivalence.

Though transition rules in Table 4.15 are defined in the flavor of single event, they can be modified into a step (a set of events within which each event is pairwise concurrent), we omit them. If we treat a single event as a step containing just one event, the proof of this soundness theorem does not exist any problem, so we use this way and still use the transition rules in Table 4.15.

- RDP. $\langle X_i|E \rangle = t_i(\langle X_1|E \rangle, \cdots, \langle X_n|E \rangle)$ $(i \in \{1, \cdots, n\})$, it is sufficient to prove that $\langle X_i|E \rangle \sim_s t_i(\langle X_1|E \rangle, \cdots, X_n|E \rangle)$ $(i \in \{1, \cdots, n\})$. By the transition rules for guarded recursion in Table 4.15, we get

$$\frac{t_i(\langle X_1|E \rangle, \cdots, \langle X_n|E \rangle) \xrightarrow{\{e_1, \cdots, e_k\}} \surd}{\langle X_i|E \rangle \xrightarrow{\{e_1, \cdots, e_k\}} \surd}$$

$$\frac{t_i(\langle X_1|E \rangle, \cdots, \langle X_n|E \rangle) \xrightarrow{\{e_1, \cdots, e_k\}} y}{\langle X_i|E \rangle \xrightarrow{\{e_1, \cdots, e_k\}} y}$$

  So, $\langle X_i|E \rangle \sim_s t_i(\langle X_1|E \rangle, \cdots, \langle X_n|E \rangle)$ $(i \in \{1, \cdots, n\})$, as desired.
- RSP. if $y_i = t_i(y_1, \cdots, y_n)$ for $i \in \{1, \cdots, n\}$, then $y_i = \langle X_i|E \rangle$ $(i \in \{1, \cdots, n\})$, it is sufficient to prove that if $y_i = t_i(y_1, \cdots, y_n)$ for $i \in \{1, \cdots, n\}$, then $y_i \sim_s \langle X_i|E \rangle$ $(i \in \{1, \cdots, n\})$. By the transition rules for guarded recursion in Table 4.15, we get

$$\frac{t_i(\langle X_1|E \rangle, \cdots, \langle X_n|E \rangle) \xrightarrow{\{e_1, \cdots, e_k\}} \surd}{\langle X_i|E \rangle \xrightarrow{\{e_1, \cdots, e_k\}} \surd}$$

$$\frac{t_i(\langle X_1|E\rangle,\cdots,\langle X_n|E\rangle) \xrightarrow{\{e_1,\cdots,e_k\}} \surd}{y_i \xrightarrow{\{e_1,\cdots,e_k\}} \surd}$$

$$\frac{t_i(\langle X_1|E\rangle,\cdots,\langle X_n|E\rangle) \xrightarrow{\{e_1,\cdots,e_k\}} y}{\langle X_i|E\rangle \xrightarrow{\{e_1,\cdots,e_k\}} y}$$

$$\frac{t_i(\langle X_1|E\rangle,\cdots,\langle X_n|E\rangle) \xrightarrow{\{e_1,\cdots,e_k\}} y}{y_i \xrightarrow{\{e_1,\cdots,e_k\}} y}$$

So, if $y_i = t_i(y_1,\cdots,y_n)$ for $i \in \{1,\cdots,n\}$, then $y_i \sim_s \langle X_i|E\rangle$ $(i \in \{1,\cdots,n\})$, as desired.

(2) Soundness of $APTC$ with guarded recursion with respect to pomset bisimulation $\sim_p$.

Since pomset bisimulation $\sim_p$ is both an equivalent and a congruent relation with respect to the guarded recursion, we only need to check if each axiom in Table 4.16 is sound modulo pomset bisimulation equivalence.

From the definition of pomset bisimulation (see Definition 2.17), we know that pomset bisimulation is defined by pomset transitions, which are labeled by pomsets. In a pomset transition, the events in the pomset are either within causality relations (defined by $\cdot$) or in concurrency (implicitly defined by $\cdot$ and $+$, and explicitly defined by $\between$), of course, they are pairwise consistent (without conflicts). In (1), we have already proven the case that all events are pairwise concurrent, so, we only need to prove the case of events in causality. Without loss of generality, we take a pomset of $P = \{e_1, e_2 : e_1 \cdot e_2\}$. Then the pomset transition labeled by the above P is just composed of one single event transition labeled by e_1 succeeded by another single event transition labeled by e_2, that is, $\xrightarrow{P} = \xrightarrow{e_1} \xrightarrow{e_2}$.

Similarly to the proof of soundness of $APTC$ with guarded recursion modulo step bisimulation equivalence (1), we can prove that each axiom in Table 4.16 is sound modulo pomset bisimulation equivalence, we omit them.

(3) Soundness of $APTC$ with guarded recursion with respect to hp-bisimulation $\sim_{hp}$.

Since hp-bisimulation $\sim_{hp}$ is both an equivalent and a congruent relation with respect to guarded recursion, we only need to check if each axiom in Table 4.16 is sound modulo hp-bisimulation equivalence.

From the definition of hp-bisimulation (see Definition 2.21), we know that hp-bisimulation is defined on the posetal product (C_1, f, C_2), $f : C_1 \to C_2$ isomorphism. Two process terms s related to C_1 and t related to C_2, and $f : C_1 \to C_2$ isomorphism. Initially, $(C_1, f, C_2) = (\emptyset, \emptyset, \emptyset)$, and $(\emptyset, \emptyset, \emptyset) \in \sim_{hp}$. When $s \xrightarrow{e} s'$ $(C_1 \xrightarrow{e} C_1')$, there will be $t \xrightarrow{e} t'$ $(C_2 \xrightarrow{e} C_2')$, and we define $f' = f[e \mapsto e]$. Then, if $(C_1, f, C_2) \in \sim_{hp}$, then $(C_1', f', C_2') \in \sim_{hp}$.

Similarly to the proof of soundness of $APTC$ with guarded recursion modulo pomset bisimulation equivalence (2), we can prove that each axiom in Table 4.16 is sound modulo hp-bisimulation equivalence, we just need additionally to check the above conditions on hp-bisimulation, we omit them. □

Theorem 4.48 (Completeness of $APTC$ with linear recursion). *Let p and q be closed $APTC$ with linear recursion terms, then,*

1. *if $p \sim_s q$ then $p = q$;*
2. *if $p \sim_p q$ then $p = q$;*
3. *if $p \sim_{hp} q$ then $p = q$.*

Proof. Firstly, by the elimination theorem of $APTC$ with guarded recursion (see Theorem 4.46), we know that each process term in $APTC$ with linear recursion is equal to a process term $\langle X_1|E \rangle$ with E a linear recursive specification.

It remains to prove the following cases.

(1) If $\langle X_1|E_1 \rangle \sim_s \langle Y_1|E_2 \rangle$ for linear recursive specification E_1 and E_2, then $\langle X_1|E_1 \rangle = \langle Y_1|E_2 \rangle$.

Let E_1 consist of recursive equations $X = t_X$ for $X \in \mathcal{X}$ and E_2 consist of recursion equations $Y = t_Y$ for $Y \in \mathcal{Y}$. Let the linear recursive specification E consist of recursion equations $Z_{XY} = t_{XY}$, and $\langle X|E_1 \rangle \sim_s \langle Y|E_2 \rangle$, and t_{XY} consist of the following summands:

1. t_{XY} contains a summand $(a_1 \parallel \cdots \parallel a_m)Z_{X'Y'}$ iff t_X contains the summand $(a_1 \parallel \cdots \parallel a_m)X'$ and t_Y contains the summand $(a_1 \parallel \cdots \parallel a_m)Y'$ such that $\langle X'|E_1 \rangle \sim_s \langle Y'|E_2 \rangle$;
2. t_{XY} contains a summand $b_1 \parallel \cdots \parallel b_n$ iff t_X contains the summand $b_1 \parallel \cdots \parallel b_n$ and t_Y contains the summand $b_1 \parallel \cdots \parallel b_n$.

Let σ map recursion variable X in E_1 to $\langle X|E_1 \rangle$, and let ψ map recursion variable Z_{XY} in E to $\langle X|E_1 \rangle$. So, $\sigma((a_1 \parallel \cdots \parallel a_m)X') \equiv (a_1 \parallel \cdots \parallel a_m)\langle X'|E_1 \rangle \equiv \psi((a_1 \parallel \cdots \parallel a_m)Z_{X'Y'})$, so by RDP, we get $\langle X|E_1 \rangle = \sigma(t_X) = \psi(t_{XY})$. Then by RSP, $\langle X|E_1 \rangle = \langle Z_{XY}|E \rangle$, particularly, $\langle X_1|E_1 \rangle = \langle Z_{X_1Y_1}|E \rangle$. Similarly, we can obtain $\langle Y_1|E_2 \rangle = \langle Z_{X_1Y_1}|E \rangle$. Finally, $\langle X_1|E_1 \rangle = \langle Z_{X_1Y_1}|E \rangle = \langle Y_1|E_2 \rangle$, as desired.

(2) If $\langle X_1|E_1 \rangle \sim_p \langle Y_1|E_2 \rangle$ for linear recursive specification E_1 and E_2, then $\langle X_1|E_1 \rangle = \langle Y_1|E_2 \rangle$.

It can be proven similarly to (1), we omit it.

(3) If $\langle X_1|E_1 \rangle \sim_{hp} \langle Y_1|E_2 \rangle$ for linear recursive specification E_1 and E_2, then $\langle X_1|E_1 \rangle = \langle Y_1|E_2 \rangle$.

It can be proven similarly to (1), we omit it. □

4.3.3 Approximation induction principle

In this subsection, we introduce approximation induction principle (AIP) and try to explain that AIP is still valid in true concurrency. AIP can be used to try and equate truly concurrent bisimilar guarded recursive specifications. AIP says that if two process terms are truly concurrent bisimilar up to any finite depth, then they are truly concurrent bisimilar.

Also, we need the auxiliary unary projection operator π_n for $n \in \mathbb{N}$ and $\mathbb{N} \triangleq \{0, 1, 2, \cdots\}$. The transition rules of π_n are expressed in Table 4.17.

Based on the transition rules for projection operator π_n in Table 4.17, we design the axioms as Table 4.18 shows.

Table 4.17   Transition rules of projection operator π_n.

$$\frac{x \xrightarrow{\{e_1,\cdots,e_k\}} \surd}{\pi_{n+1}(x) \xrightarrow{\{e_1,\cdots,e_k\}} \surd} \qquad \frac{x \xrightarrow{\{e_1,\cdots,e_k\}} x'}{\pi_{n+1}(x) \xrightarrow{\{e_1,\cdots,e_k\}} \pi_n(x')}$$

Table 4.18   Axioms of projection operator.

No.	Axiom
$PR1$	$\pi_n(x+y) = \pi_n(x) + \pi_n(y)$
$PR2$	$\pi_n(x \parallel y) = \pi_n(x) \parallel \pi_n(y)$
$PR3$	$\pi_{n+1}(e_1 \parallel \cdots \parallel e_k) = e_1 \parallel \cdots \parallel e_k$
$PR4$	$\pi_{n+1}((e_1 \parallel \cdots \parallel e_k) \cdot x) = (e_1 \parallel \cdots \parallel e_k) \cdot \pi_n(x)$
$PR5$	$\pi_0(x) = \delta$
$PR6$	$\pi_n(\delta) = \delta$

The axioms $PR1 - PR2$ say that $\pi_n(s+t)$ and $\pi_n(s \parallel t)$ can execute transitions of s and t up to depth n. $PR3$ says that $\pi_{n+1}(e_1 \parallel \cdots \parallel e_k)$ executes $\{e_1,\cdots,e_k\}$ and terminates successfully. $PR4$ says that $\pi_{n+1}((e_1 \parallel \cdots \parallel e_k) \cdot t)$ executes $\{e_1,\cdots,e_k\}$ and then executes transitions of t up to depth n. $PR5$ and $PR6$ say that $\pi_0(t)$ and $\pi_n(\delta)$ exhibit no actions.

Theorem 4.49 (Conservativity of $APTC$ with projection operator and guarded recursion). *$APTC$ with projection operator and guarded recursion is a conservative extension of $APTC$ with guarded recursion.*

Proof. It follows from the following two facts (see Theorem 2.8).

1. The transition rules of $APTC$ with guarded recursion are all source-dependent;
2. The sources of the transition rules for the projection operator contain an occurrence of π_n. □

Theorem 4.50 (Congruence theorem of projection operator π_n). *Truly concurrent bisimulation equivalences $\sim_p$, $\sim_s$, $\sim_{hp}$ and $\sim_{hhp}$ are all congruences with respect to projection operator π_n.*

Proof. (1) Case pomset bisimulation equivalence $\sim_p$.
Let x and y be $APTC$ with projection operator and guarded recursion processes, and $x \sim_p y$, it is sufficient to prove that $\pi_{n+1}(x) \sim_p \pi_{n+1}(y)$.
By the definition of pomset bisimulation $\sim_p$ (Definition 2.17), $x \sim_p y$ means that

$$x \xrightarrow{X} x' \quad y \xrightarrow{Y} y'$$

with $X \subseteq x, Y \subseteq y, X \sim Y$ and $x' \sim_p y'$.

By the pomset transition rules for projection operator π_n in Table 4.17, we can get

$$\pi_{n+1}(x) \xrightarrow{X} \surd \quad \pi_{n+1}(y) \xrightarrow{Y} \surd$$

with $X \subseteq x$, $Y \subseteq y$, and $X \sim Y$, so, we get $\pi_{n+1}(x) \sim_p \pi_{n+1}(y)$, as desired.

Or, we can get

$$\pi_{n+1}(x) \xrightarrow{X} \pi_n(x') \quad \pi_{n+1}(y) \xrightarrow{Y} \pi_n(y')$$

with $X \subseteq x$, $Y \subseteq y$, $X \sim Y$, $x' \sim_p y'$ and the assumption $\pi_n(x') \sim_p \pi_n(y')$, so, we get $\pi_{n+1}(x) \sim_p \pi_{n+1}(y)$, as desired.

(2) The cases of step bisimulation $\sim_s$, hp-bisimulation $\sim_{hp}$ and hhp-bisimulation $\sim_{hhp}$ can be proven similarly, we omit them. $\qquad\square$

Theorem 4.51 (Elimination theorem of $APTC$ with linear recursion and projection operator). *Each process term in $APTC$ with linear recursion and projection operator is equal to a process term $\langle X_1|E \rangle$ with E a linear recursive specification.*

Proof. By applying structural induction with respect to term size, each process term t_1 in $APTC$ with linear recursion and projection operator π_n generates a process can be expressed in the form of equations

$$t_i = (a_{i11} \parallel \cdots \parallel a_{i1i_1})t_{i1} + \cdots + (a_{ik_i1} \parallel \cdots \parallel a_{ik_ii_k})t_{ik_i} + (b_{i11} \parallel \cdots \parallel b_{i1i_1}) + \cdots + (b_{il_i1} \parallel \cdots \parallel b_{il_ii_l})$$

for $i \in \{1, \cdots, n\}$. Let the linear recursive specification E consist of the recursive equations

$$X_i = (a_{i11} \parallel \cdots \parallel a_{i1i_1})X_{i1} + \cdots + (a_{ik_i1} \parallel \cdots \parallel a_{ik_ii_k})X_{ik_i} + (b_{i11} \parallel \cdots \parallel b_{i1i_1}) + \cdots$$
$$+ (b_{il_i1} \parallel \cdots \parallel b_{il_ii_l})$$

for $i \in \{1, \cdots, n\}$. Replacing X_i by t_i for $i \in \{1, \cdots, n\}$ is a solution for E, RSP yields $t_1 = \langle X_1|E \rangle$.

That is, in E, there is not the occurrence of projection operator π_n. $\qquad\square$

Theorem 4.52 (Soundness of $APTC$ with projection operator and guarded recursion). *Let x and y be $APTC$ with projection operator and guarded recursion terms. If $APTC$ with projection operator and guarded recursion $\vdash x = y$, then*

1. $x \sim_s y$;
2. $x \sim_p y$;
3. $x \sim_{hp} y$.

Proof. (1) Soundness of $APTC$ with projection operator and guarded recursion with respect to step bisimulation $\sim_s$.

Since step bisimulation $\sim_s$ is both an equivalent and a congruent relation with respect to $APTC$ with projection operator and guarded recursion, we only need to check if each axiom in Table 4.18 is sound modulo step bisimulation equivalence.

Though transition rules in Table 4.17 are defined in the flavor of single event, they can be modified into a step (a set of events within which each event is pairwise concurrent), we omit them. If we treat a single event as a step containing just one event, the proof of this soundness theorem does not exist any problem, so we use this way and still use the transition rules in Table 4.17.

We only prove the soundness of the non-trivial axioms $PR1$, $PR2$ and $PR4$.

- **Axiom $PR1$.** Let p and q be $APTC$ with guarded recursion and projection operator processes. $\pi_n(p+q) = \pi_n(p) + \pi_n(q)$, it is sufficient to prove that $\pi_n(p+q) \sim_s \pi_n(p) + \pi_n(q)$. By the transition rules for projection operator π_n in Table 4.17 and $+$ in Table 4.5, we get

$$\frac{p \xrightarrow{\{e_1,\cdots,e_k\}} \surd}{\pi_{n+1}(p+q) \xrightarrow{\{e_1,\cdots,e_k\}} \surd} \qquad \frac{p \xrightarrow{\{e_1,\cdots,e_k\}} \surd}{\pi_{n+1}(p) + \pi_{n+1}(q) \xrightarrow{\{e_1,\cdots,e_k\}} \surd}$$

$$\frac{q \xrightarrow{\{e'_1,\cdots,e'_k\}} \surd}{\pi_{n+1}(p+q) \xrightarrow{\{e'_1,\cdots,e'_k\}} \surd} \qquad \frac{q \xrightarrow{\{e'_1,\cdots,e'_k\}} \surd}{\pi_{n+1}(p) + \pi_{n+1}(q) \xrightarrow{\{e'_1,\cdots,e'_k\}} \surd}$$

$$\frac{p \xrightarrow{\{e_1,\cdots,e_k\}} p'}{\pi_{n+1}(p+q) \xrightarrow{\{e_1,\cdots,e_k\}} \pi_n(p')} \qquad \frac{p \xrightarrow{\{e_1,\cdots,e_k\}} p'}{\pi_{n+1}(p) + \pi_{n+1}(q) \xrightarrow{\{e_1,\cdots,e_k\}} \pi_n(p')}$$

$$\frac{q \xrightarrow{\{e'_1,\cdots,e'_k\}} q'}{\pi_{n+1}(p+q) \xrightarrow{\{e'_1,\cdots,e'_k\}} \pi_n(q')} \qquad \frac{q \xrightarrow{\{e'_1,\cdots,e'_k\}} q'}{\pi_{n+1}(p) + \pi_{n+1}(q) \xrightarrow{\{e'_1,\cdots,e'_k\}} \pi_n(q')}$$

So, $\pi_n(p+q) \sim_s \pi_n(p) + \pi_n(q)$, as desired.

- **Axiom $PR2$.** Let p, q be $APTC$ with guarded recursion and projection operator processes, and $\pi_n(p \parallel q) = \pi_n(p) \parallel \pi_n(q)$, it is sufficient to prove that $\pi_n(p \parallel q) \sim_s \pi_n(p) \parallel \pi_n(q)$. By the transition rules for operator $\parallel$ in Table 4.7 and π_n in Table 4.17, we get

$$\frac{p \xrightarrow{e_1} \surd \quad q \xrightarrow{e_2} \surd}{\pi_{n+1}(p \parallel q) \xrightarrow{\{e_1,e_2\}} \surd} \qquad \frac{p \xrightarrow{e_1} \surd \quad q \xrightarrow{e_2} \surd}{\pi_{n+1}(p) \parallel \pi_{n+1}(q) \xrightarrow{\{e_1,e_2\}} \surd}$$

$$\frac{p \xrightarrow{e_1} p' \quad q \xrightarrow{e_2} \surd}{\pi_{n+1}(p \parallel q) \xrightarrow{\{e_1,e_2\}} \pi_n(p')} \qquad \frac{p \xrightarrow{e_1} p' \quad q \xrightarrow{e_2} \surd}{\pi_{n+1}(p) \parallel \pi_{n+1}(q) \xrightarrow{\{e_1,e_2\}} \pi_n(p')}$$

$$\frac{p \xrightarrow{e_1} \surd \quad q \xrightarrow{e_2} q'}{\pi_{n+1}(p \parallel q) \xrightarrow{\{e_1,e_2\}} \pi_n(q')} \qquad \frac{p \xrightarrow{e_1} \surd \quad q \xrightarrow{e_2} q'}{\pi_{n+1}(p) \parallel \pi_{n+1}(q) \xrightarrow{\{e_1,e_2\}} \pi_n(q')}$$

$$\frac{p \xrightarrow{e_1} p' \quad q \xrightarrow{e_2} q'}{\pi_{n+1}(p \parallel q) \xrightarrow{\{e_1,e_2\}} \pi_n(p' \between q')} \qquad \frac{p \xrightarrow{e_1} p' \quad q \xrightarrow{e_2} q'}{\pi_{n+1}(p) \parallel \pi_{n+1}(q) \xrightarrow{\{e_1,e_2\}} \pi_n(p') \between \pi_n(q')}$$

So, with the assumption $\pi_n(p' \between q') = \pi_n(p') \between \pi_n(q')$, $\pi_n(p \parallel q) \sim_s \pi_n(p) \parallel \pi_n(q)$, as desired.

- **Axiom** $PR4$. Let p be an $APTC$ with guarded recursion and projection operator process, and $\pi_{n+1}(e \cdot p) = e \cdot \pi_n(p)$, it is sufficient to prove that $\pi_{n+1}(e \cdot p) \sim_s e \cdot \pi_n(p)$. By the transition rules for operator $\cdot$ in Table 4.5 and π_n in Table 4.17, we get

$$\frac{e_1 \parallel \cdots \parallel e_k \xrightarrow{\{e_1,\cdots,e_k\}} \checkmark}{\pi_{n+1}((e_1 \parallel \cdots \parallel e_k) \cdot p) \xrightarrow{\{e_1,\cdots,e_k\}} \pi_n(p)} \qquad \frac{e_1 \parallel \cdots \parallel e_k \xrightarrow{\{e_1,\cdots,e_k\}} \checkmark}{(e_1 \parallel \cdots \parallel e_k) \cdot \pi_n(p) \xrightarrow{\{e_1,\cdots,e_k\}} \pi_n(p)}$$

So, $\pi_{n+1}(e \cdot p) \sim_s e \cdot \pi_n(p)$, as desired.

(2) Soundness of $APTC$ with guarded recursion and projection operator with respect to pomset bisimulation $\sim_p$.

Since pomset bisimulation $\sim_p$ is both an equivalent and a congruent relation with respect to $APTC$ with guarded recursion and projection operator, we only need to check if each axiom in Table 4.18 is sound modulo pomset bisimulation equivalence.

From the definition of pomset bisimulation (see Definition 2.17), we know that pomset bisimulation is defined by pomset transitions, which are labeled by pomsets. In a pomset transition, the events in the pomset are either within causality relations (defined by $\cdot$) or in concurrency (implicitly defined by $\cdot$ and $+$, and explicitly defined by $\between$), of course, they are pairwise consistent (without conflicts). In (1), we have already proven the case that all events are pairwise concurrent, so, we only need to prove the case of events in causality. Without loss of generality, we take a pomset of $P = \{e_1, e_2 : e_1 \cdot e_2\}$. Then the pomset transition labeled by the above P is just composed of one single event transition labeled by e_1 succeeded by another single event transition labeled by e_2, that is, $\xrightarrow{P} = \xrightarrow{e_1} \xrightarrow{e_2}$.

Similarly to the proof of soundness of $APTC$ with guarded recursion and projection operator modulo step bisimulation equivalence (1), we can prove that each axiom in Table 4.18 is sound modulo pomset bisimulation equivalence, we omit them.

(3) Soundness of $APTC$ with guarded recursion and projection operator with respect to hp-bisimulation $\sim_{hp}$.

Since hp-bisimulation $\sim_{hp}$ is both an equivalent and a congruent relation with respect to $APTC$ with guarded recursion and projection operator, we only need to check if each axiom in Table 4.18 is sound modulo hp-bisimulation equivalence.

From the definition of hp-bisimulation (see Definition 2.21), we know that hp-bisimulation is defined on the posetal product (C_1, f, C_2), $f : C_1 \to C_2$ isomorphism. Two process terms s related to C_1 and t related to C_2, and $f : C_1 \to C_2$ isomorphism. Initially, $(C_1, f, C_2) = (\emptyset, \emptyset, \emptyset)$, and $(\emptyset, \emptyset, \emptyset) \in \sim_{hp}$. When $s \xrightarrow{e} s'$ ($C_1 \xrightarrow{e} C_1'$), there will be $t \xrightarrow{e} t'$ ($C_2 \xrightarrow{e} C_2'$), and we define $f' = f[e \mapsto e]$. Then, if $(C_1, f, C_2) \in \sim_{hp}$, then $(C_1', f', C_2') \in \sim_{hp}$.

Similarly to the proof of soundness of $APTC$ with guarded recursion and projection operator modulo pomset bisimulation equivalence (2), we can prove that each axiom in

Table 4.19 AIP.

No.	Axiom
AIP	if $\pi_n(x) = \pi_n(y)$ for $n \in \mathbb{N}$, then $x = y$

Table 4.18 is sound modulo hp-bisimulation equivalence, we just need additionally to check the above conditions on hp-bisimulation, we omit them. □

Then AIP is given in Table 4.19.

Theorem 4.53 (Soundness of AIP). *Let x and y be $APTC$ with projection operator and guarded recursion terms.*

1. *If $\pi_n(x) \sim_s \pi_n(y)$ for $n \in \mathbb{N}$, then $x \sim_s y$;*
2. *If $\pi_n(x) \sim_p \pi_n(y)$ for $n \in \mathbb{N}$, then $x \sim_p y$;*
3. *If $\pi_n(x) \sim_{hp} \pi_n(y)$ for $n \in \mathbb{N}$, then $x \sim_{hp} y$.*

Proof. (1) If $\pi_n(x) \sim_s \pi_n(y)$ for $n \in \mathbb{N}$, then $x \sim_s y$.

Since step bisimulation $\sim_s$ is both an equivalent and a congruent relation with respect to $APTC$ with guarded recursion and projection operator, we only need to check if AIP in Table 4.19 is sound modulo step bisimulation equivalence.

Let p, p_0 and q, q_0 be closed $APTC$ with projection operator and guarded recursion terms such that $\pi_n(p_0) \sim_s \pi_n(q_0)$ for $n \in \mathbb{N}$. We define a relation R such that pRq iff $\pi_n(p) \sim_s \pi_n(q)$. Obviously, $p_0 R q_0$, next, we prove that $R \in \sim_s$.

Let pRq and $p \xrightarrow{\{e_1,\cdots,e_k\}} \surd$, then $\pi_1(p) \xrightarrow{\{e_1,\cdots,e_k\}} \surd$, $\pi_1(p) \sim_s \pi_1(q)$ yields $\pi_1(q) \xrightarrow{\{e_1,\cdots,e_k\}} \surd$. Similarly, $q \xrightarrow{\{e_1,\cdots,e_k\}} \surd$ implies $p \xrightarrow{\{e_1,\cdots,e_k\}} \surd$.

Let pRq and $p \xrightarrow{\{e_1,\cdots,e_k\}} p'$. We define the set of process terms

$$S_n \triangleq \{q' | q \xrightarrow{\{e_1,\cdots,e_k\}} q' \text{ and } \pi_n(p') \sim_s \pi_n(q')\}$$

1. Since $\pi_{n+1}(p) \sim_s \pi_{n+1}(q)$ and $\pi_{n+1}(p) \xrightarrow{\{e_1,\cdots,e_k\}} \pi_n(p')$, there exist q' such that $\pi_{n+1}(q) \xrightarrow{\{e_1,\cdots,e_k\}} \pi_n(q')$ and $\pi_n(p') \sim_s \pi_n(q')$. So, S_n is not empty.
2. There are only finitely many q' such that $q \xrightarrow{\{e_1,\cdots,e_k\}} q'$, so, S_n is finite.
3. $\pi_{n+1}(p) \sim_s \pi_{n+1}(q)$ implies $\pi_n(p') \sim_s \pi_n(q')$, so $S_n \supseteq S_{n+1}$.

So, S_n has a non-empty intersection, and let q' be in this intersection, then $q \xrightarrow{\{e_1,\cdots,e_k\}} q'$ and $\pi_n(p') \sim_s \pi_n(q')$, so $p'Rq'$. Similarly, let $p\amalg q$, we can obtain $q \xrightarrow{\{e_1,\cdots,e_k\}} q'$ implies $p \xrightarrow{\{e_1,\cdots,e_k\}} p'$ such that $p'Rq'$.

Finally, $R \in \sim_s$ and $p_0 \sim_s q_0$, as desired.

(2) If $\pi_n(x) \sim_p \pi_n(y)$ for $n \in \mathbb{N}$, then $x \sim_p y$.

Since pomset bisimulation $\sim_p$ is both an equivalent and a congruent relation with respect to $APTC$ with guarded recursion and projection operator, we only need to check if AIP in Table 4.19 is sound modulo pomset bisimulation equivalence.

From the definition of pomset bisimulation (see Definition 2.17), we know that pomset bisimulation is defined by pomset transitions, which are labeled by pomsets. In a pomset transition, the events in the pomset are either within causality relations (defined by ·) or in concurrency (implicitly defined by · and +, and explicitly defined by $\between$), of course, they are pairwise consistent (without conflicts). In (1), we have already proven the case that all events are pairwise concurrent, so, we only need to prove the case of events in causality. Without loss of generality, we take a pomset of $P = \{e_1, e_2 : e_1 \cdot e_2\}$. Then the pomset transition labeled by the above P is just composed of one single event transition labeled by e_1 succeeded by another single event transition labeled by e_2, that is, $\xrightarrow{P} = \xrightarrow{e_1} \xrightarrow{e_2}$.

Similarly to the proof of soundness of AIP modulo step bisimulation equivalence (1), we can prove that AIP in Table 4.19 is sound modulo pomset bisimulation equivalence, we omit them.

(3) If $\pi_n(x) \sim_{hp} \pi_n(y)$ for $n \in \mathbb{N}$, then $x \sim_{hp} y$.

Since hp-bisimulation $\sim_{hp}$ is both an equivalent and a congruent relation with respect to $APTC$ with guarded recursion and projection operator, we only need to check if AIP in Table 4.19 is sound modulo hp-bisimulation equivalence.

From the definition of hp-bisimulation (see Definition 2.21), we know that hp-bisimulation is defined on the posetal product (C_1, f, C_2), $f : C_1 \to C_2$ isomorphism. Two process terms s related to C_1 and t related to C_2, and $f : C_1 \to C_2$ isomorphism. Initially, $(C_1, f, C_2) = (\emptyset, \emptyset, \emptyset)$, and $(\emptyset, \emptyset, \emptyset) \in \sim_{hp}$. When $s \xrightarrow{e} s'$ $(C_1 \xrightarrow{e} C_1')$, there will be $t \xrightarrow{e} t'$ $(C_2 \xrightarrow{e} C_2')$, and we define $f' = f[e \mapsto e]$. Then, if $(C_1, f, C_2) \in \sim_{hp}$, then $(C_1', f', C_2') \in \sim_{hp}$.

Similarly to the proof of soundness of AIP modulo pomset bisimulation equivalence (2), we can prove that AIP in Table 4.19 is sound modulo hp-bisimulation equivalence, we just need additionally to check the above conditions on hp-bisimulation, we omit them.

$\square$

Theorem 4.54 (Completeness of AIP). *Let p and q be closed $APTC$ with linear recursion and projection operator terms, then,*

1. *if $p \sim_s q$ then $\pi_n(p) = \pi_n(q)$;*
2. *if $p \sim_p q$ then $\pi_n(p) = \pi_n(q)$;*
3. *if $p \sim_{hp} q$ then $\pi_n(p) = \pi_n(q)$.*

Proof. Firstly, by the elimination theorem of $APTC$ with guarded recursion and projection operator (see Theorem 4.51), we know that each process term in $APTC$ with linear recursion and projection operator is equal to a process term $\langle X_1 | E \rangle$ with E a linear recursive specification:

$$X_i = (a_{i11} \parallel \cdots \parallel a_{i1i_1}) X_{i1} + \cdots + (a_{ik_i1} \parallel \cdots \parallel a_{ik_i i_k}) X_{ik_i} + (b_{i11} \parallel \cdots \parallel b_{i1i_1}) + \cdots$$
$$+ (b_{il_i1} \parallel \cdots \parallel b_{il_i i_l})$$

for $i \in \{1, \cdots, n\}$.

It remains to prove the following cases.

Table   4.20 Transition
rule of the silent step.

$$\frac{}{\tau \xrightarrow{\tau} \surd}$$

(1) if $p \sim_s q$ then $\pi_n(p) = \pi_n(q)$.

Let $p \sim_s q$, and fix an $n \in \mathbb{N}$, there are p', q' in basic $APTC$ terms such that $p' = \pi_n(p)$ and $q' = \pi_n(q)$. Since $\sim_s$ is a congruence with respect to $APTC$, if $p \sim_s q$ then $\pi_n(p) \sim_s \pi_n(q)$. The soundness theorem yields $p' \sim_s \pi_n(p) \sim_s \pi_n(q) \sim_s q'$. Finally, the completeness of $APTC$ modulo $\sim_s$ (see Theorem 4.34) ensures $p' = q'$, and $\pi_n(p) = p' = q' = \pi_n(q)$, as desired.

(2) if $p \sim_p q$ then $\pi_n(p) = \pi_n(q)$.

Let $p \sim_p q$, and fix an $n \in \mathbb{N}$, there are p', q' in basic $APTC$ terms such that $p' = \pi_n(p)$ and $q' = \pi_n(q)$. Since $\sim_p$ is a congruence with respect to $APTC$, if $p \sim_p q$ then $\pi_n(p) \sim_p \pi_n(q)$. The soundness theorem yields $p' \sim_p \pi_n(p) \sim_p \pi_n(q) \sim_p q'$. Finally, the completeness of $APTC$ modulo $\sim_p$ (see Theorem 4.36) ensures $p' = q'$, and $\pi_n(p) = p' = q' = \pi_n(q)$, as desired.

(3) if $p \sim_{hp} q$ then $\pi_n(p) = \pi_n(q)$.

Let $p \sim_{hp} q$, and fix an $n \in \mathbb{N}$, there are p', q' in basic $APTC$ terms such that $p' = \pi_n(p)$ and $q' = \pi_n(q)$. Since $\sim_{hp}$ is a congruence with respect to $APTC$, if $p \sim_{hp} q$ then $\pi_n(p) \sim_{hp} \pi_n(q)$. The soundness theorem yields $p' \sim_{hp} \pi_n(p) \sim_{hp} \pi_n(q) \sim_{hp} q'$. Finally, the completeness of $APTC$ modulo $\sim_{hp}$ (see Theorem 4.38) ensures $p' = q'$, and $\pi_n(p) = p' = q' = \pi_n(q)$, as desired. $\qquad\square$

4.4 Abstraction

To abstract away from the internal implementations of a program, and verify that the program exhibits the desired external behaviors, the silent step τ (and making τ distinct by τ^e) and abstraction operator τ_I are introduced, where $I \subseteq \mathbb{E}$ denotes the internal events. The silent step τ represents the internal events, when we consider the external behaviors of a process, τ events can be removed, that is, τ events must keep silent. The transition rule of τ is shown in Table 4.20. In the following, let the atomic event e range over $\mathbb{E} \cup \{\delta\} \cup \{\tau\}$, and let the communication function $\gamma : \mathbb{E} \cup \{\tau\} \times \mathbb{E} \cup \{\tau\} \to \mathbb{E} \cup \{\delta\}$, with each communication involved τ resulting into δ.

The silent step τ was firstly introduced by Milner in his CCS [3], the algebraic laws about τ were introduced in [1], and finally the axiomatization of τ and τ_I formed a part of ACP [4]. Though τ has been discussed in the interleaving bisimulation background, several years ago, we introduced τ into true concurrency, called weakly true concurrency [16], and also designed its logic based on a uniform logic for true concurrency [14,15].

In this section, we try to find the algebraic laws of τ and τ_I in true concurrency, or, exactly, to what extent the laws of τ and τ_I in interleaving bisimulation fit the situation of true concurrency.

4.4.1 Rooted branching truly concurrent bisimulation equivalences

In section 2.4, we introduce τ into event structure, and also give the concept of weakly true concurrency. In this subsection, we give the concepts of rooted branching truly concurrent bisimulation equivalences, based on these concepts, we can design the axiom system of the silent step τ and the abstraction operator τ_I. Similarly to rooted branching bisimulation equivalence, rooted branching truly concurrent bisimulation equivalences are following.

Definition 4.55 (Branching pomset, step bisimulation). Assume a special termination predicate $\downarrow$, and let $\sqrt{}$ represent a state with $\sqrt{} \downarrow$. Let $\mathcal{E}_1$, $\mathcal{E}_2$ be PESs. A branching pomset bisimulation is a relation $R \subseteq \mathcal{C}(\mathcal{E}_1) \times \mathcal{C}(\mathcal{E}_2)$, such that:

1. if $(C_1, C_2) \in R$, and $C_1 \xrightarrow{X} C_1'$ then
   - either $X \equiv \tau^*$, and $(C_1', C_2) \in R$;
   - or there is a sequence of (zero or more) τ-transitions $C_2 \xrightarrow{\tau^*} C_2^0$, such that $(C_1, C_2^0) \in R$ and $C_2^0 \xrightarrow{X} C_2'$ with $(C_1', C_2') \in R$;

2. if $(C_1, C_2) \in R$, and $C_2 \xrightarrow{X} C_2'$ then
   - either $X \equiv \tau^*$, and $(C_1, C_2') \in R$;
   - or there is a sequence of (zero or more) τ-transitions $C_1 \xrightarrow{\tau^*} C_1^0$, such that $(C_1^0, C_2) \in R$ and $C_1^0 \xrightarrow{X} C_1'$ with $(C_1', C_2') \in R$;

3. if $(C_1, C_2) \in R$ and $C_1 \downarrow$, then there is a sequence of (zero or more) τ-transitions $C_2 \xrightarrow{\tau^*} C_2^0$ such that $(C_1, C_2^0) \in R$ and $C_2^0 \downarrow$;

4. if $(C_1, C_2) \in R$ and $C_2 \downarrow$, then there is a sequence of (zero or more) τ-transitions $C_1 \xrightarrow{\tau^*} C_1^0$ such that $(C_1^0, C_2) \in R$ and $C_1^0 \downarrow$.

We say that $\mathcal{E}_1$, $\mathcal{E}_2$ are branching pomset bisimilar, written $\mathcal{E}_1 \approx_{bp} \mathcal{E}_2$, if there exists a branching pomset bisimulation R, such that $(\emptyset, \emptyset) \in R$.

By replacing pomset transitions with steps, we can get the definition of branching step bisimulation. When PESs $\mathcal{E}_1$ and $\mathcal{E}_2$ are branching step bisimilar, we write $\mathcal{E}_1 \approx_{bs} \mathcal{E}_2$.

Definition 4.56 (Rooted branching pomset, step bisimulation). Assume a special termination predicate $\downarrow$, and let $\sqrt{}$ represent a state with $\sqrt{} \downarrow$. Let $\mathcal{E}_1$, $\mathcal{E}_2$ be PESs. A rooted branching pomset bisimulation is a relation $R \subseteq \mathcal{C}(\mathcal{E}_1) \times \mathcal{C}(\mathcal{E}_2)$, such that:

1. if $(C_1, C_2) \in R$, and $C_1 \xrightarrow{X} C_1'$ then $C_2 \xrightarrow{X} C_2'$ with $C_1' \approx_{bp} C_2'$;
2. if $(C_1, C_2) \in R$, and $C_2 \xrightarrow{X} C_2'$ then $C_1 \xrightarrow{X} C_1'$ with $C_1' \approx_{bp} C_2'$;

3. if $(C_1, C_2) \in R$ and $C_1 \downarrow$, then $C_2 \downarrow$;

4. if $(C_1, C_2) \in R$ and $C_2 \downarrow$, then $C_1 \downarrow$.

We say that $\mathcal{E}_1, \mathcal{E}_2$ are rooted branching pomset bisimilar, written $\mathcal{E}_1 \approx_{rbp} \mathcal{E}_2$, if there exists a rooted branching pomset bisimulation R, such that $(\emptyset, \emptyset) \in R$.

By replacing pomset transitions with steps, we can get the definition of rooted branching step bisimulation. When PESs $\mathcal{E}_1$ and $\mathcal{E}_2$ are rooted branching step bisimilar, we write $\mathcal{E}_1 \approx_{rbs} \mathcal{E}_2$.

Definition 4.57 (Branching (hereditary) history-preserving bisimulation). Assume a special termination predicate $\downarrow$, and let $\sqrt{}$ represent a state with $\sqrt{} \downarrow$. A branching history-preserving (hp-) bisimulation is a weakly posetal relation $R \subseteq \mathcal{C}(\mathcal{E}_1) \overline{\times} \mathcal{C}(\mathcal{E}_2)$ such that:

1. if $(C_1, f, C_2) \in R$, and $C_1 \xrightarrow{e_1} C_1'$ then

- either $e_1 \equiv \tau$, and $(C_1', f[e_1 \mapsto \tau^{e_1}], C_2) \in R$;
- or there is a sequence of (zero or more) τ-transitions $C_2 \xrightarrow{\tau^*} C_2^0$, such that $(C_1, f, C_2^0) \in R$ and $C_2^0 \xrightarrow{e_2} C_2'$ with $(C_1', f[e_1 \mapsto e_2], C_2') \in R$;

2. if $(C_1, f, C_2) \in R$, and $C_2 \xrightarrow{e_2} C_2'$ then

- either $e_2 \equiv \tau$, and $(C_1, f[e_2 \mapsto \tau^{e_2}], C_2') \in R$;
- or there is a sequence of (zero or more) τ-transitions $C_1 \xrightarrow{\tau^*} C_1^0$, such that $(C_1^0, f, C_2) \in R$ and $C_1^0 \xrightarrow{e_1} C_1'$ with $(C_1', f[e_2 \mapsto e_1], C_2') \in R$;

3. if $(C_1, f, C_2) \in R$ and $C_1 \downarrow$, then there is a sequence of (zero or more) τ-transitions $C_2 \xrightarrow{\tau^*} C_2^0$ such that $(C_1, f, C_2^0) \in R$ and $C_2^0 \downarrow$;

4. if $(C_1, f, C_2) \in R$ and $C_2 \downarrow$, then there is a sequence of (zero or more) τ-transitions $C_1 \xrightarrow{\tau^*} C_1^0$ such that $(C_1^0, f, C_2) \in R$ and $C_1^0 \downarrow$.

$\mathcal{E}_1, \mathcal{E}_2$ are branching history-preserving (hp-)bisimilar and are written $\mathcal{E}_1 \approx_{bhp} \mathcal{E}_2$ if there exists a branching hp-bisimulation R such that $(\emptyset, \emptyset, \emptyset) \in R$.

A branching hereditary history-preserving (hhp-)bisimulation is a downward closed branching hp-bisimulation. $\mathcal{E}_1, \mathcal{E}_2$ are branching hereditary history-preserving (hhp-)bisimilar and are written $\mathcal{E}_1 \approx_{bhhp} \mathcal{E}_2$.

Definition 4.58 (Rooted branching (hereditary) history-preserving bisimulation). Assume a special termination predicate $\downarrow$, and let $\sqrt{}$ represent a state with $\sqrt{} \downarrow$. A rooted branching history-preserving (hp-) bisimulation is a weakly posetal relation $R \subseteq \mathcal{C}(\mathcal{E}_1) \overline{\times} \mathcal{C}(\mathcal{E}_2)$ such that:

1. if $(C_1, f, C_2) \in R$, and $C_1 \xrightarrow{e_1} C_1'$, then $C_2 \xrightarrow{e_2} C_2'$ with $C_1' \approx_{bhp} C_2'$;

2. if $(C_1, f, C_2) \in R$, and $C_2 \xrightarrow{e_2} C_2'$, then $C_1 \xrightarrow{e_1} C_1'$ with $C_1' \approx_{bhp} C_2'$;

3. if $(C_1, f, C_2) \in R$ and $C_1 \downarrow$, then $C_2 \downarrow$;

4. if $(C_1, f, C_2) \in R$ and $C_2 \downarrow$, then $C_1 \downarrow$.

$\mathcal{E}_1, \mathcal{E}_2$ are rooted branching history-preserving (hp-)bisimilar and are written $\mathcal{E}_1 \approx_{rbhp} \mathcal{E}_2$ if there exists a rooted branching hp-bisimulation R such that $(\emptyset, \emptyset, \emptyset) \in R$.

A rooted branching hereditary history-preserving (hhp-)bisimulation is a downward closed rooted branching hp-bisimulation. $\mathcal{E}_1, \mathcal{E}_2$ are rooted branching hereditary history-preserving (hhp-)bisimilar and are written $\mathcal{E}_1 \approx_{rbhhp} \mathcal{E}_2$.

4.4.2 Guarded linear recursion

The silent step τ as an atomic event, is introduced into E. Considering the recursive specification $X = \tau X$, τs, $\tau \tau s$, and $\tau \cdots s$ are all its solutions, that is, the solutions make the existence of τ-loops which cause unfairness. To prevent τ-loops, we extend the definition of linear recursive specification (Definition 4.43) to the guarded one.

Definition 4.59 (Guarded linear recursive specification). A recursive specification is linear if its recursive equations are of the form

$$(a_{11} \parallel \cdots \parallel a_{1i_1})X_1 + \cdots + (a_{k1} \parallel \cdots \parallel a_{ki_k})X_k + (b_{11} \parallel \cdots \parallel b_{1j_1}) + \cdots + (b_{1j_1} \parallel \cdots \parallel b_{lj_l})$$

where $a_{11}, \cdots, a_{1i_1}, a_{k1}, \cdots, a_{ki_k}, b_{11}, \cdots, b_{1j_1}, b_{1j_1}, \cdots, b_{lj_l} \in \mathbb{E} \cup \{\tau\}$, and the sum above is allowed to be empty, in which case it represents the deadlock δ.

A linear recursive specification E is guarded if there does not exist an infinite sequence of τ-transitions $\langle X|E \rangle \xrightarrow{\tau} \langle X'|E \rangle \xrightarrow{\tau} \langle X''|E \rangle \xrightarrow{\tau} \cdots$.

Theorem 4.60 (Conservativity of $APTC$ with silent step and guarded linear recursion). *$APTC$ with silent step and guarded linear recursion is a conservative extension of $APTC$ with linear recursion.*

Proof. Since the transition rules of $APTC$ with linear recursion are source-dependent, and the transition rules for silent step in Table 4.20 contain only a fresh constant τ in their source, so the transition rules of $APTC$ with silent step and guarded linear recursion are conservative extensions of those of $APTC$ with linear recursion. $\square$

Theorem 4.61 (Congruence theorem of $APTC$ with silent step and guarded linear recursion). *Rooted branching truly concurrent bisimulation equivalences $\approx_{rbp}$, $\approx_{rbs}$ and $\approx_{rbhp}$ are all congruences with respect to $APTC$ with silent step and guarded linear recursion.*

Proof. It follows the following three facts:

1. in a guarded linear recursive specification, right-hand sides of its recursive equations can be adapted to the form by applications of the axioms in $APTC$ and replacing recursion variables by the right-hand sides of their recursive equations;
2. truly concurrent bisimulation equivalences $\sim_p$, $\sim_s$ and $\sim_{hp}$ are all congruences with respect to all operators of $APTC$, while truly concurrent bisimulation equivalences $\sim_p$, $\sim_s$ and $\sim_{hp}$ imply the corresponding rooted branching truly concurrent bisimulations $\approx_{rbp}$, $\approx_{rbs}$ and $\approx_{rbhp}$ (see Proposition 2.23), so rooted branching truly concurrent bisimulations $\approx_{rbp}$, $\approx_{rbs}$ and $\approx_{rbhp}$ are all congruences with respect to all operators of $APTC$;

Table 4.21 Axioms of silent step.

No.	Axiom
$B1$	$e \cdot \tau = e$
$B2$	$e \cdot (\tau \cdot (x + y) + x) = e \cdot (x + y)$
$B3$	$x \parallel \tau = x$

3. While $\mathbb{E}$ is extended to $\mathbb{E} \cup \{\tau\}$, it can be proved that rooted branching truly concurrent bisimulations $\approx_{rbp}$, $\approx_{rbs}$ and $\approx_{rbhp}$ are all congruences with respect to all operators of $APTC$, we omit it. $\qquad\qquad\qquad\qquad\qquad\qquad\qquad\qquad\qquad\qquad\qquad\quad$ $\square$

4.4.3 Algebraic laws for the silent step

We design the axioms for the silent step τ in Table 4.21.

The axioms $B1$, $B2$ and $B3$ are the conditions in which τ really keeps silent to act with the operators $\cdot$, $+$, and $\parallel$.

Theorem 4.62 (Elimination theorem of $APTC$ with silent step and guarded linear recursion). *Each process term in $APTC$ with silent step and guarded linear recursion is equal to a process term $\langle X_1 | E \rangle$ with E a guarded linear recursive specification.*

Proof. By applying structural induction with respect to term size, each process term t_1 in $APTC$ with silent step and guarded linear recursion generates a process can be expressed in the form of equations

$$t_i = (a_{i11} \parallel \cdots \parallel a_{i1l_1})t_{i1} + \cdots + (a_{ik_i1} \parallel \cdots \parallel a_{ik_ii_k})t_{ik_i} + (b_{i11} \parallel \cdots \parallel b_{i1l_1}) + \cdots + (b_{il_i1} \parallel \cdots \parallel b_{il_ii_l})$$

for $i \in \{1, \cdots, n\}$. Let the linear recursive specification E consist of the recursive equations

$$X_i = (a_{i11} \parallel \cdots \parallel a_{i1l_1})X_{i1} + \cdots + (a_{ik_i1} \parallel \cdots \parallel a_{ik_ii_k})X_{ik_i} + (b_{i11} \parallel \cdots \parallel b_{i1l_1}) + \cdots$$
$$+ (b_{il_i1} \parallel \cdots \parallel b_{il_ii_l})$$

for $i \in \{1, \cdots, n\}$. Replacing X_i by t_i for $i \in \{1, \cdots, n\}$ is a solution for E, RSP yields $t_1 = \langle X_1 | E \rangle$. $\qquad\qquad\qquad\qquad\qquad\qquad\qquad\qquad\qquad\qquad\qquad\qquad\qquad\quad$ $\square$

Theorem 4.63 (Soundness of $APTC$ with silent step and guarded linear recursion). *Let x and y be $APTC$ with silent step and guarded linear recursion terms. If $APTC$ with silent step and guarded linear recursion $\vdash x = y$, then*

1. $x \approx_{rbs} y$;
2. $x \approx_{rbp} y$;
3. $x \approx_{rbhp} y$.

Proof. (1) Soundness of $APTC$ with silent step and guarded linear recursion with respect to rooted branching step bisimulation $\approx_{rbs}$.

Since rooted branching step bisimulation $\approx_{rbs}$ is both an equivalent and a congruent relation with respect to $APTC$ with silent step and guarded linear recursion, we only need to check if each axiom in Table 4.21 is sound modulo rooted branching step bisimulation equivalence.

Though transition rules in Table 4.20 are defined in the flavor of single event, they can be modified into a step (a set of events within which each event is pairwise concurrent), we omit them. If we treat a single event as a step containing just one event, the proof of this soundness theorem does not exist any problem, so we use this way and still use the transition rules in Table 4.20.

- **Axiom** $B1$. Assume that $e \cdot \tau = e$, it is sufficient to prove that $e \cdot \tau \approx_{rbs} e$. By the transition rules for operator $\cdot$ in Table 4.5 and τ in Table 4.20, we get

$$\frac{e \xrightarrow{e} \surd}{e \cdot \tau \xrightarrow{e} \xrightarrow{\tau} \surd}$$

$$\frac{e \xrightarrow{e} \surd}{e \xrightarrow{e} \surd}$$

So, $e \cdot \tau \approx_{rbs} e$, as desired.

- **Axiom** $B2$. Let p and q be $APTC$ with silent step and guarded linear recursion processes, and assume that $e \cdot (\tau \cdot (p + q) + p) = e \cdot (p + q)$, it is sufficient to prove that $e \cdot (\tau \cdot (p + q) + p) \approx_{rbs} e \cdot (p + q)$. By the transition rules for operators $\cdot$ and $+$ in Table 4.5 and τ in Table 4.20, we get

$$\frac{e \xrightarrow{e} \surd \quad p \xrightarrow{e_1} \surd}{e \cdot (\tau \cdot (p + q) + p) \xrightarrow{e} \xrightarrow{e_1} \surd}$$

$$\frac{e \xrightarrow{e} \surd \quad p \xrightarrow{e_1} \surd}{e \cdot (p + q) \xrightarrow{e} \xrightarrow{e_1} \surd}$$

$$\frac{e \xrightarrow{e} \surd \quad p \xrightarrow{e_1} p'}{e \cdot (\tau \cdot (p + q) + p) \xrightarrow{e} \xrightarrow{e_1} p'}$$

$$\frac{e \xrightarrow{e} \surd \quad p \xrightarrow{e_1} p'}{e \cdot (p + q) \xrightarrow{e} \xrightarrow{e_1} p'}$$

$$\frac{e \xrightarrow{e} \surd \quad q \xrightarrow{e_2} \surd}{e \cdot (\tau \cdot (p + q) + p) \xrightarrow{e} \xrightarrow{\tau} \xrightarrow{e_2} \surd}$$

$$\frac{e \xrightarrow{e} \sqrt{\quad} \quad q \xrightarrow{e_2} \sqrt{\quad}}{e \cdot (p + q) \xrightarrow{e} \xrightarrow{e_2} \sqrt{\quad}}$$

$$\frac{e \xrightarrow{e} \sqrt{\quad} \quad q \xrightarrow{e_2} q'}{e \cdot (\tau \cdot (p + q) + p) \xrightarrow{e} \xrightarrow{\tau} \xrightarrow{e_2} q'}$$

$$\frac{e \xrightarrow{e} \sqrt{\quad} \quad q \xrightarrow{e_2} q'}{e \cdot (p + q) \xrightarrow{e} \xrightarrow{e_2} q'}$$

So, $e \cdot (\tau \cdot (p + q) + p) \approx_{rbs} e \cdot (p + q)$, as desired.

- **Axiom** $B3$. Let p be an $APTC$ with silent step and guarded linear recursion process, and assume that $p \parallel \tau = p$, it is sufficient to prove that $p \parallel \tau \approx_{rbs} p$. By the transition rules for operator $\parallel$ in Table 4.7 and τ in Table 4.20, we get

$$\frac{p \xrightarrow{e} \sqrt{\quad}}{p \parallel \tau \xRightarrow{e} \sqrt{\quad}}$$

$$\frac{p \xrightarrow{e} p'}{p \parallel \tau \xRightarrow{e} p'}$$

So, $p \parallel \tau \approx_{rbs} p$, as desired.

(2) Soundness of $APTC$ with silent step and guarded linear recursion with respect to rooted branching pomset bisimulation $\approx_{rbp}$.

Since rooted branching pomset bisimulation $\approx_{rbp}$ is both an equivalent and a congruent relation with respect to $APTC$ with silent step and guarded linear recursion, we only need to check if each axiom in Table 4.21 is sound modulo rooted branching pomset bisimulation $\approx_{rbp}$.

From the definition of rooted branching pomset bisimulation $\approx_{rbp}$ (see Definition 4.56), we know that rooted branching pomset bisimulation $\approx_{rbp}$ is defined by weak pomset transitions, which are labeled by pomsets with τ. In a weak pomset transition, the events in the pomset are either within causality relations (defined by $\cdot$) or in concurrency (implicitly defined by $\cdot$ and $+$, and explicitly defined by $\between$), of course, they are pairwise consistent (without conflicts). In (1), we have already proven the case that all events are pairwise concurrent, so, we only need to prove the case of events in causality. Without loss of generality, we take a pomset of $P = \{e_1, e_2 : e_1 \cdot e_2\}$. Then the weak pomset transition labeled by the above P is just composed of one single event transition labeled by e_1 succeeded by another single event transition labeled by e_2, that is, $\xRightarrow{P} = \xRightarrow{e_1} \xRightarrow{e_2}$.

Similarly to the proof of soundness of $APTC$ with silent step and guarded linear recursion modulo rooted branching step bisimulation $\approx_{rbs}$ (1), we can prove that each axiom in Table 4.21 is sound modulo rooted branching pomset bisimulation $\approx_{rbp}$, we omit them.

(3) Soundness of $APTC$ with silent step and guarded linear recursion with respect to rooted branching hp-bisimulation $\approx_{rbhp}$.

Since rooted branching hp-bisimulation $\approx_{rbhp}$ is both an equivalent and a congruent relation with respect to $APTC$ with silent step and guarded linear recursion, we only need to check if each axiom in Table 4.21 is sound modulo rooted branching hp-bisimulation $\approx_{rbhp}$.

From the definition of rooted branching hp-bisimulation $\approx_{rbhp}$ (see Definition 4.58), we know that rooted branching hp-bisimulation $\approx_{rbhp}$ is defined on the weakly posetal product $(C_1, f, C_2), f : \hat{C}_1 \to \hat{C}_2$ isomorphism. Two process terms s related to C_1 and t related to C_2, and $f : \hat{C}_1 \to \hat{C}_2$ isomorphism. Initially, $(C_1, f, C_2) = (\emptyset, \emptyset, \emptyset)$, and $(\emptyset, \emptyset, \emptyset) \in \approx_{rbhp}$. When $s \xrightarrow{e} s' (C_1 \xrightarrow{e} C_1')$, there will be $t \xrightarrow{e} t' (C_2 \xrightarrow{e} C_2')$, and we define $f' = f[e \mapsto e]$. Then, if $(C_1, f, C_2) \in \approx_{rbhp}$, then $(C_1', f', C_2') \in \approx_{rbhp}$.

Similarly to the proof of soundness of $APTC$ with silent step and guarded linear recursion modulo rooted branching pomset bisimulation equivalence (2), we can prove that each axiom in Table 4.21 is sound modulo rooted branching hp-bisimulation equivalence, we just need additionally to check the above conditions on rooted branching hp-bisimulation, we omit them. $\qquad\square$

Theorem 4.64 (Completeness of $APTC$ with silent step and guarded linear recursion). *Let p and q be closed $APTC$ with silent step and guarded linear recursion terms, then,*

1. *if $p \approx_{rbs} q$ then $p = q$;*
2. *if $p \approx_{rbp} q$ then $p = q$;*
3. *if $p \approx_{rbhp} q$ then $p = q$.*

Proof. Firstly, by the elimination theorem of $APTC$ with silent step and guarded linear recursion (see Theorem 4.62), we know that each process term in $APTC$ with silent step and guarded linear recursion is equal to a process term $\langle X_1|E \rangle$ with E a guarded linear recursive specification.

It remains to prove the following cases.

(1) If $\langle X_1|E_1 \rangle \approx_{rbs} \langle Y_1|E_2 \rangle$ for guarded linear recursive specification E_1 and E_2, then $\langle X_1|E_1 \rangle = \langle Y_1|E_2 \rangle$.

Firstly, the recursive equation $W = \tau + \cdots + \tau$ with $W \not\equiv X_1$ in E_1 and E_2, can be removed, and the corresponding summands aW are replaced by a, to get E_1' and E_2', by use of the axioms RDP, $A3$, and $B1$, and $\langle X|E_1 \rangle = \langle X|E_1' \rangle$, $\langle Y|E_2 \rangle = \langle Y|E_2' \rangle$.

Let E_1 consist of recursive equations $X = t_X$ for $X \in \mathcal{X}$ and E_2 consist of recursion equations $Y = t_Y$ for $Y \in \mathcal{Y}$, and are not the form $\tau + \cdots + \tau$. Let the guarded linear recursive specification E consist of recursion equations $Z_{XY} = t_{XY}$, and $\langle X|E_1 \rangle \approx_{rbs} \langle Y|E_2 \rangle$, and t_{XY} consist of the following summands:

1. t_{XY} contains a summand $(a_1 \parallel \cdots \parallel a_m)Z_{X'Y'}$ iff t_X contains the summand $(a_1 \parallel \cdots \parallel a_m)X'$ and t_Y contains the summand $(a_1 \parallel \cdots \parallel a_m)Y'$ such that $\langle X'|E_1 \rangle \approx_{rbs} \langle Y'|E_2 \rangle$;
2. t_{XY} contains a summand $b_1 \parallel \cdots \parallel b_n$ iff t_X contains the summand $b_1 \parallel \cdots \parallel b_n$ and t_Y contains the summand $b_1 \parallel \cdots \parallel b_n$;

3. t_{XY} contains a summand $\tau Z_{X'Y}$ iff $XY \not\equiv X_1 Y_1$, t_X contains the summand $\tau X'$, and $\langle X'|E_1\rangle \approx_{rbs} \langle Y|E_2\rangle$;

4. t_{XY} contains a summand $\tau Z_{XY'}$ iff $XY \not\equiv X_1 Y_1$, t_Y contains the summand $\tau Y'$, and $\langle X|E_1\rangle \approx_{rbs} \langle Y'|E_2\rangle$.

Since E_1 and E_2 are guarded, E is guarded. Constructing the process term u_{XY} consist of the following summands:

1. u_{XY} contains a summand $(a_1 \parallel \cdots \parallel a_m)\langle X'|E_1\rangle$ iff t_X contains the summand $(a_1 \parallel \cdots \parallel a_m)X'$ and t_Y contains the summand $(a_1 \parallel \cdots \parallel a_m)Y'$ such that $\langle X'|E_1\rangle \approx_{rbs} \langle Y'|E_2\rangle$;

2. u_{XY} contains a summand $b_1 \parallel \cdots \parallel b_n$ iff t_X contains the summand $b_1 \parallel \cdots \parallel b_n$ and t_Y contains the summand $b_1 \parallel \cdots \parallel b_n$;

3. u_{XY} contains a summand $\tau\langle X'|E_1\rangle$ iff $XY \not\equiv X_1 Y_1$, t_X contains the summand $\tau X'$, and $\langle X'|E_1\rangle \approx_{rbs} \langle Y|E_2\rangle$.

Let the process term s_{XY} be defined as follows:

1. $s_{XY} \triangleq \tau\langle X|E_1\rangle + u_{XY}$ iff $XY \not\equiv X_1 Y_1$, t_Y contains the summand $\tau Y'$, and $\langle X|E_1\rangle \approx_{rbs} \langle Y'|E_2\rangle$;

2. $s_{XY} \triangleq \langle X|E_1\rangle$, otherwise.

So, $\langle X|E_1\rangle = \langle X|E_1\rangle + u_{XY}$, and $(a_1 \parallel \cdots \parallel a_m)(\tau\langle X|E_1\rangle + u_{XY}) = (a_1 \parallel \cdots \parallel a_m)((\tau\langle X|E_1\rangle + u_{XY}) + u_{XY}) = (a_1 \parallel \cdots \parallel a_m)(\langle X|E_1\rangle + u_{XY}) = (a_1 \parallel \cdots \parallel a_m)\langle X|E_1\rangle$, hence, $(a_1 \parallel \cdots \parallel a_m)s_{XY} = (a_1 \parallel \cdots \parallel a_m)\langle X|E_1\rangle$.

Let σ map recursion variable X in E_1 to $\langle X|E_1\rangle$, and let ψ map recursion variable Z_{XY} in E to s_{XY}. It is sufficient to prove $s_{XY} = \psi(t_{XY})$ for recursion variables Z_{XY} in E. Either $XY \equiv X_1 Y_1$ or $XY \not\equiv X_1 Y_1$, we all can get $s_{XY} = \psi(t_{XY})$. So, $s_{XY} = \langle Z_{XY}|E\rangle$ for recursive variables Z_{XY} in E is a solution for E. Then by RSP, particularly, $\langle X_1|E_1\rangle = \langle Z_{X_1 Y_1}|E\rangle$. Similarly, we can obtain $\langle Y_1|E_2\rangle = \langle Z_{X_1 Y_1}|E\rangle$. Finally, $\langle X_1|E_1\rangle = \langle Z_{X_1 Y_1}|E\rangle = \langle Y_1|E_2\rangle$, as desired.

(2) If $\langle X_1|E_1\rangle \approx_{rbp} \langle Y_1|E_2\rangle$ for guarded linear recursive specification E_1 and E_2, then $\langle X_1|E_1\rangle = \langle Y_1|E_2\rangle$.

It can be proven similarly to (1), we omit it.

(3) If $\langle X_1|E_1\rangle \approx_{rbhb} \langle Y_1|E_2\rangle$ for guarded linear recursive specification E_1 and E_2, then $\langle X_1|E_1\rangle = \langle Y_1|E_2\rangle$.

It can be proven similarly to (1), we omit it. □

4.4.4 Abstraction

The unary abstraction operator τ_I ($I \subseteq \mathbb{E}$) renames all atomic events in I into τ. $APTC$ with silent step and abstraction operator is called $APTC_\tau$. The transition rules of operator τ_I are shown in Table 4.22.

Theorem 4.65 (Conservativity of $APTC_\tau$ with guarded linear recursion). *$APTC_\tau$ with guarded linear recursion is a conservative extension of $APTC$ with silent step and guarded linear recursion.*

Table 4.22 Transition rule of the abstraction operator.

$$\frac{x \xrightarrow{e} \checkmark}{\tau_I(x) \xrightarrow{e} \checkmark} \quad e \notin I \qquad \frac{x \xrightarrow{e} x'}{\tau_I(x) \xrightarrow{e} \tau_I(x')} \quad e \notin I$$

$$\frac{x \xrightarrow{e} \checkmark}{\tau_I(x) \xrightarrow{\tau} \checkmark} \quad e \in I \qquad \frac{x \xrightarrow{e} x'}{\tau_I(x) \xrightarrow{\tau} \tau_I(x')} \quad e \in I$$

Proof. Since the transition rules of $APTC$ with silent step and guarded linear recursion are source-dependent, and the transition rules for abstraction operator in Table 4.22 contain only a fresh operator τ_I in their source, so the transition rules of $APTC_\tau$ with guarded linear recursion are conservative extensions of those of $APTC$ with silent step and guarded linear recursion. □

Theorem 4.66 (Congruence theorem of $APTC_\tau$ with guarded linear recursion). *Rooted branching truly concurrent bisimulation equivalences $\approx_{rbp}$, $\approx_{rbs}$ and $\approx_{rbhp}$ are all congruences with respect to $APTC_\tau$ with guarded linear recursion.*

Proof. (1) Case rooted branching pomset bisimulation equivalence $\approx_{rbp}$.

Let x and y be $APTC_\tau$ with guarded linear recursion processes, and $x \approx_{rbp} y$, it is sufficient to prove that $\tau_I(x) \approx_{rbp} \tau_I(y)$.

By the transition rules for operator τ_I in Table 4.22, we can get

$$\tau_I(x) \xrightarrow{X} \checkmark (X \nsubseteq I) \quad \tau_I(y) \xrightarrow{Y} \checkmark (Y \nsubseteq I)$$

with $X \subseteq x$, $Y \subseteq y$, and $X \sim Y$.

Or, we can get

$$\tau_I(x) \xrightarrow{X} \tau_I(x')(X \nsubseteq I) \quad \tau_I(y) \xrightarrow{Y} \tau_I(y')(Y \nsubseteq I)$$

with $X \subseteq x$, $Y \subseteq y$, and $X \sim Y$ and the hypothesis $\tau_I(x') \approx_{rbp} \tau_I(y')$.

Or, we can get

$$\tau_I(x) \xrightarrow{\tau^*} \checkmark (X \subseteq I) \quad \tau_I(y) \xrightarrow{\tau^*} \checkmark (Y \subseteq I)$$

with $X \subseteq x$, $Y \subseteq y$, and $X \sim Y$.

Or, we can get

$$\tau_I(x) \xrightarrow{\tau^*} \tau_I(x')(X \subseteq I) \quad \tau_I(y) \xrightarrow{\tau^*} \tau_I(y')(Y \subseteq I)$$

with $X \subseteq x$, $Y \subseteq y$, and $X \sim Y$ and the hypothesis $\tau_I(x') \approx_{rbp} \tau_I(y')$.

So, we get $\tau_I(x) \approx_{rbp} \tau_I(y)$, as desired

Table 4.23   Axioms of abstraction operator.

No.	Axiom
$TI1$	$e \notin I \quad \tau_I(e) = e$
$TI2$	$e \in I \quad \tau_I(e) = \tau$
$TI3$	$\tau_I(\delta) = \delta$
$TI4$	$\tau_I(x + y) = \tau_I(x) + \tau_I(y)$
$TI5$	$\tau_I(x \cdot y) = \tau_I(x) \cdot \tau_I(y)$
$TI6$	$\tau_I(x \parallel y) = \tau_I(x) \parallel \tau_I(y)$

(2) The cases of rooted branching step bisimulation $\approx_{rbs}$, rooted branching hp-bisimulation $\approx_{rbhp}$ can be proven similarly, we omit them.   □

We design the axioms for the abstraction operator τ_I in Table 4.23.

The axioms $TI1 - TI3$ are the defining laws for the abstraction operator τ_I; $TI4 - TI6$ say that in process term $\tau_I(t)$, all transitions of t labeled with atomic events from I are renamed into τ.

Theorem 4.67 (Soundness of $APTC_\tau$ with guarded linear recursion). *Let x and y be $APTC_\tau$ with guarded linear recursion terms. If $APTC_\tau$ with guarded linear recursion $\vdash x = y$, then*

1. $x \approx_{rbs} y$;
2. $x \approx_{rbp} y$;
3. $x \approx_{rbhp} y$.

Proof. (1) Soundness of $APTC_\tau$ with guarded linear recursion with respect to rooted branching step bisimulation $\approx_{rbs}$.

Since rooted branching step bisimulation $\approx_{rbs}$ is both an equivalent and a congruent relation with respect to $APTC_\tau$ with guarded linear recursion, we only need to check if each axiom in Table 4.23 is sound modulo rooted branching step bisimulation equivalence.

Though transition rules in Table 4.22 are defined in the flavor of single event, they can be modified into a step (a set of events within which each event is pairwise concurrent), we omit them. If we treat a single event as a step containing just one event, the proof of this soundness theorem does not exist any problem, so we use this way and still use the transition rules in Table 4.23.

We only prove soundness of the non-trivial axioms $TI4 - TI6$, and omit the defining axioms $TI1 - TI3$.

- **Axiom $TI4$.** Let p, q be $APTC_\tau$ with guarded linear recursion processes, and $\tau_I(p + q) = \tau_I(p) + \tau_I(q)$, it is sufficient to prove that $\tau_I(p + q) \approx_{rbs} \tau_I(p) + \tau_I(q)$. By the transition rules for operator $+$ in Table 4.5 and τ_I in Table 4.22, we get

$$\frac{p \xrightarrow{e_1} \surd \quad (e_1 \notin I)}{\tau_I(p + q) \xrightarrow{e_1} \surd} \qquad \frac{p \xrightarrow{e_1} \surd \quad (e_1 \notin I)}{\tau_I(p) + \tau_I(q) \xrightarrow{e_1} \surd}$$

$$\frac{q \xrightarrow{e_2} \surd \quad (e_2 \notin I)}{\tau_I(p+q) \xrightarrow{e_2} \surd} \quad \frac{q \xrightarrow{e_2} \surd \quad (e_2 \notin I)}{\tau_I(p) + \tau_I(q) \xrightarrow{e_2} \surd}$$

$$\frac{p \xrightarrow{e_1} p' \quad (e_1 \notin I)}{\tau_I(p+q) \xrightarrow{e_1} \tau_I(p')} \quad \frac{p \xrightarrow{e_1} p' \quad (e_1 \notin I)}{\tau_I(p) + \tau_I(q) \xrightarrow{e_1} \tau_I(p')}$$

$$\frac{q \xrightarrow{e_2} q' \quad (e_2 \notin I)}{\tau_I(p+q) \xrightarrow{e_2} \tau_I(q')} \quad \frac{q \xrightarrow{e_2} q' \quad (e_2 \notin I)}{\tau_I(p) + \tau_I(q) \xrightarrow{e_2} \tau_I(q')}$$

$$\frac{p \xrightarrow{e_1} \surd \quad (e_1 \in I)}{\tau_I(p+q) \xrightarrow{\tau} \surd} \quad \frac{p \xrightarrow{e_1} \surd \quad (e_1 \in I)}{\tau_I(p) + \tau_I(q) \xrightarrow{\tau} \surd}$$

$$\frac{q \xrightarrow{e_2} \surd \quad (e_2 \in I)}{\tau_I(p+q) \xrightarrow{\tau} \surd} \quad \frac{q \xrightarrow{e_2} \surd \quad (e_2 \in I)}{\tau_I(p) + \tau_I(q) \xrightarrow{\tau} \surd}$$

$$\frac{p \xrightarrow{e_1} p' \quad (e_1 \in I)}{\tau_I(p+q) \xrightarrow{\tau} \tau_I(p')} \quad \frac{p \xrightarrow{e_1} p' \quad (e_1 \in I)}{\tau_I(p) + \tau_I(q) \xrightarrow{\tau} \tau_I(p')}$$

$$\frac{q \xrightarrow{e_2} q' \quad (e_2 \in I)}{\tau_I(p+q) \xrightarrow{\tau} \tau_I(q')} \quad \frac{q \xrightarrow{e_2} q' \quad (e_2 \in I)}{\tau_I(p) + \tau_I(q) \xrightarrow{\tau} \tau_I(q')}$$

So, $\tau_I(p+q) \approx_{rbs} \tau_I(p) + \tau_I(q)$, as desired.

- **Axiom** $TI5$. Let p, q be $APTC_\tau$ with guarded linear recursion processes, and $\tau_I(p \cdot q) = \tau_I(p) \cdot \tau_I(q)$, it is sufficient to prove that $\tau_I(p \cdot q) \approx_{rbs} \tau_I(p) \cdot \tau_I(q)$. By the transition rules for operator $\cdot$ in Table 4.5 and τ_I in Table 4.22, we get

$$\frac{p \xrightarrow{e_1} \surd \quad (e_1 \notin I)}{\tau_I(p \cdot q) \xrightarrow{e_1} \tau_I(q)} \quad \frac{p \xrightarrow{e_1} \surd \quad (e_1 \notin I)}{\tau_I(p) \cdot \tau_I(q) \xrightarrow{e_1} \tau_I(q)}$$

$$\frac{p \xrightarrow{e_1} p' \quad (e_1 \notin I)}{\tau_I(p \cdot q) \xrightarrow{e_1} \tau_I(p' \cdot q)} \quad \frac{p \xrightarrow{e_1} p' \quad (e_1 \notin I)}{\tau_I(p) \cdot \tau_I(q) \xrightarrow{e_1} \tau_I(p') \cdot \tau_I(q)}$$

$$\frac{p \xrightarrow{e_1} \surd \quad (e_1 \in I)}{\tau_I(p \cdot q) \xrightarrow{\tau} \tau_I(q)} \quad \frac{p \xrightarrow{e_1} \surd \quad (e_1 \in I)}{\tau_I(p) \cdot \tau_I(q) \xrightarrow{\tau} \tau_I(q)}$$

$$\frac{p \xrightarrow{e_1} p' \quad (e_1 \in I)}{\tau_I(p \cdot q) \xrightarrow{\tau} \tau_I(p' \cdot q)} \quad \frac{p \xrightarrow{e_1} p' \quad (e_1 \in I)}{\tau_I(p) \cdot \tau_I(q) \xrightarrow{\tau} \tau_I(p') \cdot \tau_I(q)}$$

So, with the assumption $\tau_I(p' \cdot q) = \tau_I(p') \cdot \tau_I(q)$, $\tau_I(p \cdot q) \approx_{rbs} \tau_I(p) \cdot \tau_I(q)$, as desired.

- **Axiom** $TI6$. Let p, q be $APTC_\tau$ with guarded linear recursion processes, and $\tau_I(p \parallel q) = \tau_I(p) \parallel \tau_I(q)$, it is sufficient to prove that $\tau_I(p \parallel q) \approx_{rbs} \tau_I(p) \parallel \tau_I(q)$. By the transition rules for operator $\parallel$ in Table 4.7 and τ_I in Table 4.22, we get

$$\frac{p \xrightarrow{e_1} \checkmark \quad q \xrightarrow{e_2} \checkmark \quad (e_1, e_2 \notin I)}{\tau_I(p \parallel q) \xrightarrow{\{e_1,e_2\}} \checkmark} \qquad \frac{p \xrightarrow{e_1} \checkmark \quad q \xrightarrow{e_2} \checkmark \quad (e_1, e_2 \notin I)}{\tau_I(p) \parallel \tau_I(q) \xrightarrow{\{e_1,e_2\}} \checkmark}$$

$$\frac{p \xrightarrow{e_1} p' \quad q \xrightarrow{e_2} \checkmark \quad (e_1, e_2 \notin I)}{\tau_I(p \parallel q) \xrightarrow{\{e_1,e_2\}} \tau_I(p')} \qquad \frac{p \xrightarrow{e_1} p' \quad q \xrightarrow{e_2} \checkmark \quad (e_1, e_2 \notin I)}{\tau_I(p) \parallel \tau_I(q) \xrightarrow{\{e_1,e_2\}} \tau_I(p')}$$

$$\frac{p \xrightarrow{e_1} \checkmark \quad q \xrightarrow{e_2} q' \quad (e_1, e_2 \notin I)}{\tau_I(p \parallel q) \xrightarrow{\{e_1,e_2\}} \tau_I(q')} \qquad \frac{p \xrightarrow{e_1} \checkmark \quad q \xrightarrow{e_2} q' \quad (e_1, e_2 \notin I)}{\tau_I(p) \parallel \tau_I(q) \xrightarrow{\{e_1,e_2\}} \tau_I(q')}$$

$$\frac{p \xrightarrow{e_1} p' \quad q \xrightarrow{e_2} q' \quad (e_1, e_2 \notin I)}{\tau_I(p \parallel q) \xrightarrow{\{e_1,e_2\}} \tau_I(p' \between q')} \qquad \frac{p \xrightarrow{e_1} p' \quad q \xrightarrow{e_2} q' \quad (e_1, e_2 \notin I)}{\tau_I(p) \parallel \tau_I(q) \xrightarrow{\{e_1,e_2\}} \tau_I(p') \between \tau_I(q')}$$

$$\frac{p \xrightarrow{e_1} \checkmark \quad q \xrightarrow{e_2} \checkmark \quad (e_1 \notin I, e_2 \in I)}{\tau_I(p \parallel q) \xRightarrow{e_1} \checkmark} \qquad \frac{p \xrightarrow{e_1} \checkmark \quad q \xrightarrow{e_2} \checkmark \quad (e_1 \notin I, e_2 \in I)}{\tau_I(p) \parallel \tau_I(q) \xRightarrow{e_1} \checkmark}$$

$$\frac{p \xrightarrow{e_1} p' \quad q \xrightarrow{e_2} \checkmark \quad (e_1 \notin I, e_2 \in I)}{\tau_I(p \parallel q) \xRightarrow{e_1} \tau_I(p')} \qquad \frac{p \xrightarrow{e_1} p' \quad q \xrightarrow{e_2} \checkmark \quad (e_1 \notin I, e_2 \in I)}{\tau_I(p) \parallel \tau_I(q) \xRightarrow{e_1} \tau_I(p')}$$

$$\frac{p \xrightarrow{e_1} \checkmark \quad q \xrightarrow{e_2} q' \quad (e_1 \notin I, e_2 \in I)}{\tau_I(p \parallel q) \xRightarrow{e_1} \tau_I(q')} \qquad \frac{p \xrightarrow{e_1} \checkmark \quad q \xrightarrow{e_2} q' \quad (e_1 \notin I, e_2 \in I)}{\tau_I(p) \parallel \tau_I(q) \xRightarrow{e_1} \tau_I(q')}$$

$$\frac{p \xrightarrow{e_1} p' \quad q \xrightarrow{e_2} q' \quad (e_1 \notin I, e_2 \in I)}{\tau_I(p \parallel q) \xRightarrow{e_1} \tau_I(p' \between q')} \qquad \frac{p \xrightarrow{e_1} p' \quad q \xrightarrow{e_2} q' \quad (e_1 \notin I, e_2 \in I)}{\tau_I(p) \parallel \tau_I(q) \xRightarrow{e_1} \tau_I(p') \between \tau_I(q')}$$

$$\frac{p \xrightarrow{e_1} \checkmark \quad q \xrightarrow{e_2} \checkmark \quad (e_1 \in I, e_2 \notin I)}{\tau_I(p \parallel q) \xRightarrow{e_2} \checkmark} \qquad \frac{p \xrightarrow{e_1} \checkmark \quad q \xrightarrow{e_2} \checkmark \quad (e_1 \in I, e_2 \notin I)}{\tau_I(p) \parallel \tau_I(q) \xRightarrow{e_2} \checkmark}$$

$$\frac{p \xrightarrow{e_1} p' \quad q \xrightarrow{e_2} \checkmark \quad (e_1 \in I, e_2 \notin I)}{\tau_I(p \parallel q) \xRightarrow{e_2} \tau_I(p')} \qquad \frac{p \xrightarrow{e_1} p' \quad q \xrightarrow{e_2} \checkmark \quad (e_1 \in I, e_2 \notin I)}{\tau_I(p) \parallel \tau_I(q) \xRightarrow{e_2} \tau_I(p')}$$

$$\frac{p \xrightarrow{e_1} \checkmark \quad q \xrightarrow{e_2} q' \quad (e_1 \in I, e_2 \notin I)}{\tau_I(p \parallel q) \xRightarrow{e_2} \tau_I(q')} \qquad \frac{p \xrightarrow{e_1} \checkmark \quad q \xrightarrow{e_2} q' \quad (e_1 \in I, e_2 \notin I)}{\tau_I(p) \parallel \tau_I(q) \xRightarrow{e_2} \tau_I(q')}$$

$$\frac{p \xrightarrow{e_1} p' \quad q \xrightarrow{e_2} q' \quad (e_1 \in I, e_2 \notin I)}{\tau_I(p \parallel q) \xRightarrow{e_2} \tau_I(p' \between q')} \qquad \frac{p \xrightarrow{e_1} p' \quad q \xrightarrow{e_2} q' \quad (e_1 \in I, e_2 \notin I)}{\tau_I(p) \parallel \tau_I(q) \xRightarrow{e_2} \tau_I(p') \between \tau_I(q')}$$

$$\frac{p \xrightarrow{e_1} \checkmark \quad q \xrightarrow{e_2} \checkmark \quad (e_1, e_2 \in I)}{\tau_I(p \parallel q) \xrightarrow{\tau^*} \checkmark} \qquad \frac{p \xrightarrow{e_1} \checkmark \quad q \xrightarrow{e_2} \checkmark \quad (e_1, e_2 \in I)}{\tau_I(p) \parallel \tau_I(q) \xrightarrow{\tau^*} \checkmark}$$

$$\frac{p \xrightarrow{e_1} p' \quad q \xrightarrow{e_2} \checkmark \quad (e_1, e_2 \in I)}{\tau_I(p \parallel q) \xrightarrow{\tau^*} \tau_I(p')} \qquad \frac{p \xrightarrow{e_1} p' \quad q \xrightarrow{e_2} \checkmark \quad (e_1, e_2 \in I)}{\tau_I(p) \parallel \tau_I(q) \xrightarrow{\tau^*} \tau_I(p')}$$

$$\frac{p \xrightarrow{e_1} \checkmark \quad q \xrightarrow{e_2} q' \quad (e_1, e_2 \in I)}{\tau_I(p \parallel q) \xrightarrow{\tau^*} \tau_I(q')} \qquad \frac{p \xrightarrow{e_1} \checkmark \quad q \xrightarrow{e_2} q' \quad (e_1, e_2 \in I)}{\tau_I(p) \parallel \tau_I(q) \xrightarrow{\tau^*} \tau_I(q')}$$

$$\frac{p \xrightarrow{e_1} p' \quad q \xrightarrow{e_2} q' \quad (e_1, e_2 \in I)}{\tau_I(p \parallel q) \xrightarrow{\tau^*} \tau_I(p' \between q')} \qquad \frac{p \xrightarrow{e_1} p' \quad q \xrightarrow{e_2} q' \quad (e_1, e_2 \in I)}{\tau_I(p) \parallel \tau_I(q) \xrightarrow{\tau^*} \tau_I(p' \between q')}$$

So, with the assumption $\tau_I(p' \between q') = \tau_I(p') \between \tau_I(q')$, $\tau_I(p \parallel q) \approx_{rbs} \tau_I(p) \parallel \tau_I(q)$, as desired.

(2) Soundness of $APTC_\tau$ with guarded linear recursion with respect to rooted branching pomset bisimulation $\approx_{rbp}$.

Since rooted branching pomset bisimulation $\approx_{rbp}$ is both an equivalent and a congruent relation with respect to $APTC_\tau$ with guarded linear recursion, we only need to check if each axiom in Table 4.23 is sound modulo rooted branching pomset bisimulation $\approx_{rbp}$.

From the definition of rooted branching pomset bisimulation $\approx_{rbp}$ (see Definition 4.56), we know that rooted branching pomset bisimulation $\approx_{rbp}$ is defined by weak pomset transitions, which are labeled by pomsets with τ. In a weak pomset transition, the events in the pomset are either within causality relations (defined by $\cdot$) or in concurrency (implicitly defined by $\cdot$ and $+$, and explicitly defined by $\between$), of course, they are pairwise consistent (without conflicts). In (1), we have already proven the case that all events are pairwise concurrent, so, we only need to prove the case of events in causality. Without loss of generality, we take a pomset of $P = \{e_1, e_2 : e_1 \cdot e_2\}$. Then the weak pomset transition labeled by the above P is just composed of one single event transition labeled by e_1 succeeded by another single event transition labeled by e_2, that is, $\xRightarrow{P} = \xRightarrow{e_1} \xRightarrow{e_2}$.

Similarly to the proof of soundness of $APTC_\tau$ with guarded linear recursion modulo rooted branching step bisimulation $\approx_{rbs}$ (1), we can prove that each axiom in Table 4.23 is sound modulo rooted branching pomset bisimulation $\approx_{rbp}$, we omit them.

(3) Soundness of $APTC_\tau$ with guarded linear recursion with respect to rooted branching hp-bisimulation $\approx_{rbhp}$.

Since rooted branching hp-bisimulation $\approx_{rbhp}$ is both an equivalent and a congruent relation with respect to $APTC_\tau$ with guarded linear recursion, we only need to check if each axiom in Table 4.23 is sound modulo rooted branching hp-bisimulation $\approx_{rbhp}$.

From the definition of rooted branching hp-bisimulation $\approx_{rbhp}$ (see Definition 4.58), we know that rooted branching hp-bisimulation $\approx_{rbhp}$ is defined on the weakly posetal product (C_1, f, C_2), $f : \hat{C}_1 \to \hat{C}_2$ isomorphism. Two process terms s related to C_1 and t related to C_2, and $f : \hat{C}_1 \to \hat{C}_2$ isomorphism. Initially, $(C_1, f, C_2) = (\emptyset, \emptyset, \emptyset)$, and $(\emptyset, \emptyset, \emptyset) \in \approx_{rbhp}$. When $s \xrightarrow{e} s'$ $(C_1 \xrightarrow{e} C_1')$, there will be $t \xRightarrow{e} t'$ $(C_2 \xRightarrow{e} C_2')$, and we define $f' = f[e \mapsto e]$. Then, if $(C_1, f, C_2) \in \approx_{rbhp}$, then $(C_1', f', C_2') \in \approx_{rbhp}$.

Similarly to the proof of soundness of $APTC_\tau$ with guarded linear recursion modulo rooted branching pomset bisimulation equivalence (2), we can prove that each axiom in Table 4.23 is sound modulo rooted branching hp-bisimulation equivalence, we just need additionally to check the above conditions on rooted branching hp-bisimulation, we omit them.   □

Though τ-loops are prohibited in guarded linear recursive specifications (see Definition 4.59) in a specifiable way, they can be constructed using the abstraction operator, for example, there exist τ-loops in the process term $\tau_{\{a\}}(\langle X|X = aX \rangle)$. To avoid τ-loops caused by τ_I and ensure fairness, the concept of cluster and $CFAR$ (Cluster Fair Abstraction Rule) [17] are still valid in true concurrency, we introduce them below.

Definition 4.68 (Cluster). Let E be a guarded linear recursive specification, and $I \subseteq \mathbb{E}$. Two recursion variable X and Y in E are in the same cluster for I iff there exist sequences of transitions $\langle X|E \rangle \xrightarrow{\{b_{11}, \cdots, b_{1i}\}} \cdots \xrightarrow{\{b_{m1}, \cdots, b_{mi}\}} \langle Y|E \rangle$ and $\langle Y|E \rangle \xrightarrow{\{c_{11}, \cdots, c_{1j}\}} \cdots \xrightarrow{\{c_{n1}, \cdots, c_{nj}\}} \langle X|E \rangle$, where $b_{11}, \cdots, b_{mi}, c_{11}, \cdots, c_{nj} \in I \cup \{\tau\}$.

$a_1 \parallel \cdots \parallel a_k$ or $(a_1 \parallel \cdots \parallel a_k)X$ is an exit for the cluster C iff: (1) $a_1 \parallel \cdots \parallel a_k$ or $(a_1 \parallel \cdots \parallel a_k)X$ is a summand at the right-hand side of the recursive equation for a recursion variable in C, and (2) in the case of $(a_1 \parallel \cdots \parallel a_k)X$, either $a_l \notin I \cup \{\tau\}(l \in \{1, 2, \cdots, k\})$ or $X \notin C$.

Theorem 4.69 (Soundness of $CFAR$). $CFAR$ is sound modulo rooted branching truly concurrent bisimulation equivalences $\approx_{rbs}$, $\approx_{rbp}$ and $\approx_{rbhp}$.

Proof. (1) Soundness of $CFAR$ with respect to rooted branching step bisimulation $\approx_{rbs}$.

Let X be in a cluster for I with exits $\{(a_{11} \parallel \cdots \parallel a_{1i})Y_1, \cdots, (a_{m1} \parallel \cdots \parallel a_{mi})Y_m, b_{11} \parallel \cdots \parallel b_{1j}, \cdots, b_{n1} \parallel \cdots \parallel b_{nj}\}$. Then $\langle X|E \rangle$ can execute a string of atomic events from $I \cup \{\tau\}$ inside the cluster of X, followed by an exit $(a_{i'1} \parallel \cdots \parallel a_{i'i})Y_{i'}$ for $i' \in \{1, \cdots, m\}$ or $b_{j'1} \parallel \cdots \parallel b_{j'j}$ for $j' \in \{1, \cdots, n\}$. Hence, $\tau_I(\langle X|E \rangle)$ can execute a string of τ^* inside the cluster of X, followed by an exit $\tau_I((a_{i'1} \parallel \cdots \parallel a_{i'i})\langle Y_{i'}|E \rangle)$ for $i' \in \{1, \cdots, m\}$ or $\tau_I(b_{j'1} \parallel \cdots \parallel b_{j'j})$ for $j' \in \{1, \cdots, n\}$. And these τ^* are non-initial in $\tau \tau_I(\langle X|E \rangle)$, so they are truly silent by the axiom $B1$, we obtain $\tau \tau_I(\langle X|E \rangle) \approx_{rbs} \tau \cdot \tau_I((a_{11} \parallel \cdots \parallel a_{1i})\langle Y_1|E \rangle + \cdots + (a_{m1} \parallel \cdots \parallel a_{mi})\langle Y_m|E \rangle + b_{11} \parallel \cdots \parallel b_{1j} + \cdots + b_{n1} \parallel \cdots \parallel b_{nj})$, as desired.

(2) Soundness of $CFAR$ with respect to rooted branching pomset bisimulation $\approx_{rbp}$.

From the definition of rooted branching pomset bisimulation $\approx_{rbp}$ (see Definition 4.56), we know that rooted branching pomset bisimulation $\approx_{rbp}$ is defined by weak pomset transitions, which are labeled by pomsets with τ. In a weak pomset transition, the events in the pomset are either within causality relations (defined by $\cdot$) or in concurrency (implicitly defined by $\cdot$ and $+$, and explicitly defined by $\between$), of course, they are pairwise consistent

Table 4.24   Cluster fair abstraction rule.

No.	Axiom
$CFAR$	If X is in a cluster for I with exits
	$\{(a_{11} \parallel \cdots \parallel a_{1i})Y_1, \cdots, (a_{m1} \parallel \cdots \parallel a_{mi})Y_m, b_{11} \parallel \cdots \parallel b_{1j}, \cdots, b_{n1} \parallel \cdots \parallel b_{nj}\}$,
	then $\tau \cdot \tau_I(\langle X|E\rangle) =$
	$\tau \cdot \tau_I((a_{11} \parallel \cdots \parallel a_{1i})\langle Y_1|E\rangle + \cdots + (a_{m1} \parallel \cdots \parallel a_{mi})\langle Y_m|E\rangle + b_{11} \parallel \cdots \parallel b_{1j} + \cdots$
	$+ b_{n1} \parallel \cdots \parallel b_{nj})$

(without conflicts). In (1), we have already proven the case that all events are pairwise concurrent, so, we only need to prove the case of events in causality. Without loss of generality, we take a pomset of $P = \{e_1, e_2 : e_1 \cdot e_2\}$. Then the weak pomset transition labeled by the above P is just composed of one single event transition labeled by e_1 succeeded by another single event transition labeled by e_2, that is, $\xRightarrow{P} = \xRightarrow{e_1}\xRightarrow{e_2}$.

Similarly to the proof of soundness of $CFAR$ modulo rooted branching step bisimulation $\approx_{rbs}$ (1), we can prove that $CFAR$ in Table 4.24 is sound modulo rooted branching pomset bisimulation $\approx_{rbp}$, we omit them.

(3) Soundness of $CFAR$ with respect to rooted branching hp-bisimulation $\approx_{rbhp}$.

From the definition of rooted branching hp-bisimulation $\approx_{rbhp}$ (see Definition 4.58), we know that rooted branching hp-bisimulation $\approx_{rbhp}$ is defined on the weakly posetal product (C_1, f, C_2), $f : \hat{C}_1 \to \hat{C}_2$ isomorphism. Two process terms s related to C_1 and t related to C_2, and $f : \hat{C}_1 \to \hat{C}_2$ isomorphism. Initially, $(C_1, f, C_2) = (\emptyset, \emptyset, \emptyset)$, and $(\emptyset, \emptyset, \emptyset) \in \approx_{rbhp}$. When $s \xrightarrow{e} s'$ ($C_1 \xrightarrow{e} C_1'$), there will be $t \xRightarrow{e} t'$ ($C_2 \xRightarrow{e} C_2'$), and we define $f' = f[e \mapsto e]$. Then, if $(C_1, f, C_2) \in \approx_{rbhp}$, then $(C_1', f', C_2') \in \approx_{rbhp}$.

Similarly to the proof of soundness of $CFAR$ modulo rooted branching pomset bisimulation equivalence (2), we can prove that $CFAR$ in Table 4.24 is sound modulo rooted branching hp-bisimulation equivalence, we just need additionally to check the above conditions on rooted branching hp-bisimulation, we omit them.   □

Theorem 4.70 (Completeness of $APTC_\tau$ with guarded linear recursion and $CFAR$). *Let p and q be closed $APTC_\tau$ with guarded linear recursion and $CFAR$ terms, then,*

1. *if $p \approx_{rbs} q$ then $p = q$;*
2. *if $p \approx_{rbp} q$ then $p = q$;*
3. *if $p \approx_{rbhp} q$ then $p = q$.*

Proof. (1) For the case of rooted branching step bisimulation, the proof is following.

Firstly, in the proof of Theorem 4.64, we know that each process term p in $APTC$ with silent step and guarded linear recursion is equal to a process term $\langle X_1|E\rangle$ with E a guarded linear recursive specification. And we prove if $\langle X_1|E_1\rangle \approx_{rbs} \langle Y_1|E_2\rangle$, then $\langle X_1|E_1\rangle = \langle Y_1|E_2\rangle$

The only new case is $p \equiv \tau_I(q)$. Let $q = \langle X|E\rangle$ with E a guarded linear recursive specification, so $p = \tau_I(\langle X|E\rangle)$. Then the collection of recursive variables in E can be divided into its clusters $C_1, \cdots, C_N$ for I. Let

$$(a_{1i1} \parallel \cdots \parallel a_{k_i i1})Y_{i1} + \cdots + (a_{1im_i} \parallel \cdots \parallel a_{k_{im_i} im_i})Y_{im_i} + b_{1i1} \parallel \cdots \parallel b_{l_i i1} + \cdots$$
$$+ b_{1im_i} \parallel \cdots \parallel b_{l_{im_i} im_i}$$

be the conflict composition of exits for the cluster C_i, with $i \in \{1, \cdots, N\}$.

For $Z \in C_i$ with $i \in \{1, \cdots, N\}$, we define

$s_Z \triangleq (a_{\hat{1}i1} \parallel \cdots \parallel a_{k_i\hat{i}1})\tau_I(\langle Y_{i1}|E\rangle) + \cdots + (a_{1\hat{i}m_i} \parallel \cdots \parallel a_{k_{im_i}\hat{i}m_i})\tau_I(\langle Y_{im_i}|E\rangle) + b_{\hat{1}i1} \parallel \cdots \parallel$

$b_{l_i\hat{i}1} + \cdots + b_{1\hat{i}m_i} \parallel \cdots \parallel b_{l_{im_i}im_i}$

For $Z \in C_i$ and $a_1, \cdots, a_j \in \mathbb{E} \cup \{\tau\}$ with $j \in \mathbb{N}$, we have

$(a_1 \parallel \cdots \parallel a_j)\tau_I(\langle Z|E\rangle)$

$= (a_1 \parallel \cdots \parallel a_j)\tau_I((a_{1i1} \parallel \cdots \parallel a_{k_i i1})\langle Y_{i1}|E\rangle + \cdots + (a_{1im_i} \parallel \cdots \parallel a_{k_{im_i}im_i})\langle Y_{im_i}|E\rangle + b_{1i1} \parallel$

$\cdots \parallel b_{l_i i1} + \cdots + b_{1im_i} \parallel \cdots \parallel b_{l_{im_i}im_i})$

$= (a_1 \parallel \cdots \parallel a_j)s_Z$

Let the linear recursive specification F contain the same recursive variables as E, for $Z \in C_i$, F contains the following recursive equation

$Z = (a_{\hat{1}i1} \parallel \cdots \parallel a_{k_i\hat{i}1})Y_{i1} + \cdots + (a_{1\hat{i}m_i} \parallel \cdots \parallel a_{k_{im_i}\hat{i}m_i})Y_{im_i} + b_{\hat{1}i1} \parallel \cdots \parallel b_{l_i\hat{i}1} + \cdots + b_{1\hat{i}m_i} \parallel$

$\cdots \parallel b_{l_{im_i}im_i}$

It is easy to see that there is no sequence of one or more τ-transitions from $\langle Z|F\rangle$ to itself, so F is guarded.

For

$s_Z = (a_{\hat{1}i1} \parallel \cdots \parallel a_{k_i\hat{i}1})Y_{i1} + \cdots + (a_{1\hat{i}m_i} \parallel \cdots \parallel a_{k_{im_i}\hat{i}m_i})Y_{im_i} + b_{\hat{1}i1} \parallel \cdots \parallel b_{l_i\hat{i}1} + \cdots + b_{1\hat{i}m_i} \parallel$

$\cdots \parallel b_{l_{im_i}im_i}$

is a solution for F. So, $(a_1 \parallel \cdots \parallel a_j)\tau_I(\langle Z|E\rangle) = (a_1 \parallel \cdots \parallel a_j)s_Z = (a_1 \parallel \cdots \parallel a_j)\langle Z|F\rangle$.

So,

$\langle Z|F\rangle = (a_{\hat{1}i1} \parallel \cdots \parallel a_{k_i\hat{i}1})\langle Y_{i1}|F\rangle + \cdots + (a_{1\hat{i}m_i} \parallel \cdots \parallel a_{k_{im_i}\hat{i}m_i})\langle Y_{im_i}|F\rangle + b_{\hat{1}i1} \parallel \cdots \parallel b_{l_i\hat{i}1} +$

$\cdots + b_{1\hat{i}m_i} \parallel \cdots \parallel b_{l_{im_i}im_i}$

Hence, $\tau_I(\langle X|E\rangle = \langle Z|F\rangle)$, as desired.

(2) For the case of rooted branching pomset bisimulation, it can be proven similarly to (1), we omit it.

(3) For the case of rooted branching hp-bisimulation, it can be proven similarly to (1), we omit it.   □

Finally, in section 4.2, during conflict elimination, the axioms $U25$ and $U27$ are $(\sharp(e_1, e_2))$ $e_1 \triangleleft e_2 = \tau$ and $(\sharp(e_1, e_2), e_2 \leq e_3)$ $e3 \triangleleft e_1 = \tau$. Their functions are like abstraction operator τ_I, their rigorous soundness can be proven similarly to Theorem 4.63 and Theorem 4.67, really, they are based on weakly true concurrency. We just illustrate their intuition through an example.

$P = \Theta((a \cdot b \cdot c) \parallel (d \cdot e \cdot f)$ $(\sharp(b, e)))$

$\overset{CE23}{=} (\Theta(a \cdot b \cdot c) \triangleleft (d \cdot e \cdot f)) \parallel (d \cdot e \cdot f) + (\Theta(d \cdot e \cdot f) \triangleleft (a \cdot b \cdot c)) \parallel (a \cdot b \cdot c)$ $(\sharp(b, e))$

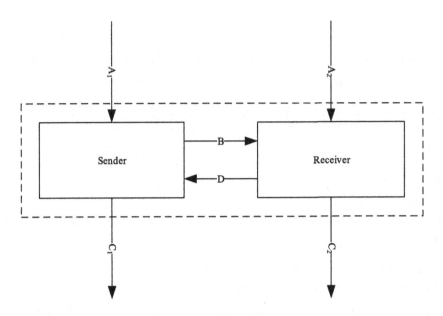

FIGURE 4.2 Alternating bit protocol.

$$\overset{CE22}{=} (a \cdot b \cdot c) \triangleleft (d \cdot e \cdot f)) \parallel (d \cdot e \cdot f) + ((d \cdot e \cdot f) \triangleleft (a \cdot b \cdot c)) \parallel (a \cdot b \cdot c) \quad (\sharp(b, e))$$

$$\overset{U31,U35}{=} (a \cdot \tau \cdot \tau) \parallel (d \cdot e \cdot f) + (d \cdot \tau \cdot \tau) \parallel (a \cdot b \cdot c)$$

$$\overset{B1}{=} a \parallel (d \cdot e \cdot f) + d \parallel (a \cdot b \cdot c)$$

We see that the conflict relation $\sharp(b, e)$ is eliminated.

4.5 Applications

$APTC$ provides a formal framework based on truly concurrent behavioral semantics, which can be used to verify the correctness of system behaviors. In this section, we tend to choose one protocol verified by ACP [4] – alternating bit protocol (ABP) [19].

The ABP protocol is used to ensure successful transmission of data through a corrupted channel. This success is based on the assumption that data can be resent an unlimited number of times, which is illustrated in Fig. 4.2, we alter it into the true concurrency situation.

1. Data elements $d_1, d_2, d_3, \cdots$ from a finite set Δ are communicated between a Sender and a Receiver.
2. If the Sender reads a datum from channel A_1, then this datum is sent to the Receiver in parallel through channel A_2.
3. The Sender processes the data in Δ, forms new data, and sends them to the Receiver through channel B.

4. And the Receiver sends the datum into channel C_2.

5. If channel B is corrupted, the message communicated through B can be turn into an error message $\perp$.

6. Every time the Receiver receives a message via channel B, it sends an acknowledgment to the Sender via channel D, which is also corrupted.

7. Finally, then Sender and the Receiver send out their outputs in parallel through channels C_1 and C_2.

In the truly concurrent ABP, the Sender sends its data to the Receiver; and the Receiver can also send its data to the Sender, for simplicity and without loss of generality, we assume that only the Sender sends its data and the Receiver only receives the data from the Sender. The Sender attaches a bit 0 to data elements d_{2k-1} and a bit 1 to data elements d_{2k}, when they are sent into channel B. When the Receiver reads a datum, it sends back the attached bit via channel D. If the Receiver receives a corrupted message, then it sends back the previous acknowledgment to the Sender.

Then the state transition of the Sender can be described by $APTC$ as follows.

$$S_b = \sum_{d\in\Delta} r_{A_1}(d) \cdot T_{db}$$

$$T_{db} = (\sum_{d'\in\Delta} (s_B(d',b) \cdot s_{C_1}(d')) + s_B(\perp)) \cdot U_{db}$$

$$U_{db} = r_D(b) \cdot S_{1-b} + (r_D(1-b) + r_D(\perp)) \cdot T_{db}$$

where s_B denotes sending data through channel B, r_D denotes receiving data through channel D, similarly, r_{A_1} means receiving data via channel A_1, s_{C_1} denotes sending data via channel C_1, and $b \in \{0,1\}$.

And the state transition of the Receiver can be described by $APTC$ as follows.

$$R_b = \sum_{d\in\Delta} r_{A_2}(d) \cdot R'_b$$

$$R'_b = \sum_{d'\in\Delta} \{r_B(d',b) \cdot s_{C_2}(d') \cdot Q_b + r_B(d',1-b) \cdot Q_{1-b}\} + r_B(\perp) \cdot Q_{1-b}$$

$$Q_b = (s_D(b) + s_D(\perp)) \cdot R_{1-b}$$

where r_{A_2} denotes receiving data via channel A_2, r_B denotes receiving data via channel B, s_{C_2} denotes sending data via channel C_2, s_D denotes sending data via channel D, and $b \in \{0,1\}$.

The send action and receive action of the same data through the same channel can communicate each other, otherwise, a deadlock δ will be caused. We define the following communication functions.

$$\gamma(s_B(d',b), r_B(d',b)) \triangleq c_B(d',b)$$

$$\gamma(s_B(\perp), r_B(\perp)) \triangleq c_B(\perp)$$

$$\gamma(s_D(b), r_D(b)) \triangleq c_D(b)$$

$$\gamma(s_D(\bot), r_D(\bot)) \triangleq c_D(\bot)$$

Let R_0 and S_0 be in parallel, then the system $R_0 S_0$ can be represented by the following process term.

$$\tau_I(\partial_H(\Theta(R_0 \between S_0))) = \tau_I(\partial_H(R_0 \between S_0))$$

where $H = \{s_B(d', b), r_B(d', b), s_D(b), r_D(b) | d' \in \Delta, b \in \{0, 1\}\}$
$\{s_B(\bot), r_B(\bot), s_D(\bot), r_D(\bot)\}$
    $I = \{c_B(d', b), c_D(b) | d' \in \Delta, b \in \{0, 1\}\} \cup \{c_B(\bot), c_D(\bot)\}$.
    Then we get the following conclusion.

Theorem 4.71 (Correctness of the ABP protocol). *The ABP protocol $\tau_I(\partial_H(R_0 \between S_0))$ exhibits desired external behaviors.*

Proof. By use of the algebraic laws of $APTC$, we have the following expansions.

$$R_0 \between S_0 \overset{P1}{=} R_0 \parallel S_0 + R_0 \mid S_0$$

$$\overset{RDP}{=} (\sum_{d \in \Delta} r_{A_2}(d) \cdot R_0') \parallel (\sum_{d \in \Delta} r_{A_1}(d) T_{d0})$$

$$+ (\sum_{d \in \Delta} r_{A_2}(d) \cdot R_0') \mid (\sum_{d \in \Delta} r_{A_1}(d) T_{d0})$$

$$\overset{P6,C14}{=} \sum_{d \in \Delta} (r_{A_2}(d) \parallel r_{A_1}(d)) R_0' \between T_{d0} + \delta \cdot R_0' \between T_{d0}$$

$$\overset{A6,A7}{=} \sum_{d \in \Delta} (r_{A_2}(d) \parallel r_{A_1}(d)) R_0' \between T_{d0}$$

$$\partial_H(R_0 \between S_0) = \partial_H(\sum_{d \in \Delta} (r_{A_2}(d) \parallel r_{A_1}(d)) R_0' \between T_{d0})$$

$$= \sum_{d \in \Delta} (r_{A_2}(d) \parallel r_{A_1}(d)) \partial_H(R_0' \between T_{d0})$$

Similarly, we can get the following equations.

$$\partial_H(R_0 \between S_0) = \sum_{d \in \Delta} (r_{A_2}(d) \parallel r_{A_1}(d)) \cdot \partial_H(T_{d0} \between R_0')$$

$$\partial_H(T_{d0} \between R_0') = c_B(d', 0) \cdot (s_{C_1}(d') \parallel s_{C_2}(d')) \cdot \partial_H(U_{d0} \between Q_0) + c_B(\bot) \cdot \partial_H(U_{d0} \between Q_1)$$

$$\partial_H(U_{d0} \between Q_1) = (c_D(1) + c_D(\bot)) \cdot \partial_H(T_{d0} \between R_0')$$

$$\partial_H(Q_0 \between U_{d0}) = c_D(0) \cdot \partial_H(R_1 \between S_1) + c_D(\bot) \cdot \partial_H(R_1' \between T_{d0})$$

$$\partial_H(R_1' \between T_{d0}) = (c_B(d', 0) + c_B(\bot)) \cdot \partial_H(Q_0 \between U_{d0})$$

$$\partial_H(R_1 \between S_1) = \sum_{d \in \Delta} (r_{A_2}(d) \parallel r_{A_1}(d)) \cdot \partial_H(T_{d1} \between R_1')$$

$$\partial_H(T_{d1} \between R_1') = c_B(d', 1) \cdot (s_{C_1}(d') \parallel s_{C_2}(d')) \cdot \partial_H(U_{d1} \between Q_1) + c_B(\bot) \cdot \partial_H(U_{d1} \between Q_0')$$
$$\partial_H(U_{d1} \between Q_0') = (c_D(0) + c_D(\bot)) \cdot \partial_H(T_{d1} \between R_1')$$
$$\partial_H(Q_1 \between U_{d1}) = c_D(1) \cdot \partial_H(R_0 \between S_0) + c_D(\bot) \cdot \partial_H(R_0' \between T_{d1})$$
$$\partial_H(R_0' \between T_{d1}) = (c_B(d', 1) + c_B(\bot)) \cdot \partial_H(Q_1 \between U_{d1})$$

Let $\partial_H(R_0 \between S_0) = \langle X_1 | E \rangle$, where E is the following guarded linear recursion specification:

$$\{ X_1 = \sum_{d \in \Delta} (r_{A_2}(d) \parallel r_{A_1}(d)) \cdot X_{2d}, \quad Y_1 = \sum_{d \in \Delta} (r_{A_2}(d) \parallel r_{A_1}(d)) \cdot Y_{2d},$$

$$X_{2d} = c_B(d', 0) \cdot X_{4d} + c_B(\bot) \cdot X_{3d}, \quad Y_{2d} = c_B(d', 1) \cdot Y_{4d} + c_B(\bot) \cdot Y_{3d},$$

$$X_{3d} = (c_D(1) + c_D(\bot)) \cdot X_{2d}, \quad Y_{3d} = (c_D(0) + c_D(\bot)) \cdot Y_{2d},$$

$$X_{4d} = (s_{C_1}(d') \parallel s_{C_2}(d')) \cdot X_{5d}, \quad Y_{4d} = (s_{C_1}(d') \parallel s_{C_2}(d')) \cdot Y_{5d},$$

$$X_{5d} = c_D(0) \cdot Y_1 + c_D(\bot) \cdot X_{6d}, \quad Y_{5d} = c_D(1) \cdot X_1 + c_D(\bot) \cdot Y_{6d},$$

$$X_{6d} = (c_B(d, 0) + c_B(\bot)) \cdot X_{5d}, \quad Y_{6d} = (c_B(d, 1) + c_B(\bot)) \cdot Y_{5d}$$

$$|d, d' \in \Delta \}$$

Then we apply abstraction operator τ_I into $\langle X_1 | E \rangle$.

$$\tau_I(\langle X_1 | E \rangle) = \sum_{d \in \Delta} (r_{A_1}(d) \parallel r_{A_2}(d)) \cdot \tau_I(\langle X_{2d} | E \rangle)$$

$$= \sum_{d \in \Delta} (r_{A_1}(d) \parallel r_{A_2}(d)) \cdot \tau_I(\langle X_{4d} | E \rangle)$$

$$= \sum_{d, d' \in \Delta} (r_{A_1}(d) \parallel r_{A_2}(d)) \cdot (s_{C_1}(d') \parallel s_{C_2}(d')) \cdot \tau_I(\langle X_{5d} | E \rangle)$$

$$= \sum_{d, d' \in \Delta} (r_{A_1}(d) \parallel r_{A_2}(d)) \cdot (s_{C_1}(d') \parallel s_{C_2}(d')) \cdot \tau_I(\langle Y_1 | E \rangle)$$

Similarly, we can get $\tau_I(\langle Y_1 | E \rangle) = \sum_{d, d' \in \Delta} (r_{A_1}(d) \parallel r_{A_2}(d)) \cdot (s_{C_1}(d') \parallel s_{C_2}(d')) \cdot \tau_I(\langle X_1 | E \rangle)$.

We get $\tau_I(\partial_H(R_0 \between S_0)) = \sum_{d, d' \in \Delta} (r_{A_1}(d) \parallel r_{A_2}(d)) \cdot (s_{C_1}(d') \parallel s_{C_2}(d')) \cdot \tau_I(\partial_H(R_0 \between S_0))$. So, the ABP protocol $\tau_I(\partial_H(R_0 \between S_0))$ exhibits desired external behaviors.   □

4.6 Extensions

APTC also has the modularity as *ACP*, so, *APTC* can be extended easily. By introducing new operators or new constants, *APTC* can have more properties, modularity provides *APTC* an elegant fashion to express a new property. In this section, we take examples of placeholder which maybe capture the nature of true concurrency, renaming operator which is used to rename the atomic events and firstly introduced by Milner in his CCS [3], state operator which can explicitly define states, and a more complex extension called guards which can express conditionals.

Table 4.25 Transition rule of the shadow constant.

$$\overline{\text{\textcircled{s}} \to \surd}$$

4.6.1 Placeholder

Through verification of ABP protocol [19] in section 4.5, we see that the verification is in a structural symmetric way. Let see the following example.

$$\partial_H((a \cdot r_b) \between w_b) = \partial_H((a \parallel w_b) \cdot r_b + \gamma(a, w_b) \cdot r_b)$$
$$= \delta \cdot r_b + \delta \cdot r_b$$
$$= \delta + \delta$$
$$= \delta$$

With $H = \{r_b, w_b\}$ and $\gamma(r_b, w_b) \triangleq c_b$. The communication c_b does not occur and the above equation should be able to be equal to $a \cdot c_b$. How to deal this situation?

It is caused that the two communicating actions are not at the same causal depth. That is, we must pad something in hole of $(a \cdot r_b) \between ([-] \cdot w_b)$ to make r_b and w_b in the same causal depth.

Can we pad τ into that hole? No. Because $\tau \cdot w_b \neq w_b$. We must pad something new to that hole.

We introduce a constant called shadow constant $\text{\textcircled{s}}$ to act for the placeholder that we ever used to deal entanglement in quantum process algebra. The transition rule of the shadow constant $\text{\textcircled{s}}$ is shown in Table 4.25. The rule says that $\text{\textcircled{s}}$ can terminate successfully without executing any action.

We need to adjust the definition of guarded linear recursive specification to the following one.

Definition 4.72 (Guarded linear recursive specification). A linear recursive specification E is guarded if there does not exist an infinite sequence of τ-transitions $\langle X|E \rangle \xrightarrow{\tau} \langle X'|E \rangle \xrightarrow{\tau} \langle X''|E \rangle \xrightarrow{\tau} \cdots$, and there does not exist an infinite sequence of $\text{\textcircled{s}}$-transitions $\langle X|E \rangle \to \langle X'|E \rangle \to \langle X''|E \rangle \to \cdots$.

Theorem 4.73 (Conservativity of $APTC$ with respect to the shadow constant). $APTC_\tau$ *with guarded linear recursion and shadow constant is a conservative extension of* $APTC_\tau$ *with guarded linear recursion.*

Proof. It follows from the following two facts (see Theorem 2.8).

1. The transition rules of $APTC_\tau$ with guarded linear recursion in section 4.4 are all source-dependent;

Table 4.26 Axioms of shadow constant.

No.	Axiom
$SC1$	$\circledS \cdot x = x$
$SC2$	$x \cdot \circledS = x$
$SC3$	$\circledS^e \parallel e = e$
$SC4$	$e \parallel (\circledS^e \cdot y) = e \cdot y$
$SC5$	$\circledS^e \parallel (e \cdot y) = e \cdot y$
$SC6$	$(e \cdot x) \parallel \circledS^e = e \cdot x$
$SC7$	$(\circledS^e \cdot x) \parallel e = e \cdot x$
$SC8$	$(e \cdot x) \parallel (\circledS^e \cdot y) = e \cdot (x \between y)$
$SC9$	$(\circledS^e \cdot x) \parallel (e \cdot y) = e \cdot (x \between y)$

2. The sources of the transition rules for the shadow constant contain an occurrence of $\circledS$. □

We design the axioms for the shadow constant $\circledS$ in Table 4.26. And for $\circledS_i^e$, we add superscript e to denote $\circledS$ is belonging to e and subscript i to denote that it is the i-th shadow of e. And we extend the set $\mathbb{E}$ to the set $\mathbb{E} \cup \{\tau\} \cup \{\delta\} \cup \{\circledS_i^e\}$.

The mismatch of action and its shadows in parallelism will cause deadlock, that is, $e \parallel \circledS^{e'} = \delta$ with $e \neq e'$. We must make all shadows $\circledS_i^e$ are distinct, to ensure f in hp-bisimulation is an isomorphism.

Theorem 4.74 (Soundness of the shadow constant). *Let x and y be $APTC_\tau$ with guarded linear recursion and the shadow constant terms. If $APTC_\tau$ with guarded linear recursion and the shadow constant $\vdash x = y$, then*

1. $x \approx_{rbs} y$;
2. $x \approx_{rbp} y$;
3. $x \approx_{rbhp} y$.

Proof. (1) Soundness of $APTC_\tau$ with guarded linear recursion and the shadow constant with respect to rooted branching step bisimulation $\approx_{rbs}$.

Since rooted branching step bisimulation $\approx_{rbs}$ is both an equivalent and a congruent relation with respect to $APTC_\tau$ with guarded linear recursion and the shadow constant, we only need to check if each axiom in Table 4.26 is sound modulo rooted branching step bisimulation equivalence.

Though transition rules in Table 4.25 are defined in the flavor of single event, they can be modified into a step (a set of events within which each event is pairwise concurrent), we omit them. If we treat a single event as a step containing just one event, the proof of this soundness theorem does not exist any problem, so we use this way and still use the transition rules in Table 4.26.

The proof of soundness of $SC1 - SC9$ modulo rooted branching step bisimulation is trivial, and we omit it.

(2) Soundness of $APTC_\tau$ with guarded linear recursion and the shadow constant with respect to rooted branching pomset bisimulation $\approx_{rbp}$.

Since rooted branching pomset bisimulation $\approx_{rbp}$ is both an equivalent and a congruent relation with respect to $APTC_\tau$ with guarded linear recursion and the shadow constant, we only need to check if each axiom in Table 4.26 is sound modulo rooted branching pomset bisimulation $\approx_{rbp}$.

From the definition of rooted branching pomset bisimulation $\approx_{rbp}$ (see Definition 4.56), we know that rooted branching pomset bisimulation $\approx_{rbp}$ is defined by weak pomset transitions, which are labeled by pomsets with τ. In a weak pomset transition, the events in the pomset are either within causality relations (defined by $\cdot$) or in concurrency (implicitly defined by $\cdot$ and $+$, and explicitly defined by $\between$), of course, they are pairwise consistent (without conflicts). In (1), we have already proven the case that all events are pairwise concurrent, so, we only need to prove the case of events in causality. Without loss of generality, we take a pomset of $P = \{e_1, e_2 : e_1 \cdot e_2\}$. Then the weak pomset transition labeled by the above P is just composed of one single event transition labeled by e_1 succeeded by another single event transition labeled by e_2, that is, $\xRightarrow{P} = \xRightarrow{e_1} \xRightarrow{e_2}$.

Similarly to the proof of soundness of $APTC_\tau$ with guarded linear recursion and the shadow constant modulo rooted branching step bisimulation $\approx_{rbs}$ (1), we can prove that each axiom in Table 4.26 is sound modulo rooted branching pomset bisimulation $\approx_{rbp}$, we omit them.

(3) Soundness of $APTC_\tau$ with guarded linear recursion and the shadow constant with respect to rooted branching hp-bisimulation $\approx_{rbhp}$.

Since rooted branching hp-bisimulation $\approx_{rbhp}$ is both an equivalent and a congruent relation with respect to $APTC_\tau$ with guarded linear recursion and the shadow constant, we only need to check if each axiom in Table 4.26 is sound modulo rooted branching hp-bisimulation $\approx_{rbhp}$.

From the definition of rooted branching hp-bisimulation $\approx_{rbhp}$ (see Definition 4.58), we know that rooted branching hp-bisimulation $\approx_{rbhp}$ is defined on the weakly posetal product $(C_1, f, C_2), f : \hat{C_1} \to \hat{C_2}$ isomorphism. Two process terms s related to C_1 and t related to C_2, and $f : \hat{C_1} \to \hat{C_2}$ isomorphism. Initially, $(C_1, f, C_2) = (\emptyset, \emptyset, \emptyset)$, and $(\emptyset, \emptyset, \emptyset) \in \approx_{rbhp}$. When $s \xrightarrow{e} s'$ $(C_1 \xrightarrow{e} C_1')$, there will be $t \xrightarrow{e} t'$ $(C_2 \xrightarrow{e} C_2')$, and we define $f' = f[e \mapsto e]$. Then, if $(C_1, f, C_2) \in \approx_{rbhp}$, then $(C_1', f', C_2') \in \approx_{rbhp}$.

Similarly to the proof of soundness of $APTC_\tau$ with guarded linear recursion and the shadow constant modulo rooted branching pomset bisimulation equivalence (2), we can prove that each axiom in Table 4.26 is sound modulo rooted branching hp-bisimulation equivalence, we just need additionally to check the above conditions on rooted branching hp-bisimulation, we omit them. □

Theorem 4.75 (Completeness of the shadow constant). *Let p and q be closed $APTC_\tau$ with guarded linear recursion and $CFAR$ and the shadow constant terms, then,*

1. *if $p \approx_{rbs} q$ then $p = q$;*
2. *if $p \approx_{rbp} q$ then $p = q$;*
3. *if $p \approx_{rbhp} q$ then $p = q$.*

Proof. (1) For the case of rooted branching step bisimulation, the proof is following.

Firstly, in the proof of Theorem 4.70, we know that each process term p in $APTC_\tau$ with guarded linear recursion is equal to a process term $\langle X_1 | E \rangle$ with E a guarded linear recursive specification. And we prove if $\langle X_1 | E_1 \rangle \approx_{rbs} \langle Y_1 | E_2 \rangle$, then $\langle X_1 | E_1 \rangle = \langle Y_1 | E_2 \rangle$

It is no necessary to induct with respect to the structure of process term p, because there are no new cases. The only new situation is that now the set $\mathbb{E}$ contains some new constants $\circledS_i^e$ for $e \in \mathbb{E}$ and $i \in \mathbb{N}$. Since $\circledS_i^e$ does not do anything, so, naturally, if $\langle X_1 | E_1 \rangle \approx_{rbs} \langle Y_1 | E_2 \rangle$, the only thing is that we should prevent $\circledS$-loops in the recursion in a specific way as same as preventing τ-loops, then $\langle X_1 | E_1 \rangle = \langle Y_1 | E_2 \rangle$.

(2) For the case of rooted branching pomset bisimulation, it can be proven similarly to (1), we omit it.

(3) For the case of rooted branching hp-bisimulation, it can be proven similarly to (1), we omit it. $\qquad\square$

With the shadow constant, we have

$$\partial_H((a \cdot r_b) \between w_b) = \partial_H((a \cdot r_b) \between (\circledS_1^a \cdot w_b))$$
$$= a \cdot c_b$$

with $H = \{r_b, w_b\}$ and $\gamma(r_b, w_b) \triangleq c_b$.

And we see the following example:

$$a \between b = a \parallel b + a \mid b$$
$$= a \parallel b + a \parallel b + a \parallel b + a \mid b$$
$$= a \parallel (\circledS_1^a \cdot b) + (\circledS_1^b \cdot a) \parallel b + a \parallel b + a \mid b$$
$$= (a \parallel \circledS_1^a) \cdot b + (\circledS_1^b \parallel b) \cdot a + a \parallel b + a \mid b$$
$$= a \cdot b + b \cdot a + a \parallel b + a \mid b$$

What do we see? Yes. The parallelism contains both interleaving and true concurrency. This may be why true concurrency is called ***true*** concurrency.

With the help of shadow constant, now we can verify the traditional alternating bit protocol (ABP) [19].

The ABP protocol is used to ensure successful transmission of data through a corrupted channel. This success is based on the assumption that data can be resent an unlimited number of times, which is illustrated in Fig. 4.3, we alter it into the true concurrency situation.

1. Data elements $d_1, d_2, d_3, \cdots$ from a finite set Δ are communicated between a Sender and a Receiver.

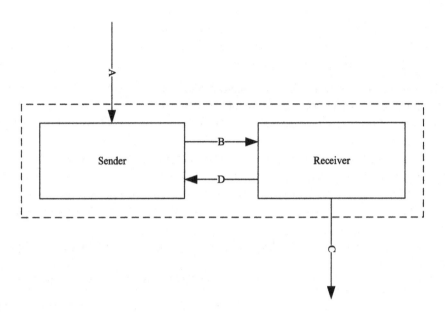

FIGURE 4.3 Alternating bit protocol.

2. If the Sender reads a datum from channel A.
3. The Sender processes the data in Δ, forms new data, and sends them to the Receiver through channel B.
4. And the Receiver sends the datum into channel C.
5. If channel B is corrupted, the message communicated through B can be turn into an error message $\perp$.
6. Every time the Receiver receives a message via channel B, it sends an acknowledgment to the Sender via channel D, which is also corrupted.

The Sender attaches a bit 0 to data elements d_{2k-1} and a bit 1 to data elements d_{2k}, when they are sent into channel B. When the Receiver reads a datum, it sends back the attached bit via channel D. If the Receiver receives a corrupted message, then it sends back the previous acknowledgment to the Sender.

Then the state transition of the Sender can be described by $APTC$ as follows.

$$S_b = \sum_{d \in \Delta} r_A(d) \cdot T_{db}$$

$$T_{db} = (\sum_{d' \in \Delta} (s_B(d', b) \cdot \mathbb{S}^{sc(d')}) + s_B(\perp)) \cdot U_{db}$$

$$U_{db} = r_D(b) \cdot S_{1-b} + (r_D(1-b) + r_D(\perp)) \cdot T_{db}$$

where s_B denotes sending data through channel B, r_D denotes receiving data through channel D, similarly, r_A means receiving data via channel A, $\circledS^{s_C(d')}$ denotes the shadow of $s_C(d')$.

And the state transition of the Receiver can be described by $APTC$ as follows.

$$R_b = \sum_{d \in \Delta} \circledS^{r_A(d)} \cdot R_b'$$

$$R_b' = \sum_{d' \in \Delta} \{r_B(d', b) \cdot s_C(d') \cdot Q_b + r_B(d', 1-b) \cdot Q_{1-b}\} + r_B(\bot) \cdot Q_{1-b}$$

$$Q_b = (s_D(b) + s_D(\bot)) \cdot R_{1-b}$$

where $\circledS^{r_A(d)}$ denotes the shadow of $r_A(d)$, r_B denotes receiving data via channel B, s_C denotes sending data via channel C, s_D denotes sending data via channel D, and $b \in \{0, 1\}$.

The send action and receive action of the same data through the same channel can communicate each other, otherwise, a deadlock δ will be caused. We define the following communication functions.

$$\gamma(s_B(d', b), r_B(d', b)) \triangleq c_B(d', b)$$

$$\gamma(s_B(\bot), r_B(\bot)) \triangleq c_B(\bot)$$

$$\gamma(s_D(b), r_D(b)) \triangleq c_D(b)$$

$$\gamma(s_D(\bot), r_D(\bot)) \triangleq c_D(\bot)$$

Let R_0 and S_0 be in parallel, then the system $R_0 S_0$ can be represented by the following process term.

$$\tau_I(\partial_H(\Theta(R_0 \between S_0))) = \tau_I(\partial_H(R_0 \between S_0))$$

where $H = \{s_B(d', b), r_B(d', b), s_D(b), r_D(b) | d' \in \Delta, b \in \{0, 1\}\}$
$\{s_B(\bot), r_B(\bot), s_D(\bot), r_D(\bot)\}$
   $I = \{c_B(d', b), c_D(b) | d' \in \Delta, b \in \{0, 1\}\} \cup \{c_B(\bot), c_D(\bot)\}$.
   Then we get the following conclusion.

Theorem 4.76 (Correctness of the ABP protocol). *The ABP protocol $\tau_I(\partial_H(R_0 \between S_0))$ exhibits desired external behaviors.*

Proof. Similarly, we can get $\tau_I(\langle X_1 | E \rangle) = \sum_{d, d' \in \Delta} r_A(d) \cdot s_C(d') \cdot \tau_I(\langle Y_1 | E \rangle)$ and $\tau_I(\langle Y_1 | E \rangle) = \sum_{d, d' \in \Delta} r_A(d) \cdot s_C(d') \cdot \tau_I(\langle X_1 | E \rangle)$.

So, the ABP protocol $\tau_I(\partial_H(R_0 \between S_0))$ exhibits desired external behaviors. $\square$

4.6.2 Renaming

Renaming operator $\rho_f(t)$ renames all actions in process term t, and assumes a renaming function $f : \mathbb{E} \cup \{\tau\} \to \mathbb{E} \cup \{\tau\}$ with $f(\tau) \triangleq \tau$, which is expressed by the following two transition rules in Table 4.27.

Table 4.27   Transition rule of the renaming operator.

$$\frac{x \xrightarrow{e} \surd}{\rho_f(x) \xrightarrow{f(e)} \surd} \qquad \frac{x \xrightarrow{e} x'}{\rho_f(x) \xrightarrow{f(e)} \rho_f(x')}$$

Theorem 4.77 (Conservativity of $APTC$ with respect to the renaming operator). *$APTC_\tau$ with guarded linear recursion and renaming operator is a conservative extension of $APTC_\tau$ with guarded linear recursion.*

Proof. It follows from the following two facts (see Theorem 2.8).

1. The transition rules of $APTC_\tau$ with guarded linear recursion in section 4.4 are all source-dependent;
2. The sources of the transition rules for the renaming operator contain an occurrence of ρ_f. $\qquad \square$

Theorem 4.78 (Congruence theorem of the renaming operator). *Rooted branching truly concurrent bisimulation equivalences $\approx_{rbp}$, $\approx_{rbs}$ and $\approx_{rbhp}$ are all congruences with respect to $APTC_\tau$ with guarded linear recursion and the renaming operator.*

Proof. (1) Case rooted branching pomset bisimulation equivalence $\approx_{rbp}$.

Let x and y be $APTC_\tau$ with guarded linear recursion and the renaming operator processes, and $x \approx_{rbp} y$, it is sufficient to prove that $\rho_f(x) \approx_{rbp} \rho_f(y)$.

By the transition rules for operator ρ_f in Table 4.27, we can get

$$\rho_f(x) \xrightarrow{f(X)} \surd \quad \rho_f(y) \xrightarrow{f(Y)} \surd$$

with $X \subseteq x$, $Y \subseteq y$, and $X \sim Y$.

Or, we can get

$$\rho_f(x) \xrightarrow{f(X)} \rho_f(x') \quad \rho_f(y) \xrightarrow{f(Y)} \rho_f(y')$$

with $X \subseteq x$, $Y \subseteq y$, and $X \sim Y$ and the hypothesis $\rho_f(x') \approx_{rbp} \rho_f(y')$.

So, we get $\rho_f(x) \approx_{rbp} \rho_f(y)$, as desired

(2) The cases of rooted branching step bisimulation $\approx_{rbs}$, rooted branching hp-bisimulation $\approx_{rbhp}$ can be proven similarly, we omit them. $\qquad \square$

We design the axioms for the renaming operator ρ_f in Table 4.28.

$RN1 - RN2$ are the defining equations for the renaming operator ρ_f; $RN3 - RN5$ say that in $\rho_f(t)$, the labels of all transitions of t are renamed by means of the mapping f.

Theorem 4.79 (Soundness of the renaming operator). *Let x and y be $APTC_\tau$ with guarded linear recursion and the renaming operator terms. If $APTC_\tau$ with guarded linear recursion and the renaming operator $\vdash x = y$, then*

Table 4.28 Axioms of renaming operator.

No.	Axiom
$RN1$	$\rho_f(e) = f(e)$
$RN2$	$\rho_f(\delta) = \delta$
$RN3$	$\rho_f(x + y) = \rho_f(x) + \rho_f(y)$
$RN4$	$\rho_f(x \cdot y) = \rho_f(x) \cdot \rho_f(y)$
$RN5$	$\rho_f(x \parallel y) = \rho_f(x) \parallel \rho_f(y)$

1. $x \approx_{rbs} y$;

2. $x \approx_{rbp} y$;

3. $x \approx_{rbhp} y$.

Proof. (1) Soundness of $APTC_\tau$ with guarded linear recursion and the renaming operator with respect to rooted branching step bisimulation $\approx_{rbs}$.

Since rooted branching step bisimulation $\approx_{rbs}$ is both an equivalent and a congruent relation with respect to $APTC_\tau$ with guarded linear recursion and the renaming operator, we only need to check if each axiom in Table 4.28 is sound modulo rooted branching step bisimulation equivalence.

Though transition rules in Table 4.27 are defined in the flavor of single event, they can be modified into a step (a set of events within which each event is pairwise concurrent), we omit them. If we treat a single event as a step containing just one event, the proof of this soundness theorem does not exist any problem, so we use this way and still use the transition rules in Table 4.28.

We only prove soundness of the non-trivial axioms $RN3 - RN5$, and omit the defining axioms $RN1 - RN2$.

- **Axiom** $RN3$. Let p, q be $APTC_\tau$ with guarded linear recursion and the renaming opera-tor processes, and $\rho_f(p + q) = \rho_f(p) + \rho_f(q)$, it is sufficient to prove that $\rho_f(p + q) \approx_{rbs} \rho_f(p) + \rho_f(q)$. By the transition rules for operator $+$ in Table 4.5 and ρ_f in Table 4.27, we get

$$\frac{p \xrightarrow{e_1} \surd}{\rho_f(p + q) \xrightarrow{f(e_1)} \surd} \qquad \frac{p \xrightarrow{e_1} \surd}{\rho_f(p) + \rho_f(q) \xrightarrow{f(e_1)} \surd}$$

$$\frac{q \xrightarrow{e_2} \surd}{\rho_f(p + q) \xrightarrow{f(e_2)} \surd} \qquad \frac{q \xrightarrow{e_2} \surd}{\rho_f(p) + \rho_f(q) \xrightarrow{f(e_2)} \surd}$$

$$\frac{p \xrightarrow{e_1} p'}{\rho_f(p + q) \xrightarrow{f(e_1)} \rho_f(p')} \qquad \frac{p \xrightarrow{e_1} p'}{\rho_f(p) + \rho_f(q) \xrightarrow{f(e_1)} \rho_f(p')}$$

$$\frac{q \xrightarrow{e_2} q'}{\rho_f(p+q) \xrightarrow{f(e_2)} \rho_f(q')} \qquad \frac{q \xrightarrow{e_2} q'}{\rho_f(p)+\rho_f(q) \xrightarrow{f(e_2)} \rho_f(q')}$$

So, $\rho_f(p+q) \approx_{rbs} \rho_f(p) + \rho_f(q)$, as desired.

- **Axiom** $RN4$. Let p, q be $APTC_\tau$ with guarded linear recursion and the renaming operator processes, and $\rho_f(p \cdot q) = \rho_f(p) \cdot \rho_f(q)$, it is sufficient to prove that $\rho_f(p \cdot q) \approx_{rbs} \rho_f(p) \cdot \rho_f(q)$. By the transition rules for operator $\cdot$ in Table 4.5 and ρ_f in Table 4.27, we get

$$\frac{p \xrightarrow{e_1} \surd}{\rho_f(p \cdot q) \xrightarrow{f(e_1)} \rho_f(q)} \qquad \frac{p \xrightarrow{e_1} \surd}{\rho_f(p) \cdot \rho_f(q) \xrightarrow{f(e_1)} \rho_f(q)}$$

$$\frac{p \xrightarrow{e_1} p'}{\rho_f(p \cdot q) \xrightarrow{f(e_1)} \rho_f(p' \cdot q)} \qquad \frac{p \xrightarrow{e_1} p'}{\rho_f(p) \cdot \rho_f(q) \xrightarrow{f(e_1)} \rho_f(p') \cdot \rho_f(q)}$$

So, with the assumption $\rho_f(p' \cdot q) = \rho_f(p') \cdot \rho_f(q)$, $\rho_f(p \cdot q) \approx_{rbs} \rho_f(p) \cdot \rho_f(q)$, as desired.

- **Axiom** $RN5$. Let p, q be $APTC_\tau$ with guarded linear recursion and the renaming operator processes, and $\rho_f(p \parallel q) = \rho_f(p) \parallel \rho_f(q)$, it is sufficient to prove that $\rho_f(p \parallel q) \approx_{rbs} \rho_f(p) \parallel \rho_f(q)$. By the transition rules for operator $\parallel$ in Table 4.7 and ρ_f in Table 4.27, we get

$$\frac{p \xrightarrow{e_1} \surd \quad q \xrightarrow{e_2} \surd}{\rho_f(p \parallel q) \xrightarrow{\{f(e_1),f(e_2)\}} \surd} \qquad \frac{p \xrightarrow{e_1} \surd \quad q \xrightarrow{e_2} \surd}{\rho_f(p) \parallel \rho_f(q) \xrightarrow{\{f(e_1),f(e_2)\}} \surd}$$

$$\frac{p \xrightarrow{e_1} p' \quad q \xrightarrow{e_2} \surd}{\rho_f(p \parallel q) \xrightarrow{\{f(e_1),f(e_2)\}} \rho_f(p')} \qquad \frac{p \xrightarrow{e_1} p' \quad q \xrightarrow{e_2} \surd}{\rho_f(p) \parallel \rho_f(q) \xrightarrow{\{f(e_1),f(e_2)\}} \rho_f(p')}$$

$$\frac{p \xrightarrow{e_1} \surd \quad q \xrightarrow{e_2} q'}{\rho_f(p \parallel q) \xrightarrow{\{f(e_1),f(e_2)\}} \rho_f(q')} \qquad \frac{p \xrightarrow{e_1} \surd \quad q \xrightarrow{e_2} q'}{\rho_f(p) \parallel \rho_f(q) \xrightarrow{\{f(e_1),f(e_2)\}} \rho_f(q')}$$

$$\frac{p \xrightarrow{e_1} p' \quad q \xrightarrow{e_2} q'}{\rho_f(p \parallel q) \xrightarrow{\{f(e_1),f(e_2)\}} \rho_f(p' \between q')} \qquad \frac{p \xrightarrow{e_1} p' \quad q \xrightarrow{e_2} q'}{\rho_f(p) \parallel \rho_f(q) \xrightarrow{\{f(e_1),f(e_2)\}} \rho_f(p') \between \rho_f(q')}$$

So, with the assumption $\rho_f(p' \between q') = \rho_f(p') \between \rho_f(q')$, $\rho_f(p \parallel q) \approx_{rbs} \rho_f(p) \parallel \rho_f(q)$, as desired.

(2) Soundness of $APTC_\tau$ with guarded linear recursion and the renaming operator with respect to rooted branching pomset bisimulation $\approx_{rbp}$.

Since rooted branching pomset bisimulation $\approx_{rbp}$ is both an equivalent and a congruent relation with respect to $APTC_\tau$ with guarded linear recursion and the renaming operator, we only need to check if each axiom in Table 4.28 is sound modulo rooted branching pomset bisimulation $\approx_{rbp}$.

From the definition of rooted branching pomset bisimulation $\approx_{rbp}$ (see Definition 4.56), we know that rooted branching pomset bisimulation $\approx_{rbp}$ is defined by weak pomset transitions, which are labeled by pomsets with τ. In a weak pomset transition, the events in the pomset are either within causality relations (defined by $\cdot$) or in concurrency (implicitly defined by $\cdot$ and $+$, and explicitly defined by $\between$), of course, they are pairwise consistent (without conflicts). In (1), we have already proven the case that all events are pairwise concurrent, so, we only need to prove the case of events in causality. Without loss of generality, we take a pomset of $P = \{e_1, e_2 : e_1 \cdot e_2\}$. Then the weak pomset transition labeled by the above P is just composed of one single event transition labeled by e_1 succeeded by another single event transition labeled by e_2, that is, $\xRightarrow{P} = \xRightarrow{e_1}\xRightarrow{e_2}$.

Similarly to the proof of soundness of $APTC_\tau$ with guarded linear recursion and the renaming operator modulo rooted branching step bisimulation $\approx_{rbs}$ (1), we can prove that each axiom in Table 4.28 is sound modulo rooted branching pomset bisimulation $\approx_{rbp}$, we omit them.

(3) Soundness of $APTC_\tau$ with guarded linear recursion and the renaming operator with respect to rooted branching hp-bisimulation $\approx_{rbhp}$.

Since rooted branching hp-bisimulation $\approx_{rbhp}$ is both an equivalent and a congruent relation with respect to $APTC_\tau$ with guarded linear recursion and the renaming operator, we only need to check if each axiom in Table 4.28 is sound modulo rooted branching hp-bisimulation $\approx_{rbhp}$.

From the definition of rooted branching hp-bisimulation $\approx_{rbhp}$ (see Definition 4.58), we know that rooted branching hp-bisimulation $\approx_{rbhp}$ is defined on the weakly posetal product (C_1, f, C_2), $f : \hat{C_1} \to \hat{C_2}$ isomorphism. Two process terms s related to C_1 and t related to C_2, and $f : \hat{C_1} \to \hat{C_2}$ isomorphism. Initially, $(C_1, f, C_2) = (\emptyset, \emptyset, \emptyset)$, and $(\emptyset, \emptyset, \emptyset) \in \approx_{rbhp}$. When $s \xrightarrow{e} s'$ $(C_1 \xrightarrow{e} C_1')$, there will be $t \xrightarrow{e} t'$ $(C_2 \xrightarrow{e} C_2')$, and we define $f' = f[e \mapsto e]$. Then, if $(C_1, f, C_2) \in \approx_{rbhp}$, then $(C_1', f', C_2') \in \approx_{rbhp}$.

Similarly to the proof of soundness of $APTC_\tau$ with guarded linear recursion and the renaming operator modulo rooted branching pomset bisimulation equivalence (2), we can prove that each axiom in Table 4.28 is sound modulo rooted branching hp-bisimulation equivalence, we just need additionally to check the above conditions on rooted branching hp-bisimulation, we omit them. □

Theorem 4.80 (Completeness of the renaming operator). *Let p and q be closed $APTC_\tau$ with guarded linear recursion and $CFAR$ and the renaming operator terms, then,*

1. *if $p \approx_{rbs} q$ then $p = q$;*
2. *if $p \approx_{rbp} q$ then $p = q$;*
3. *if $p \approx_{rbhp} q$ then $p = q$.*

Proof. (1) For the case of rooted branching step bisimulation, the proof is following.

Firstly, in the proof of Theorem 4.70, we know that each process term p in $APTC_\tau$ with guarded linear recursion is equal to a process term $\langle X_1|E \rangle$ with E a guarded linear recursive specification. And we prove if $\langle X_1|E_1 \rangle \approx_{rbs} \langle Y_1|E_2 \rangle$, then $\langle X_1|E_1 \rangle = \langle Y_1|E_2 \rangle$

Structural induction with respect to process term p can be applied. The only new case (where $RN1 - RN5$ are needed) is $p \equiv \rho_f(q)$. First assuming $q = \langle X_1|E \rangle$ with a guarded linear recursive specification E, we prove the case of $p = \rho_f(\langle X_1|E \rangle)$. Let E consist of guarded linear recursive equations

$$X_i = (a_{1i1} \parallel \cdots \parallel a_{k_i i1})X_{i1} + ... + (a_{1im_i} \parallel \cdots \parallel a_{k_{im_i} im_i})X_{im_i} + b_{1i1} \parallel \cdots \parallel b_{l_i i1} + ...$$
$$+ b_{1im_i} \parallel \cdots \parallel b_{l_{im_i} im_i}$$

for $i \in 1, ..., n$. Let F consist of guarded linear recursive equations

$$Y_i = (f(a_{1i1}) \parallel \cdots \parallel f(a_{k_i i1}))Y_{i1} + ... + (f(a_{1im_i}) \parallel \cdots \parallel f(a_{k_{im_i} im_i}))Y_{im_i}$$
$$+ f(b_{1i1}) \parallel \cdots \parallel f(b_{l_i i1}) + ... + f(b_{1im_i}) \parallel \cdots \parallel f(b_{l_{im_i} im_i})$$

for $i \in 1, ..., n$.

$\rho_f(\langle X_i|E \rangle)$
$\stackrel{RDP}{=} \rho_f((a_{1i1} \parallel \cdots \parallel a_{k_i i1})X_{i1} + ... + (a_{1im_i} \parallel \cdots \parallel a_{k_{im_i} im_i})X_{im_i}$
$+ b_{1i1} \parallel \cdots \parallel b_{l_i i1} + ... + b_{1im_i} \parallel \cdots \parallel b_{l_{im_i} im_i})$
$\stackrel{RN1\text{-}RN5}{=} (f(a_{1i1}) \parallel \cdots \parallel f(a_{k_i i1}))\rho_f(X_{i1}) + ... + (f(a_{1im_i}) \parallel \cdots \parallel f(a_{k_{im_i} im_i}))\rho_f(X_{im_i})$
$+ f(b_{1i1}) \parallel \cdots \parallel f(b_{l_i i1}) + ... + f(b_{1im_i}) \parallel \cdots \parallel f(b_{l_{im_i} im_i})$

Replacing Y_i by $\rho_f(\langle X_i|E \rangle)$ for $i \in \{1, ..., n\}$ is a solution for F. So by RSP, $\rho_f(\langle X_1|E \rangle) = \langle Y_1|F \rangle$, as desired.

(2) For the case of rooted branching pomset bisimulation, it can be proven similarly to (1), we omit it.

(3) For the case of rooted branching hp-bisimulation, it can be proven similarly to (1), we omit it.   □

4.6.3 States

State operator permits explicitly to describe states, where S denotes a finite set of states, $action(s, e)$ denotes the visible behavior of e in state s with $action : S \times \mathbb{E} \to \mathbb{E}$, $effect(s, e)$ represents the state that results if e is executed in s with $effect : S \times \mathbb{E} \to S$. State operator $\lambda_s(t)$ which denotes process term t in s, is expressed by the following transition rules in Table 4.29. Note that $action$ and $effect$ are extended to $\mathbb{E} \cup \{\tau\}$ by defining $action(s, \tau) \triangleq \tau$ and $effect(s, \tau) \triangleq s$. We use $e_1 \% e_2$ to denote that e_1 and e_2 are in race condition.

Theorem 4.81 (Conservativity of $APTC$ with respect to the state operator). *$APTC_\tau$ with guarded linear recursion and state operator is a conservative extension of $APTC_\tau$ with guarded linear recursion.*

Table 4.29   Transition rule of the state operator.

$$\frac{x \xrightarrow{e} \surd}{\lambda_s(x) \xrightarrow{action(s,e)} \surd} \qquad \frac{x \xrightarrow{e} x'}{\lambda_s(x) \xrightarrow{action(s,e)} \lambda_{effect(s,e)}(x')}$$

$$\frac{x \xrightarrow{e_1} \surd \quad y \xnrightarrow{e_2} \quad (e_1 \% e_2)}{\lambda_s(x \parallel y) \xrightarrow{action(s,e_1)} \lambda_{effect(s,e_1)}(y)} \qquad \frac{x \xrightarrow{e_1} x' \quad y \xnrightarrow{e_2} \quad (e_1 \% e_2)}{\lambda_s(x \parallel y) \xrightarrow{action(s,e_1)} \lambda_{effect(s,e_1)}(x' \between y)}$$

$$\frac{x \xnrightarrow{e_1} \quad y \xrightarrow{e_2} \surd \quad (e_1 \% e_2)}{\lambda_s(x \parallel y) \xrightarrow{action(s,e_2)} \lambda_{effect(s,e_2)}(x)} \qquad \frac{x \xnrightarrow{e_1} \quad y \xrightarrow{e_2} y' \quad (e_1 \% e_2)}{\lambda_s(x \parallel y) \xrightarrow{action(s,e_2)} \lambda_{effect(s,e_2)}(x \between y')}$$

$$\frac{x \xrightarrow{e_1} \surd \quad y \xrightarrow{e_2} \surd}{\lambda_s(x \parallel y) \xrightarrow{\{action(s,e_1),action(s,e_2)\}} \surd}$$

$$\frac{x \xrightarrow{e_1} x' \quad y \xrightarrow{e_2} \surd}{\lambda_s(x \parallel y) \xrightarrow{\{action(s,e_1),action(s,e_2)\}} \lambda_{effect(s,e_1) \cup effect(s,e_2)}(x')}$$

$$\frac{x \xrightarrow{e_1} \surd \quad y \xrightarrow{e_2} y'}{\lambda_s(x \parallel y) \xrightarrow{\{action(s,e_1),action(s,e_2)\}} \lambda_{effect(s,e_1) \cup effect(s,e_2)}(y')}$$

$$\frac{x \xrightarrow{e_1} x' \quad y \xrightarrow{e_2} y'}{\lambda_s(x \parallel y) \xrightarrow{\{action(s,e_1),action(s,e_2)\}} \lambda_{effect(s,e_1) \cup effect(s,e_2)}(x' \between y')}$$

Proof. It follows from the following two facts.

1. The transition rules of $APTC_\tau$ with guarded linear recursion are all source-dependent;
2. The sources of the transition rules for the state operator contain an occurrence of λ_s.

□

Theorem 4.82 (Congruence theorem of the state operator). *Rooted branching truly concurrent bisimulation equivalences $\approx_{rbp}$, $\approx_{rbs}$ and $\approx_{rbhp}$ are all congruences with respect to $APTC_\tau$ with guarded linear recursion and the state operator.*

Proof. (1) Case rooted branching pomset bisimulation equivalence $\approx_{rbp}$.

Let x and y be $APTC_\tau$ with guarded linear recursion and the state operator processes, and $x \approx_{rbp} y$, it is sufficient to prove that $\lambda_s(x) \approx_{rbp} \lambda_s(y)$.

By the transition rules for operator λ_s in Table 4.29, we can get

$$\lambda_s(x) \xrightarrow{action(s,X)} \surd \quad \lambda_s(y) \xrightarrow{action(s,Y)} \surd$$

Table 4.30   Axioms of state operator.

No.	Axiom
$SO1$	$\lambda_s(e) = action(s, e)$
$SO2$	$\lambda_s(\delta) = \delta$
$SO3$	$\lambda_s(x + y) = \lambda_s(x) + \lambda_s(y)$
$SO4$	$\lambda_s(e \cdot y) = action(s, e) \cdot \lambda_{effect(s,e)}(y)$
$SO5$	$\lambda_s(x \parallel y) = \lambda_s(x) \parallel \lambda_s(y)$

with $X \subseteq x$, $Y \subseteq y$, and $X \sim Y$.

Or, we can get

$$\lambda_s(x) \xrightarrow{action(s,X)} \lambda_{effect(s,X)}(x') \quad \lambda_s(y) \xrightarrow{action(s,Y)} \lambda_{effect(s,Y)}(y')$$

with $X \subseteq x$, $Y \subseteq y$, and $X \sim Y$ and the hypothesis $\lambda_{effect(s,X)}(x') \approx_{rbp} \lambda_{effect(s,Y)}(y')$.

So, we get $\lambda_s(x) \approx_{rbp} \lambda_s(y)$, as desired

(2) The cases of rooted branching step bisimulation $\approx_{rbs}$, rooted branching hp-bisimulation $\approx_{rbhp}$ can be proven similarly, we omit them. $\qquad\square$

We design the axioms for the state operator λ_s in Table 4.30.

Theorem 4.83 (Soundness of the state operator). *Let x and y be $APTC_\tau$ with guarded linear recursion and the state operator terms. If $APTC_\tau$ with guarded linear recursion and the state operator $\vdash x = y$, then*

1. $x \approx_{rbs} y$;
2. $x \approx_{rbp} y$;
3. $x \approx_{rbhp} y$.

Proof. (1) Soundness of $APTC_\tau$ with guarded linear recursion and the state operator with respect to rooted branching step bisimulation $\approx_{rbs}$.

Since rooted branching step bisimulation $\approx_{rbs}$ is both an equivalent and a congruent relation with respect to $APTC_\tau$ with guarded linear recursion and the state operator, we only need to check if each axiom in Table 4.30 is sound modulo rooted branching step bisimulation equivalence.

Though transition rules in Table 4.29 are defined in the flavor of single event, they can be modified into a step (a set of events within which each event is pairwise concurrent), we omit them. If we treat a single event as a step containing just one event, the proof of this soundness theorem does not exist any problem, so we use this way and still use the transition rules in Table 4.30.

We only prove soundness of the non-trivial axioms $SO3 - SO5$, and omit the defining axioms $SO1 - SO2$.

- **Axiom** $SO3$. Let p, q be $APTC_\tau$ with guarded linear recursion and the state operator processes, and $\lambda_s(p + q) = \lambda_s(p) + \lambda_s(q)$, it is sufficient to prove that $\lambda_s(p + q) \approx_{rbs}$

$\lambda_s(p) + \lambda_s(q)$. By the transition rules for operator $+$ and λ_s in Table 4.29, we get

$$\frac{p \xrightarrow{e_1} \surd}{\lambda_s(p+q) \xrightarrow{action(s,e_1)} \surd} \qquad \frac{p \xrightarrow{e_1} \surd}{\lambda_s(p) + \lambda_s(q) \xrightarrow{action(s,e_1)} \surd}$$

$$\frac{q \xrightarrow{e_2} \surd}{\lambda_s(p+q) \xrightarrow{action(s,e_2)} \surd} \qquad \frac{q \xrightarrow{e_2} \surd}{\lambda_s(p) + \lambda_s(q) \xrightarrow{action(s,e_2)} \surd}$$

$$\frac{p \xrightarrow{e_1} p'}{\lambda_s(p+q) \xrightarrow{action(s,e_1)} \lambda_{effect(s,e_1)}(p')} \qquad \frac{p \xrightarrow{e_1} p'}{\lambda_s(p) + \lambda_s(q) \xrightarrow{action(s,e_1)} \lambda_{effect(s,e_1)}(p')}$$

$$\frac{q \xrightarrow{e_2} q'}{\lambda_s(p+q) \xrightarrow{action(s,e_2)} \lambda_{effect(s,e_2)}(q')} \qquad \frac{q \xrightarrow{e_2} q'}{\lambda_s(p) + \lambda_s(q) \xrightarrow{action(s,e_2)} \lambda_{effect(s,e_2)}(q')}$$

So, $\lambda_s(p+q) \approx_{rbs} \lambda_s(p) + \lambda_s(q)$, as desired.

- **Axiom** $SO4$. Let q be $APTC_\tau$ with guarded linear recursion and the state operator processes, and $\lambda_s(e \cdot q) = action(s, e) \cdot \lambda_{effect(s,e)}(q)$, it is sufficient to prove that $\lambda_s(e \cdot q) \approx_{rbs} action(s, e) \cdot \lambda_{effect(s,e)}(q)$. By the transition rules for operator $\cdot$ and λ_s in Table 4.29, we get

$$\frac{e \xrightarrow{e} \surd}{\lambda_s(e \cdot q) \xrightarrow{action(s,e)} \lambda_{effect(s,e)}(q)} \qquad \frac{action(s, e) \xrightarrow{action(s,e)} \surd}{action(s, e) \cdot \lambda_{effect(s,e)}(q) \xrightarrow{action(s,e)} \lambda_{effect(s,e)}(q)}$$

So, $\lambda_s(e \cdot q) \approx_{rbs} action(s, e) \cdot \lambda_{effect(s,e)}(q)$, as desired.

- **Axiom** $SO5$. Let p, q be $APTC_\tau$ with guarded linear recursion and the state operator processes, and $\lambda_s(p \parallel q) = \lambda_s(p) \parallel \lambda_s(q)$, it is sufficient to prove that $\lambda_s(p \parallel q) \approx_{rbs} \lambda_s(p) \parallel \lambda_s(q)$. By the transition rules for operator $\parallel$ and λ_s in Table 4.29, we get for the case $\neg(e_1 \% e_2)$

$$\frac{p \xrightarrow{e_1} \surd \quad q \xrightarrow{e_2} \surd}{\lambda_s(p \parallel q) \xrightarrow{\{action(s,e_1),action(s,e_2)\}} \surd}$$

$$\frac{p \xrightarrow{e_1} \surd \quad q \xrightarrow{e_2} \surd}{\lambda_s(p) \parallel \lambda_s(q) \xrightarrow{\{action(s,e_1),action(s,e_2)\}} \surd}$$

$$\frac{p \xrightarrow{e_1} p' \quad q \xrightarrow{e_2} \surd}{\lambda_s(p \parallel q) \xrightarrow{\{action(s,e_1),action(s,e_2)\}} \lambda_{effect(s,e_1) \cup effect(s,e_2)}(p')}$$

$$\frac{p \xrightarrow{e_1} p' \quad q \xrightarrow{e_2} \checkmark}{\lambda_s(p) \parallel \lambda_s(q) \xrightarrow{\{action(s,e_1),action(s,e_2)\}} \lambda_{effect(s,e_1) \cup effect(s,e_2)}(p')}$$

$$\frac{p \xrightarrow{e_1} \checkmark \quad q \xrightarrow{e_2} q'}{\lambda_s(p \parallel q) \xrightarrow{\{action(s,e_1),action(s,e_2)\}} \lambda_{effect(s,e_1) \cup effect(s,e_2)}(q')}$$

$$\frac{p \xrightarrow{e_1} \checkmark \quad q \xrightarrow{e_2} q'}{\lambda_s(p) \parallel \lambda_s(q) \xrightarrow{\{action(s,e_1),action(s,e_2)\}} \lambda_{effect(s,e_1) \cup effect(s,e_2)}(q')}$$

$$\frac{p \xrightarrow{e_1} p' \quad q \xrightarrow{e_2} q'}{\lambda_s(p \parallel q) \xrightarrow{\{action(s,e_1),action(s,e_2)\}} \lambda_{effect(s,e_1) \cup effect(s,e_2)}(p' \between q')}$$

$$\frac{p \xrightarrow{e_1} p' \quad q \xrightarrow{e_2} q'}{\lambda_s(p) \parallel \lambda_s(q) \xrightarrow{\{action(s,e_1),action(s,e_2)\}} \lambda_{effect(s,e_1) \cup effect(s,e_2)}(p') \between \lambda_{effect(s,e_1) \cup effect(s,e_2)}(q')}$$

So, with the assumption $\lambda_{effect(s,e_1) \cup effect(s,e_2)}(p' \between q') = \lambda_{effect(s,e_1) \cup effect(s,e_2)}(p') \between \lambda_{effect(s,e_1) \cup effect(s,e_2)}(q')$, $\lambda_s(p \parallel q) \approx_{rbs} \lambda_s(p) \parallel \lambda_s(q)$, as desired. For the case $e_1 \% e_2$, we get

$$\frac{p \xrightarrow{e_1} \checkmark \quad q \not\xrightarrow{e_2}}{\lambda_s(p \parallel q) \xrightarrow{action(s,e_1)} \lambda_{effect(s,e_1)}(q)}$$

$$\frac{p \xrightarrow{e_1} \checkmark \quad q \not\xrightarrow{e_2}}{\lambda_s(p) \parallel \lambda_s(q) \xrightarrow{action(s,e_1)} \lambda_{effect(s,e_1)}(q)}$$

$$\frac{p \xrightarrow{e_1} p' \quad q \not\xrightarrow{e_2}}{\lambda_s(p \parallel q) \xrightarrow{action(s,e_1)} \lambda_{effect(s,e_1)}(p' \between q)}$$

$$\frac{p \xrightarrow{e_1} p' \quad q \not\xrightarrow{e_2}}{\lambda_s(p) \parallel \lambda_s(q) \xrightarrow{action(s,e_1)} \lambda_{effect(s,e_1)}(p') \between \lambda_{effect(s,e_1)}(q)}$$

$$\frac{p \not\xrightarrow{e_1} \quad q \xrightarrow{e_2} \checkmark}{\lambda_s(p \parallel q) \xrightarrow{action(s,e_2)} \lambda_{effect(s,e_2)}(p)}$$

$$\frac{p \not\xrightarrow{e_1} \quad q \xrightarrow{e_2} \checkmark}{\lambda_s(p) \parallel \lambda_s(q) \xrightarrow{action(s,e_2)} \lambda_{effect(s,e_2)}(p)}$$

$$\frac{p \overset{e_1}{\nrightarrow} \quad q \overset{e_2}{\rightarrow} q'}{\lambda_s(p \parallel q) \xrightarrow{action(s,e_2)} \lambda_{effect(s,e_2)}(p \between q')}$$

$$\frac{p \overset{e_1}{\nrightarrow} \quad q \overset{e_2}{\rightarrow} q'}{\lambda_s(p) \parallel \lambda_s(q) \xrightarrow{action(s,e_2)} \lambda_{effect(s,e_2)}(p) \between \lambda_{effect(s,e_2)}(q')}$$

So, with the assumption $\lambda_{effect(s,e_1)}(p' \between q) = \lambda_{effect(s,e_1)}(p') \between \lambda_{effect(s,e_1)}(q)$ and $\lambda_{effect(s,e_2)}(p \between q') = \lambda_{effect(s,e_2)}(p) \between \lambda_{effect(s,e_2)}(q')$, $\lambda_s(p \parallel q) \approx_{rbs} \lambda_s(p) \parallel \lambda_s(q)$, as desired.

(2) Soundness of $APTC_\tau$ with guarded linear recursion and the state operator with respect to rooted branching pomset bisimulation $\approx_{rbp}$.

Since rooted branching pomset bisimulation $\approx_{rbp}$ is both an equivalent and a congruent relation with respect to $APTC_\tau$ with guarded linear recursion and the state operator, we only need to check if each axiom in Table 4.30 is sound modulo rooted branching pomset bisimulation $\approx_{rbp}$.

From the definition of rooted branching pomset bisimulation $\approx_{rbp}$ (see Definition 4.56), we know that rooted branching pomset bisimulation $\approx_{rbp}$ is defined by weak pomset transitions, which are labeled by pomsets with τ. In a weak pomset transition, the events in the pomset are either within causality relations (defined by ·) or in concurrency (implicitly defined by · and +, and explicitly defined by $\between$), of course, they are pairwise consistent (without conflicts). In (1), we have already proven the case that all events are pairwise concurrent, so, we only need to prove the case of events in causality. Without loss of generality, we take a pomset of $P = \{e_1, e_2 : e_1 \cdot e_2\}$. Then the weak pomset transition labeled by the above P is just composed of one single event transition labeled by e_1 succeeded by another single event transition labeled by e_2, that is, $\overset{P}{\Rightarrow} = \overset{e_1}{\Rightarrow} \overset{e_2}{\Rightarrow}$.

Similarly to the proof of soundness of $APTC_\tau$ with guarded linear recursion and the state operator modulo rooted branching step bisimulation $\approx_{rbs}$ (1), we can prove that each axiom in Table 4.30 is sound modulo rooted branching pomset bisimulation $\approx_{rbp}$, we omit them.

(3) Soundness of $APTC_\tau$ with guarded linear recursion and the state operator with respect to rooted branching hp-bisimulation $\approx_{rbhp}$.

Since rooted branching hp-bisimulation $\approx_{rbhp}$ is both an equivalent and a congruent relation with respect to $APTC_\tau$ with guarded linear recursion and the state operator, we only need to check if each axiom in Table 4.30 is sound modulo rooted branching hp-bisimulation $\approx_{rbhp}$.

From the definition of rooted branching hp-bisimulation $\approx_{rbhp}$ (see Definition 4.58), we know that rooted branching hp-bisimulation $\approx_{rbhp}$ is defined on the weakly posetal product (C_1, f, C_2), $f : \hat{C}_1 \rightarrow \hat{C}_2$ isomorphism. Two process terms s related to C_1 and t related to C_2, and $f : \hat{C}_1 \rightarrow \hat{C}_2$ isomorphism. Initially, $(C_1, f, C_2) = (\emptyset, \emptyset, \emptyset)$, and $(\emptyset, \emptyset, \emptyset) \in \approx_{rbhp}$. When $s \overset{e}{\rightarrow} s'$ $(C_1 \overset{e}{\rightarrow} C_1')$, there will be $t \overset{e}{\Rightarrow} t'$ $(C_2 \overset{e}{\Rightarrow} C_2')$, and we define $f' = f[e \mapsto e]$. Then, if $(C_1, f, C_2) \in \approx_{rbhp}$, then $(C_1', f', C_2') \in \approx_{rbhp}$.

Similarly to the proof of soundness of $APTC_\tau$ with guarded linear recursion and the state operator modulo rooted branching pomset bisimulation equivalence (2), we can prove that each axiom in Table 4.30 is sound modulo rooted branching hp-bisimulation equivalence, we just need additionally to check the above conditions on rooted branching hp-bisimulation, we omit them. $\square$

Theorem 4.84 (Completeness of the state operator). *Let p and q be closed $APTC_\tau$ with guarded linear recursion and $CFAR$ and the state operator terms, then,*

1. *if $p \approx_{rbs} q$ then $p = q$;*
2. *if $p \approx_{rbp} q$ then $p = q$;*
3. *if $p \approx_{rbhp} q$ then $p = q$.*

Proof. (1) For the case of rooted branching step bisimulation, the proof is following.

Firstly, we know that each process term p in $APTC_\tau$ with guarded linear recursion is equal to a process term $\langle X_1|E \rangle$ with E a guarded linear recursive specification. And we prove if $\langle X_1|E_1 \rangle \approx_{rbs} \langle Y_1|E_2 \rangle$, then $\langle X_1|E_1 \rangle = \langle Y_1|E_2 \rangle$

Structural induction with respect to process term p can be applied. The only new case (where $SO1 - SO5$ are needed) is $p \equiv \lambda_{s_0}(q)$. First assuming $q = \langle X_1|E \rangle$ with a guarded linear recursive specification E, we prove the case of $p = \lambda_{s_0}(\langle X_1|E \rangle)$. Let E consist of guarded linear recursive equations

$$X_i = (a_{1i1} \parallel \cdots \parallel a_{k_i1 i1})X_{i1} + \ldots + (a_{1im_i} \parallel \cdots \parallel a_{k_{im_i} im_i})X_{im_i} + b_{1i1} \parallel \cdots \parallel b_{l_i1 i1} + \ldots$$
$$+ b_{1im_i} \parallel \cdots \parallel b_{l_{im_i} im_i}$$

for $i \in 1, \ldots, n$. Let F consist of guarded linear recursive equations

$$Y_i(s) = (action(s, a_{1i1}) \parallel \cdots \parallel action(s, a_{k_i1 i1}))Y_{i1}(effect(s, a_{1i1})$$
$$\cup \cdots \cup effect(s, a_{k_i1 i1}))$$
$$+ \ldots + (action(s, a_{1im_i}) \parallel \cdots \parallel action(s, a_{k_{im_i} im_i}))Y_{im_i}(effect(s, a_{1im_i})$$
$$\cup \cdots \cup effect(s, a_{k_{im_i} im_i}))$$
$$+ action(s, b_{1i1}) \parallel \cdots \parallel action(s, b_{l_i1 i1}) + \ldots + action(s, b_{1im_i}) \parallel \cdots \parallel action(s, b_{l_{im_i} im_i})$$

for $i \in 1, \ldots, n$.

$$\lambda_s(\langle X_i|E \rangle)$$
$$\overset{RDP}{=} \lambda_s((a_{1i1} \parallel \cdots \parallel a_{k_i1 i1})X_{i1} + \ldots + (a_{1im_i} \parallel \cdots \parallel a_{k_{im_i} im_i})X_{im_i}$$
$$+ b_{1i1} \parallel \cdots \parallel b_{l_i1 i1} + \ldots + b_{1im_i} \parallel \cdots \parallel b_{l_{im_i} im_i})$$
$$\overset{SO1\text{-}SO5}{=} (action(s, a_{1i1}) \parallel \cdots \parallel action(s, a_{k_i1 i1}))\lambda_{effect(s,a_{1i1})\cup\cdots\cup effect(s,a_{k_i1 i1})}(X_{i1})$$
$$+ \ldots + (action(s, a_{1im_i}) \parallel \cdots \parallel action(s, a_{k_{im_i} im_i}))\lambda_{effect(s,a_{1im_i})\cup\cdots\cup effect(s,a_{k_{im_i} im_i})}(X_{im_i})$$
$$+ action(s, b_{1i1}) \parallel \cdots \parallel action(s, b_{l_i1 i1}) + \ldots + action(s, b_{1im_i}) \parallel \cdots \parallel action(s, b_{l_{im_i} im_i})$$

Table 4.31 Transition rules of left parallel operator $\parallel$.

$$\frac{x \xrightarrow{e_1} \surd \quad y \xrightarrow{e_2} \surd \quad (e_1 \leq e_2)}{x \parallel y \xrightarrow{\{e_1,e_2\}} \surd} \qquad \frac{x \xrightarrow{e_1} x' \quad y \xrightarrow{e_2} \surd \quad (e_1 \leq e_2)}{x \parallel y \xrightarrow{\{e_1,e_2\}} x'}$$

$$\frac{x \xrightarrow{e_1} \surd \quad y \xrightarrow{e_2} y' \quad (e_1 \leq e_2)}{x \parallel y \xrightarrow{\{e_1,e_2\}} y'} \qquad \frac{x \xrightarrow{e_1} x' \quad y \xrightarrow{e_2} y' \quad (e_1 \leq e_2)}{x \parallel y \xrightarrow{\{e_1,e_2\}} x' \between y'}$$

Replacing $Y_i(s)$ by $\lambda_s(\langle X_i | E \rangle)$ for $i \in \{1, ..., n\}$ is a solution for F. So by RSP, $\lambda_{s_0}(\langle X_1 | E \rangle) = \langle Y_1(s_0) | F \rangle$, as desired.

(2) For the case of rooted branching pomset bisimulation, it can be proven similarly to (1), we omit it.

(3) For the case of rooted branching hp-bisimulation, it can be proven similarly to (1), we omit it. $\qquad\square$

4.7 Axiomatization for hhp-bisimilarity

Since hhp-bisimilarity is a downward closed hp-bisimilarity and can be downward closed to single atomic event, which implies bisimilarity. As Moller [22] proven, there is not a finite sound and complete axiomatization for parallelism $\parallel$ modulo bisimulation equivalence, so there is not a finite sound and complete axiomatization for parallelism $\parallel$ modulo hhp-bisimulation equivalence either. Inspired by the way of left merge to modeling the full merge for bisimilarity, we introduce a left parallel composition $\parallel$ to model the full parallelism $\parallel$ for hhp-bisimilarity.

In the following subsection, we add left parallel composition $\parallel$ to the whole theory. Because the resulting theory is similar to the former, we only list the significant differences, and all proofs of the conclusions are left to the reader.

4.7.1 *APTC* with left parallel composition

The transition rules of left parallel composition $\parallel$ are shown in Table 4.31. With a little abuse, we extend the causal order relation $\leq$ on $\mathbb{E}$ to include the original partial order (denoted by $<$) and concurrency (denoted by $=$).

The new axioms for parallelism are listed in Table 4.32.

Definition 4.85 (Basic terms of *APTC* with left parallel composition). The set of basic terms of *APTC*, $\mathcal{B}(APTC)$, is inductively defined as follows:

1. $\mathbb{E} \subset \mathcal{B}(APTC)$;
2. if $e \in \mathbb{E}, t \in \mathcal{B}(APTC)$ then $e \cdot t \in \mathcal{B}(APTC)$;
3. if $t, s \in \mathcal{B}(APTC)$ then $t + s \in \mathcal{B}(APTC)$;
4. if $t, s \in \mathcal{B}(APTC)$ then $t \parallel s \in \mathcal{B}(APTC)$.

Table 4.32 Axioms of parallelism with left parallel composition.

No.	Axiom
A6	$x + \delta = x$
A7	$\delta \cdot x = \delta$
P1	$x \between y = x \parallel y + x \mid y$
P2	$x \parallel y = y \parallel x$
P3	$(x \parallel y) \parallel z = x \parallel (y \parallel z)$
P4	$x \parallel y = x \between y + y \between x$
P5	$(e_1 \le e_2) \quad e_1 \between (e_2 \cdot y) = (e_1 \between e_2) \cdot y$
P6	$(e_1 \le e_2) \quad (e_1 \cdot x) \between e_2 = (e_1 \between e_2) \cdot x$
P7	$(e_1 \le e_2) \quad (e_1 \cdot x) \between (e_2 \cdot y) = (e_1 \between e_2) \cdot (x \between y)$
P8	$(x + y) \between z = (x \between z) + (y \between z)$
P9	$\delta \between x = \delta$
C10	$e_1 \mid e_2 = \gamma(e_1, e_2)$
C11	$e_1 \mid (e_2 \cdot y) = \gamma(e_1, e_2) \cdot y$
C12	$(e_1 \cdot x) \mid e_2 = \gamma(e_1, e_2) \cdot x$
C13	$(e_1 \cdot x) \mid (e_2 \cdot y) = \gamma(e_1, e_2) \cdot (x \between y)$
C14	$(x + y) \mid z = (x \mid z) + (y \mid z)$
C15	$x \mid (y + z) = (x \mid y) + (x \mid z)$
C16	$\delta \mid x = \delta$
C17	$x \mid \delta = \delta$
CE18	$\Theta(e) = e$
CE19	$\Theta(\delta) = \delta$
CE20	$\Theta(x + y) = \Theta(x) + \Theta(y)$
CE21	$\Theta(x \cdot y) = \Theta(x) \cdot \Theta(y)$
CE22	$\Theta(x \between y) = ((\Theta(x) \triangleleft y) \between y) + ((\Theta(y) \triangleleft x) \between x)$
CE23	$\Theta(x \mid y) = ((\Theta(x) \triangleleft y) \mid y) + ((\Theta(y) \triangleleft x) \mid x)$
U24	$(\sharp(e_1, e_2)) \quad e_1 \triangleleft e_2 = \tau$
U25	$(\sharp(e_1, e_2), e_2 \le e_3) \quad e_1 \triangleleft e_3 = \tau$
U26	$(\sharp(e_1, e_2), e_2 \le e_3) \quad e_3 \triangleleft e_1 = \tau$
U27	$e \triangleleft \delta = e$
U28	$\delta \triangleleft e = \delta$
U29	$(x + y) \triangleleft z = (x \triangleleft z) + (y \triangleleft z)$
U30	$(x \cdot y) \triangleleft z = (x \triangleleft z) \cdot (y \triangleleft z)$
U31	$(x \between y) \triangleleft z = (x \triangleleft z) \between (y \triangleleft z)$
U32	$(x \mid y) \triangleleft z = (x \triangleleft z) \mid (y \triangleleft z)$
U33	$x \triangleleft (y + z) = (x \triangleleft y) \triangleleft z$
U34	$x \triangleleft (y \cdot z) = (x \triangleleft y) \triangleleft z$
U35	$x \triangleleft (y \between z) = (x \triangleleft y) \triangleleft z$
U36	$x \triangleleft (y \mid z) = (x \triangleleft y) \triangleleft z$

Table 4.33 Axioms of encapsulation operator with left parallel composition.

No.	Axiom
D1	$e \notin H$ $\partial_H(e) = e$
D2	$e \in H$ $\partial_H(e) = \delta$
D3	$\partial_H(\delta) = \delta$
D4	$\partial_H(x + y) = \partial_H(x) + \partial_H(y)$
D5	$\partial_H(x \cdot y) = \partial_H(x) \cdot \partial_H(y)$
D6	$\partial_H(x \parallel\!\!\!\!\perp y) = \partial_H(x) \parallel\!\!\!\!\perp \partial_H(y)$

Theorem 4.86 (Generalization of the algebra for left parallelism with respect to $BATC$). *The algebra for left parallelism is a generalization of $BATC$.*

Theorem 4.87 (Congruence theorem of $APTC$ with left parallel composition). *Truly concurrent bisimulation equivalences $\sim_p$, $\sim_s$, $\sim_{hp}$ and $\sim_{hhp}$ are all congruences with respect to $APTC$ with left parallel composition.*

Theorem 4.88 (Elimination theorem of parallelism with left parallel composition). *Let p be a closed $APTC$ with left parallel composition term. Then there is a basic $APTC$ term q such that $APTC \vdash p = q$.*

Theorem 4.89 (Soundness of parallelism with left parallel composition modulo truly concurrent bisimulation equivalences). *Let x and y be $APTC$ with left parallel composition terms. If $APTC \vdash x = y$, then*

1. $x \sim_s y$;
2. $x \sim_p y$;
3. $x \sim_{hp} y$;
4. $x \sim_{hhp} y$.

Theorem 4.90 (Completeness of parallelism with left parallel composition modulo truly concurrent bisimulation equivalences). *Let x and y be $APTC$ terms.*

1. *If $x \sim_s y$, then $APTC \vdash x = y$;*
2. *if $x \sim_p y$, then $APTC \vdash x = y$;*
3. *if $x \sim_{hp} y$, then $APTC \vdash x = y$;*
4. *if $x \sim_{hhp} y$, then $APTC \vdash x = y$.*

The transition rules of encapsulation operator are the same, and the axioms are shown in Table 4.33.

Theorem 4.91 (Conservativity of $APTC$ with respect to the algebra for parallelism with left parallel composition). *$APTC$ is a conservative extension of the algebra for parallelism with left parallel composition.*

Theorem 4.92 (Congruence theorem of encapsulation operator ∂_H). *Truly concurrent bisimulation equivalences $\sim_p$, $\sim_s$, $\sim_{hp}$, and $\sim_{hhp}$ are all congruences with respect to encapsulation operator ∂_H.*

Theorem 4.93 (Elimination theorem of $APTC$). *Let p be a closed $APTC$ term including the encapsulation operator ∂_H. Then there is a basic $APTC$ term q such that $APTC \vdash p = q$.*

Theorem 4.94 (Soundness of $APTC$ modulo truly concurrent bisimulation equivalences). *Let x and y be $APTC$ terms including encapsulation operator ∂_H. If $APTC \vdash x = y$, then*

1. $x \sim_s y$;
2. $x \sim_p y$;
3. $x \sim_{hp} y$;
4. $x \sim_{hhp} y$.

Theorem 4.95 (Completeness of $APTC$ modulo truly concurrent bisimulation equivalences). *Let p and q be closed $APTC$ terms including encapsulation operator ∂_H,*

1. *if $p \sim_s q$ then $p = q$;*
2. *if $p \sim_p q$ then $p = q$;*
3. *if $p \sim_{hp} q$ then $p = q$;*
4. *if $p \sim_{hhp} q$ then $p = q$.*

4.7.2 Recursion

Definition 4.96 (Recursive specification). A recursive specification is a finite set of recursive equations

$$X_1 = t_1(X_1, \cdots, X_n)$$

$$\cdots$$

$$X_n = t_n(X_1, \cdots, X_n)$$

where the left-hand sides of X_i are called recursion variables, and the right-hand sides $t_i(X_1, \cdots, X_n)$ are process terms in $APTC$ with possible occurrences of the recursion variables $X_1, \cdots, X_n$.

Definition 4.97 (Solution). Processes $p_1, \cdots, p_n$ are solutions for a recursive specification $\{X_i = t_i(X_1, \cdots, X_n) | i \in \{1, \cdots, n\}\}$ (with respect to truly concurrent bisimulation equivalences $\sim_s (\sim_p, \sim_{hp}, \sim_{hhp})$) if $p_i \sim_s (\sim_p, \sim_{hp}, \sim_{hhp}) t_i(p_1, \cdots, p_n)$ for $i \in \{1, \cdots, n\}$.

Definition 4.98 (Guarded recursive specification). A recursive specification

$$X_1 = t_1(X_1, \cdots, X_n)$$

$$\cdots$$

$$X_n = t_n(X_1, \cdots, X_n)$$

is guarded if the right-hand sides of its recursive equations can be adapted to the form by applications of the axioms in $APTC$ and replacing recursion variables by the right-hand sides of their recursive equations,

$$(a_{11} \parallel \cdots \parallel a_{1i_1}) \cdot s_1(X_1, \cdots, X_n) + \cdots + (a_{k1} \parallel \cdots \parallel a_{ki_k}) \cdot s_k(X_1, \cdots, X_n)$$
$$+ (b_{11} \parallel \cdots \parallel b_{1j_1}) + \cdots + (b_{1j_1} \parallel \cdots \parallel b_{lj_l})$$

where $a_{11}, \cdots, a_{1i_1}, a_{k1}, \cdots, a_{ki_k}, b_{11}, \cdots, b_{1j_1}, b_{1j_1}, \cdots, b_{lj_l} \in \mathbb{E}$, and the sum above is allowed to be empty, in which case it represents the deadlock δ.

Definition 4.99 (Linear recursive specification). A recursive specification is linear if its recursive equations are of the form

$$(a_{11} \parallel \cdots \parallel a_{1i_1})X_1 + \cdots + (a_{k1} \parallel \cdots \parallel a_{ki_k})X_k + (b_{11} \parallel \cdots \parallel b_{1j_1}) + \cdots + (b_{1j_1} \parallel \cdots \parallel b_{lj_l})$$

where $a_{11}, \cdots, a_{1i_1}, a_{k1}, \cdots, a_{ki_k}, b_{11}, \cdots, b_{1j_1}, b_{1j_1}, \cdots, b_{lj_l} \in \mathbb{E}$, and the sum above is allowed to be empty, in which case it represents the deadlock δ.

Theorem 4.100 (Conservativity of $APTC$ with guarded recursion). *$APTC$ with guarded recursion is a conservative extension of $APTC$.*

Theorem 4.101 (Congruence theorem of $APTC$ with guarded recursion). *Truly concurrent bisimulation equivalences $\sim_p$, $\sim_s$, $\sim_{hp}$, $\sim_{hhp}$ are all congruences with respect to $APTC$ with guarded recursion.*

Theorem 4.102 (Elimination theorem of $APTC$ with linear recursion). *Each process term in $APTC$ with linear recursion is equal to a process term $\langle X_1|E \rangle$ with E a linear recursive specification.*

Theorem 4.103 (Soundness of $APTC$ with guarded recursion). *Let x and y be $APTC$ with guarded recursion terms. If $APTC$ with guarded recursion $\vdash x = y$, then*

1. $x \sim_s y$;
2. $x \sim_p y$;
3. $x \sim_{hp} y$;
4. $x \sim_{hhp} y$.

Theorem 4.104 (Completeness of $APTC$ with linear recursion). *Let p and q be closed $APTC$ with linear recursion terms, then,*

1. *if $p \sim_s q$ then $p = q$;*
2. *if $p \sim_p q$ then $p = q$;*
3. *if $p \sim_{hp} q$ then $p = q$;*
4. *if $p \sim_{hhp} q$ then $p = q$.*

Table 4.34 Axioms of silent step.

No.	Axiom
B1	$e \cdot \tau = e$
B2	$e \cdot (\tau \cdot (x + y) + x) = e \cdot (x + y)$
B3	$x \parallel \tau = x$

4.7.3 Abstraction

Definition 4.105 (Guarded linear recursive specification). A recursive specification is linear if its recursive equations are of the form

$$(a_{11} \parallel \cdots \parallel a_{1i_1})X_1 + \cdots + (a_{k1} \parallel \cdots \parallel a_{ki_k})X_k + (b_{11} \parallel \cdots \parallel b_{1j_1}) + \cdots + (b_{1j_1} \parallel \cdots \parallel b_{lj_l})$$

where $a_{11}, \cdots, a_{1i_1}, a_{k1}, \cdots, a_{ki_k}, b_{11}, \cdots, b_{1j_1}, b_{1j_1}, \cdots, b_{lj_l} \in \mathbb{E} \cup \{\tau\}$, and the sum above is allowed to be empty, in which case it represents the deadlock δ.

A linear recursive specification E is guarded if there does not exist an infinite sequence of τ-transitions $\langle X|E \rangle \xrightarrow{\tau} \langle X'|E \rangle \xrightarrow{\tau} \langle X''|E \rangle \xrightarrow{\tau} \cdots$.

The transition rules of τ are the same, and axioms of τ are as Table 4.34 shows.

Theorem 4.106 (Conservativity of $APTC$ with silent step and guarded linear recursion). *$APTC$ with silent step and guarded linear recursion is a conservative extension of $APTC$ with linear recursion.*

Theorem 4.107 (Congruence theorem of $APTC$ with silent step and guarded linear recursion). *Rooted branching truly concurrent bisimulation equivalences $\approx_{rbp}$, $\approx_{rbs}$, $\approx_{rbhp}$, and $\approx_{rbhhp}$ are all congruences with respect to $APTC$ with silent step and guarded linear recursion.*

Theorem 4.108 (Elimination theorem of $APTC$ with silent step and guarded linear recursion). *Each process term in $APTC$ with silent step and guarded linear recursion is equal to a process term $\langle X_1|E \rangle$ with E a guarded linear recursive specification.*

Theorem 4.109 (Soundness of $APTC$ with silent step and guarded linear recursion). *Let x and y be $APTC$ with silent step and guarded linear recursion terms. If $APTC$ with silent step and guarded linear recursion $\vdash x = y$, then*

1. $x \approx_{rbs} y$;
2. $x \approx_{rbp} y$;
3. $x \approx_{rbhp} y$;
4. $x \approx_{rbhhp} y$.

Theorem 4.110 (Completeness of $APTC$ with silent step and guarded linear recursion). *Let p and q be closed $APTC$ with silent step and guarded linear recursion terms, then,*

Table 4.35   Axioms of abstraction operator.

No.	Axiom
$TI1$	$e \notin I \quad \tau_I(e) = e$
$TI2$	$e \in I \quad \tau_I(e) = \tau$
$TI3$	$\tau_I(\delta) = \delta$
$TI4$	$\tau_I(x + y) = \tau_I(x) + \tau_I(y)$
$TI5$	$\tau_I(x \cdot y) = \tau_I(x) \cdot \tau_I(y)$
$TI6$	$\tau_I(x \between y) = \tau_I(x) \between \tau_I(y)$

1. *if* $p \approx_{rbs} q$ *then* $p = q$;
2. *if* $p \approx_{rbp} q$ *then* $p = q$;
3. *if* $p \approx_{rbhp} q$ *then* $p = q$;
4. *if* $p \approx_{rbhhp} q$ *then* $p = q$.

The transition rules of τ_I are the same, and the axioms are shown in Table 4.35.

Theorem 4.111 (Conservativity of $APTC_\tau$ with guarded linear recursion). *$APTC_\tau$ with guarded linear recursion is a conservative extension of $APTC$ with silent step and guarded linear recursion.*

Theorem 4.112 (Congruence theorem of $APTC_\tau$ with guarded linear recursion). *Rooted branching truly concurrent bisimulation equivalences $\approx_{rbp}$, $\approx_{rbs}$, $\approx_{rbhp}$ and $\approx_{rbhhp}$ are all congruences with respect to $APTC_\tau$ with guarded linear recursion.*

Theorem 4.113 (Soundness of $APTC_\tau$ with guarded linear recursion). *Let x and y be $APTC_\tau$ with guarded linear recursion terms. If $APTC_\tau$ with guarded linear recursion $\vdash x = y$, then*

1. $x \approx_{rbs} y$;
2. $x \approx_{rbp} y$;
3. $x \approx_{rbhp} y$;
4. $x \approx_{rbhhp} y$.

Definition 4.114 (Cluster). Let E be a guarded linear recursive specification, and $I \subseteq \mathbb{E}$. Two recursion variable X and Y in E are in the same cluster for I iff there exist sequences of transitions $\langle X|E \rangle \xrightarrow{\{b_{11}, \cdots, b_{1i}\}} \cdots \xrightarrow{\{b_{m1}, \cdots, b_{mi}\}} \langle Y|E \rangle$ and $\langle Y|E \rangle \xrightarrow{\{c_{11}, \cdots, c_{1j}\}} \cdots \xrightarrow{\{c_{n1}, \cdots, c_{nj}\}} \langle X|E \rangle$, where $b_{11}, \cdots, b_{mi}, c_{11}, \cdots, c_{nj} \in I \cup \{\tau\}$.

$a_1 \between \cdots \between a_k$ or $(a_1 \between \cdots \between a_k)X$ is an exit for the cluster C iff: (1) $a_1 \between \cdots \between a_k$ or $(a_1 \between \cdots \between a_k)X$ is a summand at the right-hand side of the recursive equation for a recursion variable in C, and (2) in the case of $(a_1 \between \cdots \between a_k)X$, either $a_l \notin I \cup \{\tau\}(l \in \{1, 2, \cdots, k\})$ or $X \notin C$.

The Cluster Fair Abstraction Rule (CFAR) is shown in Table 4.36.

Table 4.36 Cluster fair abstraction rule.

No.	Axiom
$CFAR$	If X is in a cluster for I with exits
	$\{(a_{11} \between \cdots \between a_{1i})Y_1, \cdots, (a_{m1} \between \cdots \between a_{mi})Y_m, b_{11} \between \cdots \between b_{1j}, \cdots, b_{n1} \between \cdots \between b_{nj}\}$,
	then $\tau \cdot \tau_I(\langle X|E \rangle) =$
	$\tau \cdot \tau_I((a_{11} \between \cdots \between a_{1i})\langle Y_1|E \rangle + \cdots + (a_{m1} \between \cdots \between a_{mi})\langle Y_m|E \rangle + b_{11} \between \cdots \between b_{1j} + \cdots$
	$+ b_{n1} \between \cdots \between b_{nj})$

Theorem 4.115 (Soundness of $CFAR$). *$CFAR$ is sound modulo rooted branching truly concurrent bisimulation equivalences $\approx_{rbs}$, $\approx_{rbp}$, $\approx_{rbhp}$ and $\approx_{rbhhp}$.*

Theorem 4.116 (Completeness of $APTC_\tau$ with guarded linear recursion and $CFAR$). *Let p and q be closed $APTC_\tau$ with guarded linear recursion and $CFAR$ terms, then,*

1. *if $p \approx_{rbs} q$ then $p = q$;*
2. *if $p \approx_{rbp} q$ then $p = q$;*
3. *if $p \approx_{rbhp} q$ then $p = q$;*
4. *if $p \approx_{rbhhp} q$ then $p = q$.*

4.8 Conclusions

Now, let us conclude this chapter. We try to find the algebraic laws for true concurrency, as a uniform logic for true concurrency [14] [15] already existed. There are simple comparisons between Hennessy and Milner (HM) logic for bisimulation equivalence and the uniform logic for truly concurrent bisimulation equivalences, the algebraic laws [1], ACP [4] for bisimulation equivalence, and *what* for truly concurrent bisimulation equivalences, which is still missing.

Following the above idea, we find the algebraic laws for true concurrency, which is called $APTC$, an algebra for true concurrency. Like ACP, $APTC$ also has four modules: $BATC$ (Basic Algebra for True Concurrency), $APTC$ (Algebra for Parallelism in True Concurrency), recursion and abstraction, and we prove the soundness and completeness of their algebraic laws modulo truly concurrent bisimulation equivalences. And we show its applications in verification of behaviors of system in a truly concurrent flavor, and its modularity by extending new computational properties into it.

Unlike ACP, in $APTC$, the parallelism is a fundamental computational pattern, and cannot be steadied by other computational patterns. We establish a whole theory which has correspondences to ACP.

5

Mobility

In this chapter, we design a calculus of truly concurrent mobile processes (π_{tc}) following the way paved by π-calculus for bisimulation and our previous work on truly concurrent process algebra CTC and APTC. This chapter is organized as follows. We introduce the syntax and operational semantics of π_{tc} in section 5.1, its properties for strongly truly concurrent bisimulations in section 5.2, its axiomatization in section 5.3. Finally, in section 5.4, we conclude this chapter.

5.1 Syntax and operational semantics

We assume an infinite set $\mathcal{N}$ of (action or event) names, and use $a, b, c, \cdots$ to range over $\mathcal{N}$, use x, y, z, w, u, v as meta-variables over names. We denote by $\overline{\mathcal{N}}$ the set of co-names and let $\overline{a}, \overline{b}, \overline{c}, \cdots$ range over $\overline{\mathcal{N}}$. Then we set $\mathcal{L} = \mathcal{N} \cup \overline{\mathcal{N}}$ as the set of labels, and use $l, \overline{l}$ to range over $\mathcal{L}$. We extend complementation to $\mathcal{L}$ such that $\overline{\overline{a}} = a$. Let τ denote the silent step (internal action or event) and define $Act = \mathcal{L} \cup \{\tau\}$ to be the set of actions, α, β range over Act. And K, L are used to stand for subsets of $\mathcal{L}$ and $\overline{L}$ is used for the set of complements of labels in L.

Further, we introduce a set $\mathcal{X}$ of process variables, and a set $\mathcal{K}$ of process constants, and let $X, Y, \cdots$ range over $\mathcal{X}$, and $A, B, \cdots$ range over $\mathcal{K}$. For each process constant A, a nonnegative arity $ar(A)$ is assigned to it. Let $\widetilde{x} = x_1, \cdots, x_{ar(A)}$ be a tuple of distinct name variables, then $A(\widetilde{x})$ is called a process constant. $\widetilde{X}$ is a tuple of distinct process variables, and also $E, F, \cdots$ range over the recursive expressions. We write $\mathcal{P}$ for the set of processes. Sometimes, we use I, J to stand for an indexing set, and we write $E_i : i \in I$ for a family of expressions indexed by I. Id_D is the identity function or relation over set D. The symbol $\equiv_\alpha$ denotes equality under standard alpha-convertibility, note that the subscript α has no relation to the action α.

5.1.1 Syntax

We use the Prefix . to model the causality relation $\leq$ in true concurrency, the Summation $+$ to model the conflict relation $\sharp$ in true concurrency, and the Composition $\parallel$ to explicitly model concurrent relation in true concurrency. And we follow the conventions of process algebra.

Definition 5.1 (Syntax). A truly concurrent process P is defined inductively by the following formation rules:

1. $A(\widetilde{x}) \in \mathcal{P}$;

2. nil $\in \mathcal{P}$;

3. if $P \in \mathcal{P}$, then the Prefix $\tau.P \in \mathcal{P}$, for $\tau \in Act$ is the silent action;

4. if $P \in \mathcal{P}$, then the Output $\bar{x}y.P \in \mathcal{P}$, for $x, y \in Act$;

5. if $P \in \mathcal{P}$, then the Input $x(y).P \in \mathcal{P}$, for $x, y \in Act$;

6. if $P \in \mathcal{P}$, then the Restriction $(x)P \in \mathcal{P}$, for $x \in Act$;

7. if $P, Q \in \mathcal{P}$, then the Summation $P + Q \in \mathcal{P}$;

8. if $P, Q \in \mathcal{P}$, then the Composition $P \parallel Q \in \mathcal{P}$;

The standard BNF grammar of syntax of π_{tc} can be summarized as follows:

$$P ::= A(\tilde{x}) \mid \mathbf{nil} \mid \tau.P \mid \bar{x}y.P \mid x(y).P \mid (x)P \mid P + P \mid P \parallel P$$

In $\bar{x}y$, $x(y)$ and $\bar{x}(y)$, x is called the subject, y is called the object and it may be free or bound.

Definition 5.2 (Free variables). The free names of a process P, $fn(P)$, are defined as follows.

1. $fn(A(\tilde{x})) \subseteq \{\tilde{x}\}$;

2. $fn(\mathbf{nil}) = \emptyset$;

3. $fn(\tau.P) = fn(P)$;

4. $fn(\bar{x}y.P) = fn(P) \cup \{x\} \cup \{y\}$;

5. $fn(x(y).P) = fn(P) \cup \{x\} - \{y\}$;

6. $fn((x)P) = fn(P) - \{x\}$;

7. $fn(P + Q) = fn(P) \cup fn(Q)$;

8. $fn(P \parallel Q) = fn(P) \cup fn(Q)$.

Definition 5.3 (Bound variables). Let $n(P)$ be the names of a process P, then the bound names $bn(P) = n(P) - fn(P)$.

For each process constant schema $A(\tilde{x})$, a defining equation of the form

$$A(\tilde{x}) \stackrel{\text{def}}{=} P$$

is assumed, where P is a process with $fn(P) \subseteq \{\tilde{x}\}$.

Definition 5.4 (Substitutions). A substitution is a function $\sigma : \mathcal{N} \to \mathcal{N}$. For $x_i\sigma = y_i$ with $1 \le i \le n$, we write $\{y_1/x_1, \cdots, y_n/x_n\}$ or $\{\tilde{y}/\tilde{x}\}$ for σ. For a process $P \in \mathcal{P}$, $P\sigma$ is defined inductively as follows:

1. if P is a process constant $A(\tilde{x}) = A(x_1, \cdots, x_n)$, then $P\sigma = A(x_1\sigma, \cdots, x_n\sigma)$;

2. if $P = \mathbf{nil}$, then $P\sigma = \mathbf{nil}$;

3. if $P = \tau.P'$, then $P\sigma = \tau.P'\sigma$;

4. if $P = \bar{x}y.P'$, then $P\sigma = \overline{x\sigma}\,y\sigma.P'\sigma$;

5. if $P = x(y).P'$, then $P\sigma = x\sigma(y).P'\sigma$;

6. if $P = (x)P'$, then $P\sigma = (x\sigma)P'\sigma$;

7. if $P = P_1 + P_2$, then $P\sigma = P_1\sigma + P_2\sigma$;

8. if $P = P_1 \parallel P_2$, then $P\sigma = P_1\sigma \parallel P_2\sigma$.

5.1.2 Operational semantics

The operational semantics is defined by LTSs (labeled transition systems), and it is detailed by the following definition.

Definition 5.5 (Semantics). The operational semantics of π_{tc} corresponding to the syntax in Definition 5.1 is defined by a series of transition rules, named **ACT**, **SUM**, **IDE**, **PAR**, **COM** and **CLOSE**, **RES** and **OPEN** indicate that the rules are associated respectively with Prefix, Summation, Match, Identity, Parallel Composition, Communication, and Restriction in Definition 5.1. They are shown in Table 5.1.

5.1.3 Properties of transitions

Proposition 5.6. 1. *If $P \xrightarrow{\alpha} P'$ then*
   a. $fn(\alpha) \subseteq fn(P)$;
   b. $fn(P') \subseteq fn(P) \cup bn(\alpha)$;
2. *If $P \xrightarrow{\{\alpha_1,\cdots,\alpha_n\}} P'$ then*
   a. $fn(\alpha_1) \cup \cdots \cup fn(\alpha_n) \subseteq fn(P)$;
   b. $fn(P') \subseteq fn(P) \cup bn(\alpha_1) \cup \cdots \cup bn(\alpha_n)$.

Proof. By induction on the depth of inference. □

Proposition 5.7. *Suppose that $P \xrightarrow{\alpha(y)} P'$, where $\alpha = x$ or $\alpha = \bar{x}$, and $x \notin n(P)$, then there exists some $P'' \equiv_\alpha P'\{z/y\}$, $P \xrightarrow{\alpha(z)} P''$.*

Proof. By induction on the depth of inference. □

Proposition 5.8. *If $P \rightarrow P'$, $bn(\alpha) \cap fn(P'\sigma) = \emptyset$, and $\sigma \lceil bn(\alpha) = id$, then there exists some $P'' \equiv_\alpha P'\sigma$, $P\sigma \xrightarrow{\alpha\sigma} P''$.*

Proof. By the definition of substitution (Definition 5.4) and induction on the depth of inference. □

Proposition 5.9. 1. *If $P\{w/z\} \xrightarrow{\alpha} P'$, where $w \notin fn(P)$ and $bn(\alpha) \cap fn(P, w) = \emptyset$, then there exist some Q and β with $Q\{w/z\} \equiv_\alpha P'$ and $\beta\sigma = \alpha$, $P \xrightarrow{\beta} Q$;*
2. *If $P\{w/z\} \xrightarrow{\{\alpha_1,\cdots,\alpha_n\}} P'$, where $w \notin fn(P)$ and $bn(\alpha_1) \cap \cdots \cap bn(\alpha_n) \cap fn(P, w) = \emptyset$, then there exist some Q and $\beta_1, \cdots, \beta_n$ with $Q\{w/z\} \equiv_\alpha P'$ and $\beta_1\sigma = \alpha_1, \cdots, \beta_n\sigma = \alpha_n$, $P \xrightarrow{\{\beta_1,\cdots,\beta_n\}} Q$.*

Proof. By the definition of substitution (Definition 5.4) and induction on the depth of inference. □

Table 5.1 Transition rules of π_{tc}.

$$\textbf{TAU-ACT} \quad \frac{}{\tau.P \xrightarrow{\tau} P} \qquad \textbf{OUTPUT-ACT} \quad \frac{}{\overline{x}y.P \xrightarrow{\overline{x}y} P}$$

$$\textbf{INPUT-ACT} \quad \frac{}{x(z).P \xrightarrow{x(w)} P\{w/z\}} \quad (w \notin fn((z)P))$$

$$\textbf{PAR}_1 \quad \frac{P \xrightarrow{\alpha} P' \quad Q \nrightarrow}{P \parallel Q \xrightarrow{\alpha} P' \parallel Q} \quad (bn(\alpha) \cap fn(Q) = \emptyset) \qquad \textbf{PAR}_2 \quad \frac{Q \xrightarrow{\alpha} Q' \quad P \nrightarrow}{P \parallel Q \xrightarrow{\alpha} P \parallel Q'} \quad (bn(\alpha) \cap fn(P) = \emptyset)$$

$$\textbf{PAR}_3 \quad \frac{P \xrightarrow{\alpha} P' \quad Q \xrightarrow{\beta} Q'}{P \parallel Q \xrightarrow{\{\alpha,\beta\}} P' \parallel Q'} \quad (\beta \neq \overline{\alpha}, bn(\alpha) \cap bn(\beta) = \emptyset, bn(\alpha) \cap fn(Q) = \emptyset, bn(\beta) \cap fn(P) = \emptyset)$$

$$\textbf{PAR}_4 \quad \frac{P \xrightarrow{x_1(z)} P' \quad Q \xrightarrow{x_2(z)} Q'}{P \parallel Q \xrightarrow{\{x_1(w),x_2(w)\}} P'\{w/z\} \parallel Q'\{w/z\}} \quad (w \notin fn((z)P) \cup fn((z)Q))$$

$$\textbf{COM} \quad \frac{P \xrightarrow{\overline{x}y} P' \quad Q \xrightarrow{x(z)} Q'}{P \parallel Q \xrightarrow{\tau} P' \parallel Q'\{y/z\}}$$

$$\textbf{CLOSE} \quad \frac{P \xrightarrow{\overline{x}(w)} P' \quad Q \xrightarrow{x(w)} Q'}{P \parallel Q \xrightarrow{\tau} (w)(P' \parallel Q')}$$

$$\textbf{SUM}_1 \quad \frac{P \xrightarrow{\alpha} P'}{P + Q \xrightarrow{\alpha} P'} \qquad \textbf{SUM}_2 \quad \frac{P \xrightarrow{\{\alpha_1,\cdots,\alpha_n\}} P'}{P + Q \xrightarrow{\{\alpha_1,\cdots,\alpha_n\}} P'}$$

$$\textbf{IDE}_1 \quad \frac{P\{\widetilde{y}/\widetilde{x}\} \xrightarrow{\alpha} P'}{A(\widetilde{y}) \xrightarrow{\alpha} P'} \quad (A(\widetilde{x}) \overset{\text{def}}{=} P) \qquad \textbf{IDE}_2 \quad \frac{P\{\widetilde{y}/\widetilde{x}\} \xrightarrow{\{\alpha_1,\cdots,\alpha_n\}} P'}{A(\widetilde{y}) \xrightarrow{\{\alpha_1,\cdots,\alpha_n\}} P'} \quad (A(\widetilde{x}) \overset{\text{def}}{=} P)$$

$$\textbf{RES}_1 \quad \frac{P \xrightarrow{\alpha} P'}{(y)P \xrightarrow{\alpha} (y)P'} \quad (y \notin n(\alpha)) \qquad \textbf{RES}_2 \quad \frac{P \xrightarrow{\{\alpha_1,\cdots,\alpha_n\}} P'}{(y)P \xrightarrow{\{\alpha_1,\cdots,\alpha_n\}} (y)P'} \quad (y \notin n(\alpha_1) \cup \cdots \cup n(\alpha_n))$$

$$\textbf{OPEN}_1 \quad \frac{P \xrightarrow{\overline{x}y} P'}{(y)P \xrightarrow{\overline{x}(w)} P'\{w/y\}} \quad (y \neq x, w \notin fn((y)P'))$$

$$\textbf{OPEN}_2 \quad \frac{P \xrightarrow{\{\overline{x}_1 y,\cdots,\overline{x}_n y\}} P'}{(y)P \xrightarrow{\{\overline{x}_1(w),\cdots,\overline{x}_n(w)\}} P'\{w/y\}} \quad (y \neq x_1 \neq \cdots \neq x_n, w \notin fn((y)P'))$$

5.2 Strongly truly concurrent bisimilarities

5.2.1 Basic definitions

Firstly, in this subsection, we introduce concepts of (strongly) truly concurrent bisimilarities, including pomset bisimilarity, step bisimilarity, history-preserving (hp-)bisimilarity and hereditary history-preserving (hhp-)bisimilarity. In contrast to traditional truly concurrent bisimilarities in CTC and APTC, these versions in π_{tc} must take care of actions with bound objects. Note that, these truly concurrent bisimilarities are defined as late bisimilarities, but not early bisimilarities, as defined in π-calculus [23,24]. Note that, here, a PES $\mathcal{E}$ is deemed as a process.

Definition 5.10 (Pomset transitions and step). Let $\mathcal{E}$ be a PES and let $C \in \mathcal{C}(\mathcal{E})$, and $\emptyset \neq X \subseteq \mathbb{E}$, if $C \cap X = \emptyset$ and $C' = C \cup X \in \mathcal{C}(\mathcal{E})$, then $C \xrightarrow{X} C'$ is called a pomset transition from C to C'. When the events in X are pairwise concurrent, we say that $C \xrightarrow{X} C'$ is a step.

Definition 5.11 (Strong pomset, step bisimilarity). Let $\mathcal{E}_1$, $\mathcal{E}_2$ be PESs. A strong pomset bisimulation is a relation $R \subseteq \mathcal{C}(\mathcal{E}_1) \times \mathcal{C}(\mathcal{E}_2)$, such that if $(C_1, C_2) \in R$, and $C_1 \xrightarrow{X_1} C_1'$ (with $\mathcal{E}_1 \xrightarrow{X_1} \mathcal{E}_1'$) then $C_2 \xrightarrow{X_2} C_2'$ (with $\mathcal{E}_2 \xrightarrow{X_2} \mathcal{E}_2'$), with $X_1 \subseteq \mathbb{E}_1$, $X_2 \subseteq \mathbb{E}_2$, $X_1 \sim X_2$ and $(C_1', C_2') \in R$:

1. for each fresh action $\alpha \in X_1$, if $C_1'' \xrightarrow{\alpha} C_1'''$ (with $\mathcal{E}_1'' \xrightarrow{\alpha} \mathcal{E}_1'''$), then for some C_2'' and C_2''', $C_2'' \xrightarrow{\alpha} C_2'''$ (with $\mathcal{E}_2'' \xrightarrow{\alpha} \mathcal{E}_2'''$), such that if $(C_1'', C_2'') \in R$ then $(C_1''', C_2''') \in R$;

2. for each $x(y) \in X_1$ with $(y \notin n(\mathcal{E}_1, \mathcal{E}_2))$, if $C_1'' \xrightarrow{x(y)} C_1'''$ (with $\mathcal{E}_1'' \xrightarrow{x(y)} \mathcal{E}_1'''\{w/y\}$) for all w, then for some C_2'' and C_2''', $C_2'' \xrightarrow{x(y)} C_2'''$ (with $\mathcal{E}_2'' \xrightarrow{x(y)} \mathcal{E}_2'''\{w/y\}$) for all w, such that if $(C_1'', C_2'') \in R$ then $(C_1''', C_2''') \in R$;

3. for each two $x_1(y), x_2(y) \in X_1$ with $(y \notin n(\mathcal{E}_1, \mathcal{E}_2))$, if $C_1'' \xrightarrow{\{x_1(y), x_2(y)\}} C_1'''$ (with $\mathcal{E}_1'' \xrightarrow{\{x_1(y), x_2(y)\}} \mathcal{E}_1'''\{w/y\}$) for all w, then for some C_2'' and C_2''', $C_2'' \xrightarrow{\{x_1(y), x_2(y)\}} C_2'''$ (with $\mathcal{E}_2'' \xrightarrow{\{x_1(y), x_2(y)\}} \mathcal{E}_2'''\{w/y\}$) for all w, such that if $(C_1'', C_2'') \in R$ then $(C_1''', C_2''') \in R$;

4. for each $\overline{x}(y) \in X_1$ with $y \notin n(\mathcal{E}_1, \mathcal{E}_2)$, if $C_1'' \xrightarrow{\overline{x}(y)} C_1'''$ (with $\mathcal{E}_1'' \xrightarrow{\overline{x}(y)} \mathcal{E}_1'''$), then for some C_2'' and C_2''', $C_2'' \xrightarrow{\overline{x}(y)} C_2'''$ (with $\mathcal{E}_2'' \xrightarrow{\overline{x}(y)} \mathcal{E}_2'''$), such that if $(C_1'', C_2'') \in R$ then $(C_1''', C_2''') \in R$.

and vice-versa.

We say that $\mathcal{E}_1$, $\mathcal{E}_2$ are strong pomset bisimilar, written $\mathcal{E}_1 \sim_p \mathcal{E}_2$, if there exists a strong pomset bisimulation R, such that $(\emptyset, \emptyset) \in R$. By replacing pomset transitions with steps, we can get the definition of strong step bisimulation. When PESs $\mathcal{E}_1$ and $\mathcal{E}_2$ are strong step bisimilar, we write $\mathcal{E}_1 \sim_s \mathcal{E}_2$.

Definition 5.12 (Posetal product). Given two PESs $\mathcal{E}_1$, $\mathcal{E}_2$, the posetal product of their configurations, denoted $\mathcal{C}(\mathcal{E}_1)\overline{\times}\mathcal{C}(\mathcal{E}_2)$, is defined as

$$\{(C_1, f, C_2) | C_1 \in \mathcal{C}(\mathcal{E}_1), C_2 \in \mathcal{C}(\mathcal{E}_2), f : C_1 \to C_2 \text{ isomorphism}\}.$$

A subset $R \subseteq C(\mathcal{E}_1) \overline{\times} C(\mathcal{E}_2)$ is called a posetal relation. We say that R is downward closed when for any $(C_1, f, C_2), (C_1', f', C_2') \in C(\mathcal{E}_1) \overline{\times} C(\mathcal{E}_2)$, if $(C_1, f, C_2) \subseteq (C_1', f', C_2')$ pointwise and $(C_1', f', C_2') \in R$, then $(C_1, f, C_2) \in R$.

For $f : X_1 \to X_2$, we define $f[x_1 \mapsto x_2] : X_1 \cup \{x_1\} \to X_2 \cup \{x_2\}$, $z \in X_1 \cup \{x_1\}$, (1) $f[x_1 \mapsto x_2](z) = x_2$, if $z = x_1$; (2) $f[x_1 \mapsto x_2](z) = f(z)$, otherwise. Where $X_1 \subseteq \mathbb{E}_1$, $X_2 \subseteq \mathbb{E}_2$, $x_1 \in \mathbb{E}_1$, $x_2 \in \mathbb{E}_2$.

Definition 5.13 (Strong (hereditary) history-preserving bisimilarity). A strong history-preserving (hp-) bisimulation is a posetal relation $R \subseteq C(\mathcal{E}_1) \overline{\times} C(\mathcal{E}_2)$ such that if $(C_1, f, C_2) \in R$, and

1. for $e_1 = \alpha$ a fresh action, if $C_1 \xrightarrow{\alpha} C_1'$ (with $\mathcal{E}_1 \xrightarrow{\alpha} \mathcal{E}_1'$), then for some C_2' and $e_2 = \alpha$, $C_2 \xrightarrow{\alpha} C_2'$ (with $\mathcal{E}_2 \xrightarrow{\alpha} \mathcal{E}_2'$), such that $(C_1', f[e_1 \mapsto e_2], C_2') \in R$;

2. for $e_1 = x(y)$ with $(y \notin n(\mathcal{E}_1, \mathcal{E}_2))$, if $C_1 \xrightarrow{x(y)} C_1'$ (with $\mathcal{E}_1 \xrightarrow{x(y)} \mathcal{E}_1'\{w/y\}$) for all w, then for some C_2' and $e_2 = x(y)$, $C_2 \xrightarrow{x(y)} C_2'$ (with $\mathcal{E}_2 \xrightarrow{x(y)} \mathcal{E}_2'\{w/y\}$) for all w, such that $(C_1', f[e_1 \mapsto e_2], C_2') \in R$;

3. for $e_1 = \overline{x}(y)$ with $y \notin n(\mathcal{E}_1, \mathcal{E}_2)$, if $C_1 \xrightarrow{\overline{x}(y)} C_1'$ (with $\mathcal{E}_1 \xrightarrow{\overline{x}(y)} \mathcal{E}_1'$), then for some C_2' and $e_2 = \overline{x}(y)$, $C_2 \xrightarrow{\overline{x}(y)} C_2'$ (with $\mathcal{E}_2 \xrightarrow{\overline{x}(y)} \mathcal{E}_2'$), such that $(C_1', f[e_1 \mapsto e_2], C_2') \in R$.

and vice-versa. $\mathcal{E}_1, \mathcal{E}_2$ are strong history-preserving (hp-)bisimilar and are written $\mathcal{E}_1 \sim_{hp} \mathcal{E}_2$ if there exists a strong hp-bisimulation R such that $(\emptyset, \emptyset, \emptyset) \in R$.

A strongly hereditary history-preserving (hhp-)bisimulation is a downward closed strong hp-bisimulation. $\mathcal{E}_1, \mathcal{E}_2$ are strongly hereditary history-preserving (hhp-)bisimilar and are written $\mathcal{E}_1 \sim_{hhp} \mathcal{E}_2$.

Since the Parallel composition $\parallel$ is a fundamental computational pattern in CTC and APTC, and also it is fundamental in π_{tc} as defined in Table 5.1, and cannot be instead of other operators. So, the above truly concurrent bisimilarities are preserved by substitutions of names as defined in Definition 5.4. We illustrate it by an example. We assume $P \equiv \overline{x}v$, abbreviated to $\overline{x}$; and $Q \equiv y(u)$, abbreviated to y. Then the following equations are true when $x \neq y$ and $u \neq v$:

$$P \parallel Q \sim_p \overline{x} \parallel y$$

$$P \parallel Q \sim_s \overline{x} \parallel y$$

$$P \parallel Q \sim_{hp} \overline{x} \parallel y$$

$$P \parallel Q \sim_{hhp} \overline{x} \parallel y.$$

By substituting y to x, the following equations still hold:

$$P \parallel Q\{x/y\} \sim_p \overline{x} \parallel x$$

$$P \parallel Q\{x/y\} \sim_s \overline{x} \parallel x$$

$$P \parallel Q\{x/y\} \sim_{hp} \overline{x} \parallel x$$

$$P \parallel Q\{x/y\} \sim_{hhp} \overline{x} \parallel x.$$

Theorem 5.14. $\equiv_\alpha$ *are strongly truly concurrent bisimulations. That is, if $P \equiv_\alpha Q$, then,*

1. $P \sim_p Q$;
2. $P \sim_s Q$;
3. $P \sim_{hp} Q$;
4. $P \sim_{hhp} Q$.

Proof. By induction on the depth of inference (see Table 5.1), we can get the following facts:

1. If α is a free action and $P \xrightarrow{\alpha} P'$, then equally for some Q' with $P' \equiv_\alpha Q'$, $Q \xrightarrow{\alpha} Q'$;
2. If $P \xrightarrow{a(y)} P'$ with $a = x$ or $a = \overline{x}$ and $z \notin n(Q)$, then equally for some Q' with $P'\{z/y\} \equiv_\alpha Q'$, $Q \xrightarrow{a(z)} Q'$.

   Then, we can get:

1. by the definition of strong pomset bisimilarity (Definition 5.11), $P \sim_p Q$;
2. by the definition of strong step bisimilarity (Definition 5.11), $P \sim_s Q$;
3. by the definition of strong hp-bisimilarity (Definition 5.13), $P \sim_{hp} Q$;
4. by the definition of strongly hhp-bisimilarity (Definition 5.13), $P \sim_{hhp} Q$. □

5.2.2 Laws and congruence

Similarly to CTC, we can obtain the following laws with respect to truly concurrent bisimilarities.

Theorem 5.15 (Summation laws for strong pomset bisimilarity). *The summation laws for strong pomset bisimilarity are as follows.*

1. $P + \boldsymbol{nil} \sim_p P$;
2. $P + P \sim_p P$;
3. $P_1 + P_2 \sim_p P_2 + P_1$;
4. $P_1 + (P_2 + P_3) \sim_p (P_1 + P_2) + P_3$.

Proof. **1.** It is sufficient to prove the relation $R = \{(P + \boldsymbol{nil}, P)\} \cup \textbf{Id}$ is a strong pomset bisimulation. It can be proved similarly to the proof of Monoid laws for strong pomset bisimulation in CTC, we omit it;

2. It is sufficient to prove the relation $R = \{(P + P, P)\} \cup \textbf{Id}$ is a strong pomset bisimulation. It can be proved similarly to the proof of Monoid laws for strong pomset bisimulation in CTC, we omit it;

3. It is sufficient to prove the relation $R = \{(P_1 + P_2, P_2 + P_1)\} \cup \mathbf{Id}$ is a strong pomset bisimulation. It can be proved similarly to the proof of Monoid laws for strong pomset bisimulation in CTC, we omit it;

4. It is sufficient to prove the relation $R = \{(P_1 + (P_2 + P_3), (P_1 + P_2) + P_3)\} \cup \mathbf{Id}$ is a strong pomset bisimulation. It can be proved similarly to the proof of Monoid laws for strong pomset bisimulation in CTC, we omit it. □

Theorem 5.16 (Summation laws for strong step bisimilarity). *The summation laws for strong step bisimilarity are as follows.*

1. $P + \mathbf{nil} \sim_s P$;
2. $P + P \sim_s P$;
3. $P_1 + P_2 \sim_s P_2 + P_1$;
4. $P_1 + (P_2 + P_3) \sim_s (P_1 + P_2) + P_3$.

Proof. **1.** It is sufficient to prove the relation $R = \{(P + \mathbf{nil}, P)\} \cup \mathbf{Id}$ is a strong step bisimulation. It can be proved similarly to the proof of Monoid laws for strong step bisimulation in CTC, we omit it;

2. It is sufficient to prove the relation $R = \{(P + P, P)\} \cup \mathbf{Id}$ is a strong step bisimulation. It can be proved similarly to the proof of Monoid laws for strong step bisimulation in CTC, we omit it;

3. It is sufficient to prove the relation $R = \{(P_1 + P_2, P_2 + P_1)\} \cup \mathbf{Id}$ is a strong step bisimulation. It can be proved similarly to the proof of Monoid laws for strong step bisimulation in CTC, we omit it;

4. It is sufficient to prove the relation $R = \{(P_1 + (P_2 + P_3), (P_1 + P_2) + P_3)\} \cup \mathbf{Id}$ is a strong step bisimulation. It can be proved similarly to the proof of Monoid laws for strong step bisimulation in CTC, we omit it. □

Theorem 5.17 (Summation laws for strong hp-bisimilarity). *The summation laws for strong hp-bisimilarity are as follows.*

1. $P + \mathbf{nil} \sim_{hp} P$;
2. $P + P \sim_{hp} P$;
3. $P_1 + P_2 \sim_{hp} P_2 + P_1$;
4. $P_1 + (P_2 + P_3) \sim_{hp} (P_1 + P_2) + P_3$.

Proof. **1.** It is sufficient to prove the relation $R = \{(P + \mathbf{nil}, P)\} \cup \mathbf{Id}$ is a strong hp-bisimulation. It can be proved similarly to the proof of Monoid laws for strong hp-bisimulation in CTC, we omit it;

2. It is sufficient to prove the relation $R = \{(P + P, P)\} \cup \mathbf{Id}$ is a strong hp-bisimulation. It can be proved similarly to the proof of Monoid laws for strong hp-bisimulation in CTC, we omit it;

3. It is sufficient to prove the relation $R = \{(P_1 + P_2, P_2 + P_1)\} \cup \mathbf{Id}$ is a strong hp-bisimulation. It can be proved similarly to the proof of Monoid laws for strong hp-bisimulation in CTC, we omit it;

4. It is sufficient to prove the relation $R = \{(P_1 + (P_2 + P_3), (P_1 + P_2) + P_3)\} \cup \mathbf{Id}$ is a strong hp-bisimulation. It can be proved similarly to the proof of Monoid laws for strong hp-bisimulation in CTC, we omit it. □

Theorem 5.18 (Summation laws for strongly hhp-bisimilarity). *The summation laws for strongly hhp-bisimilarity are as follows.*

1. $P + \mathbf{nil} \sim_{hhp} P$;
2. $P + P \sim_{hhp} P$;
3. $P_1 + P_2 \sim_{hhp} P_2 + P_1$;
4. $P_1 + (P_2 + P_3) \sim_{hhp} (P_1 + P_2) + P_3$.

Proof. **1.** It is sufficient to prove the relation $R = \{(P + \mathbf{nil}, P)\} \cup \mathbf{Id}$ is a strongly hhp-bisimulation. It can be proved similarly to the proof of Monoid laws for strongly hhp-bisimulation in CTC, we omit it;

2. It is sufficient to prove the relation $R = \{(P + P, P)\} \cup \mathbf{Id}$ is a strongly hhp-bisimulation. It can be proved similarly to the proof of Monoid laws for strongly hhp-bisimulation in CTC, we omit it;

3. It is sufficient to prove the relation $R = \{(P_1 + P_2, P_2 + P_1)\} \cup \mathbf{Id}$ is a strongly hhp-bisimulation. It can be proved similarly to the proof of Monoid laws for strongly hhp-bisimulation in CTC, we omit it;

4. It is sufficient to prove the relation $R = \{(P_1 + (P_2 + P_3), (P_1 + P_2) + P_3)\} \cup \mathbf{Id}$ is a strongly hhp-bisimulation. It can be proved similarly to the proof of Monoid laws for strongly hhp-bisimulation in CTC, we omit it. □

Theorem 5.19 (Identity law for truly concurrent bisimilarities). *If* $A(\tilde{x}) \stackrel{def}{=} P$, *then*

1. $A(\tilde{y}) \sim_p P\{\tilde{y}/\tilde{x}\}$;
2. $A(\tilde{y}) \sim_s P\{\tilde{y}/\tilde{x}\}$;
3. $A(\tilde{y}) \sim_{hp} P\{\tilde{y}/\tilde{x}\}$;
4. $A(\tilde{y}) \sim_{hhp} P\{\tilde{y}/\tilde{x}\}$.

Proof. **1.** It is straightforward to see that $R = \{A(\tilde{y}, P\{\tilde{y}/\tilde{x}\})\} \cup \mathbf{Id}$ is a strong pomset bisimulation;

2. It is straightforward to see that $R = \{A(\tilde{y}, P\{\tilde{y}/\tilde{x}\})\} \cup \mathbf{Id}$ is a strong step bisimulation;

3. It is straightforward to see that $R = \{A(\tilde{y}, P\{\tilde{y}/\tilde{x}\})\} \cup \mathbf{Id}$ is a strong hp-bisimulation;

4. It is straightforward to see that $R = \{A(\tilde{y}, P\{\tilde{y}/\tilde{x}\})\} \cup \mathbf{Id}$ is a strongly hhp-bisimulation. □

Theorem 5.20 (Restriction Laws for strong pomset bisimilarity). *The restriction laws for strong pomset bisimilarity are as follows.*

1. $(y)P \sim_p P$, *if* $y \notin fn(P)$;
2. $(y)(z)P \sim_p (z)(y)P$;
3. $(y)(P + Q) \sim_p (y)P + (y)Q$;

4. $(y)\alpha.P \sim_p \alpha.(y)P$ if $y \notin n(\alpha)$;
5. $(y)\alpha.P \sim_p$ **nil** if y is the subject of α.

Proof. **1.** It is sufficient to prove the relation $R = \{((y)P, P)|$ if $y \notin fn(P)\} \cup$ **Id** is a strong pomset bisimulation. It can be proved similarly to the proof of Static laws about restriction \ for strong pomset bisimulation in CTC, we omit it;

2. It is sufficient to prove the relation $R = \{((y)(z)P, (z)(y)P)\} \cup$ **Id** is a strong pomset bisimulation. It can be proved similarly to the proof of Static laws about restriction \ for strong pomset bisimulation in CTC, we omit it;

3. It is sufficient to prove the relation $R = \{((y)(P+Q), (y)P+(y)Q)\} \cup$ **Id** is a strong pomset bisimulation. It can be proved similarly to the proof of Static laws about restriction \ for strong pomset bisimulation in CTC, we omit it;

4. It is sufficient to prove the relation $R = \{((y)\alpha.P, \alpha.(y)P)|$ if $y \notin n(\alpha)\} \cup$ **Id** is a strong pomset bisimulation. It can be proved similarly to the proof of Static laws about restriction \ for strong pomset bisimulation in CTC, we omit it;

5. It is sufficient to prove the relation $R = \{((y)\alpha.P, \textbf{nil})|$ if y is the subject of $\alpha\} \cup$ **Id** is a strong pomset bisimulation. It can be proved similarly to the proof of Static laws about restriction \ for strong pomset bisimulation in CTC, we omit it. $\square$

Theorem 5.21 (Restriction Laws for strong step bisimilarity). *The restriction laws for strong step bisimilarity are as follows.*

1. $(y)P \sim_s P$, if $y \notin fn(P)$;
2. $(y)(z)P \sim_s (z)(y)P$;
3. $(y)(P+Q) \sim_s (y)P + (y)Q$;
4. $(y)\alpha.P \sim_s \alpha.(y)P$ if $y \notin n(\alpha)$;
5. $(y)\alpha.P \sim_s$ **nil** if y is the subject of α.

Proof. **1.** It is sufficient to prove the relation $R = \{((y)P, P)|$ if $y \notin fn(P)\} \cup$ **Id** is a strong step bisimulation. It can be proved similarly to the proof of Static laws about restriction \ for strong step bisimulation in CTC, we omit it;

2. It is sufficient to prove the relation $R = \{((y)(z)P, (z)(y)P)\} \cup$ **Id** is a strong step bisimulation. It can be proved similarly to the proof of Static laws about restriction \ for strong step bisimulation in CTC, we omit it;

3. It is sufficient to prove the relation $R = \{((y)(P+Q), (y)P+(y)Q)\} \cup$ **Id** is a strong step bisimulation. It can be proved similarly to the proof of Static laws about restriction \ for strong step bisimulation in CTC, we omit it;

4. It is sufficient to prove the relation $R = \{((y)\alpha.P, \alpha.(y)P)|$ if $y \notin n(\alpha)\} \cup$ **Id** is a strong step bisimulation. It can be proved similarly to the proof of Static laws about restriction \ for strong step bisimulation in CTC, we omit it;

5. It is sufficient to prove the relation $R = \{((y)\alpha.P, \textbf{nil})|$ if y is the subject of $\alpha\} \cup$ **Id** is a strong step bisimulation. It can be proved similarly to the proof of Static laws about restriction \ for strong step bisimulation in CTC, we omit it. $\square$

Theorem 5.22 (Restriction Laws for strong hp-bisimilarity). *The restriction laws for strong hp-bisimilarity are as follows.*

1. $(y)P \sim_{hp} P$, *if* $y \notin fn(P)$;
2. $(y)(z)P \sim_{hp} (z)(y)P$;
3. $(y)(P + Q) \sim_{hp} (y)P + (y)Q$;
4. $(y)\alpha.P \sim_{hp} \alpha.(y)P$ *if* $y \notin n(\alpha)$;
5. $(y)\alpha.P \sim_{hp}$ **nil** *if* y *is the subject of* α.

Proof. **1.** It is sufficient to prove the relation $R = \{((y)P, P)| \text{ if } y \notin fn(P)\} \cup$ **Id** is a strong hp-bisimulation. It can be proved similarly to the proof of Static laws about restriction \ for strong hp-bisimulation in CTC, we omit it;

2. It is sufficient to prove the relation $R = \{((y)(z)P, (z)(y)P)\} \cup$ **Id** is a strong hp-bisimulation. It can be proved similarly to the proof of Static laws about restriction \ for strong hp-bisimulation in CTC, we omit it;

3. It is sufficient to prove the relation $R = \{((y)(P + Q), (y)P + (y)Q)\} \cup$ **Id** is a strong hp-bisimulation. It can be proved similarly to the proof of Static laws about restriction \ for strong hp-bisimulation in CTC, we omit it;

4. It is sufficient to prove the relation $R = \{((y)\alpha.P, \alpha.(y)P)| \text{ if } y \notin n(\alpha)\} \cup$ **Id** is a strong hp-bisimulation. It can be proved similarly to the proof of Static laws about restriction \ for strong hp-bisimulation in CTC, we omit it;

5. It is sufficient to prove the relation $R = \{((y)\alpha.P, \textbf{nil})| \text{ if } y \text{ is the subject of } \alpha\} \cup$ **Id** is a strong hp-bisimulation. It can be proved similarly to the proof of Static laws about restriction \ for strong hp-bisimulation in CTC, we omit it. $\square$

Theorem 5.23 (Restriction Laws for strongly hhp-bisimilarity). *The restriction laws for strongly hhp-bisimilarity are as follows.*

1. $(y)P \sim_{hhp} P$, *if* $y \notin fn(P)$;
2. $(y)(z)P \sim_{hhp} (z)(y)P$;
3. $(y)(P + Q) \sim_{hhp} (y)P + (y)Q$;
4. $(y)\alpha.P \sim_{hhp} \alpha.(y)P$ *if* $y \notin n(\alpha)$;
5. $(y)\alpha.P \sim_{hhp}$ **nil** *if* y *is the subject of* α.

Proof. **1.** It is sufficient to prove the relation $R = \{((y)P, P)| \text{ if } y \notin fn(P)\} \cup$ **Id** is a strongly hhp-bisimulation. It can be proved similarly to the proof of Static laws about restriction \ for strongly hhp-bisimulation in CTC, we omit it;

2. It is sufficient to prove the relation $R = \{((y)(z)P, (z)(y)P)\} \cup$ **Id** is a strongly hhp-bisimulation. It can be proved similarly to the proof of Static laws about restriction \ for strongly hhp-bisimulation in CTC, we omit it;

3. It is sufficient to prove the relation $R = \{((y)(P + Q), (y)P + (y)Q)\} \cup$ **Id** is a strongly hhp-bisimulation. It can be proved similarly to the proof of Static laws about restriction \ for strongly hhp-bisimulation in CTC, we omit it;

4. It is sufficient to prove the relation $R = \{((y)\alpha.P, \alpha.(y)P)|$ if $y \notin n(\alpha)\} \cup \mathbf{Id}$ is a strongly hhp-bisimulation. It can be proved similarly to the proof of Static laws about restriction \ for strongly hhp-bisimulation in CTC, we omit it;

5. It is sufficient to prove the relation $R = \{((y)\alpha.P, \mathbf{nil})|$ if y is the subject of $\alpha\} \cup \mathbf{Id}$ is a strongly hhp-bisimulation. It can be proved similarly to the proof of Static laws about restriction \ for strongly hhp-bisimulation in CTC, we omit it. $\qquad\square$

Theorem 5.24 (Parallel laws for strong pomset bisimilarity). *The parallel laws for strong pomset bisimilarity are as follows.*

1. $P \parallel \mathbf{nil} \sim_p P;$
2. $P_1 \parallel P_2 \sim_p P_2 \parallel P_1;$
3. $(y)P_1 \parallel P_2 \sim_p (y)(P_1 \parallel P_2)$
4. $(P_1 \parallel P_2) \parallel P_3 \sim_p P_1 \parallel (P_2 \parallel P_3);$
5. $(y)(P_1 \parallel P_2) \sim_p (y)P_1 \parallel (y)P_2,$ *if* $y \notin fn(P_1) \cap fn(P_2).$

Proof. **1.** It is sufficient to prove the relation $R = \{(P \parallel \mathbf{nil}, P)\} \cup \mathbf{Id}$ is a strong pomset bisimulation. It can be proved similarly to the proof of Static laws about parallel $\parallel$ for strong pomset bisimulation in CTC, we omit it;

2. It is sufficient to prove the relation $R = \{(P_1 \parallel P_2, P_2 \parallel P_1)\} \cup \mathbf{Id}$ is a strong pomset bisimulation. It can be proved similarly to the proof of Static laws about parallel $\parallel$ for strong pomset bisimulation in CTC, we omit it;

3. It is sufficient to prove the relation $R = \{((y)P_1 \parallel P_2, (y)(P_1 \parallel P_2))\} \cup \mathbf{Id}$ is a strong pomset bisimulation. It can be proved similarly to the proof of Static laws about parallel $\parallel$ for strong pomset bisimulation in CTC, we omit it;

4. It is sufficient to prove the relation $R = \{((P_1 \parallel P_2) \parallel P_3, P_1 \parallel (P_2 \parallel P_3))\} \cup \mathbf{Id}$ is a strong pomset bisimulation. It can be proved similarly to the proof of Static laws about parallel $\parallel$ for strong pomset bisimulation in CTC, we omit it;

5. It is sufficient to prove the relation $R = \{(y)(P_1 \parallel P_2), (y)P_1 \parallel (y)P_2)|$ if $y \notin fn(P_1) \cap fn(P_2)\} \cup \mathbf{Id}$ is a strong pomset bisimulation. It can be proved similarly to the proof of Static laws about parallel $\parallel$ for strong pomset bisimulation in CTC, we omit it. $\qquad\square$

Theorem 5.25 (Parallel laws for strong step bisimilarity). *The parallel laws for strong step bisimilarity are as follows.*

1. $P \parallel \mathbf{nil} \sim_s P;$
2. $P_1 \parallel P_2 \sim_s P_2 \parallel P_1;$
3. $(y)P_1 \parallel P_2 \sim_s (y)(P_1 \parallel P_2)$
4. $(P_1 \parallel P_2) \parallel P_3 \sim_s P_1 \parallel (P_2 \parallel P_3);$
5. $(y)(P_1 \parallel P_2) \sim_s (y)P_1 \parallel (y)P_2,$ *if* $y \notin fn(P_1) \cap fn(P_2).$

Proof. **1.** It is sufficient to prove the relation $R = \{(P \parallel \mathbf{nil}, P)\} \cup \mathbf{Id}$ is a strong step bisimulation. It can be proved similarly to the proof of Static laws about parallel $\parallel$ for strong step bisimulation in CTC, we omit it;

2. It is sufficient to prove the relation $R = \{(P_1 \parallel P_2, P_2 \parallel P_1)\} \cup \mathbf{Id}$ is a strong step bisimulation. It can be proved similarly to the proof of Static laws about parallel $\parallel$ for strong step bisimulation in CTC, we omit it;

3. It is sufficient to prove the relation $R = \{((y)P_1 \parallel P_2, (y)(P_1 \parallel P_2))\} \cup \mathbf{Id}$ is a strong step bisimulation. It can be proved similarly to the proof of Static laws about parallel $\parallel$ for strong step bisimulation in CTC, we omit it;

4. It is sufficient to prove the relation $R = \{((P_1 \parallel P_2) \parallel P_3, P_1 \parallel (P_2 \parallel P_3))\} \cup \mathbf{Id}$ is a strong step bisimulation. It can be proved similarly to the proof of Static laws about parallel $\parallel$ for strong step bisimulation in CTC, we omit it;

5. It is sufficient to prove the relation $R = \{(y)(P_1 \parallel P_2), (y)P_1 \parallel (y)P_2)|$ if $y \notin fn(P_1) \cap fn(P_2)\} \cup \mathbf{Id}$ is a strong step bisimulation. It can be proved similarly to the proof of Static laws about parallel $\parallel$ for strong step bisimulation in CTC, we omit it. $\square$

Theorem 5.26 (Parallel laws for strong hp-bisimilarity). *The parallel laws for strong hp-bisimilarity are as follows.*

1. $P \parallel \mathbf{nil} \sim_{hp} P;$
2. $P_1 \parallel P_2 \sim_{hp} P_2 \parallel P_1;$
3. $(y)P_1 \parallel P_2 \sim_{hp} (y)(P_1 \parallel P_2)$
4. $(P_1 \parallel P_2) \parallel P_3 \sim_{hp} P_1 \parallel (P_2 \parallel P_3);$
5. $(y)(P_1 \parallel P_2) \sim_{hp} (y)P_1 \parallel (y)P_2, if y \notin fn(P_1) \cap fn(P_2).$

Proof. **1.** It is sufficient to prove the relation $R = \{(P \parallel \mathbf{nil}, P)\} \cup \mathbf{Id}$ is a strong hp-bisimulation. It can be proved similarly to the proof of Static laws about parallel $\parallel$ for strong hp-bisimulation in CTC, we omit it;

2. It is sufficient to prove the relation $R = \{(P_1 \parallel P_2, P_2 \parallel P_1)\} \cup \mathbf{Id}$ is a strong hp-bisimulation. It can be proved similarly to the proof of Static laws about parallel $\parallel$ for strong hp-bisimulation in CTC, we omit it;

3. It is sufficient to prove the relation $R = \{((y)P_1 \parallel P_2, (y)(P_1 \parallel P_2))\} \cup \mathbf{Id}$ is a strong hp-bisimulation. It can be proved similarly to the proof of Static laws about parallel $\parallel$ for strong hp-bisimulation in CTC, we omit it;

4. It is sufficient to prove the relation $R = \{((P_1 \parallel P_2) \parallel P_3, P_1 \parallel (P_2 \parallel P_3))\} \cup \mathbf{Id}$ is a strong hp-bisimulation. It can be proved similarly to the proof of Static laws about parallel $\parallel$ for strong hp-bisimulation in CTC, we omit it;

5. It is sufficient to prove the relation $R = \{(y)(P_1 \parallel P_2), (y)P_1 \parallel (y)P_2)|$ if $y \notin fn(P_1) \cap fn(P_2)\} \cup \mathbf{Id}$ is a strong hp-bisimulation. It can be proved similarly to the proof of Static laws about parallel $\parallel$ for strong hp-bisimulation in CTC, we omit it. $\square$

Theorem 5.27 (Parallel laws for strongly hhp-bisimilarity). *The parallel laws for strongly hhp-bisimilarity are as follows.*

1. $P \parallel \mathbf{nil} \sim_{hhp} P;$
2. $P_1 \parallel P_2 \sim_{hhp} P_2 \parallel P_1;$
3. $(y)P_1 \parallel P_2 \sim_{hhp} (y)(P_1 \parallel P_2)$

4. $(P_1 \parallel P_2) \parallel P_3 \sim_{hhp} P_1 \parallel (P_2 \parallel P_3)$;
5. $(y)(P_1 \parallel P_2) \sim_{hhp} (y)P_1 \parallel (y)P_2$, if $y \notin fn(P_1) \cap fn(P_2)$.

Proof. **1.** It is sufficient to prove the relation $R = \{(P \parallel \mathbf{nil}, P)\} \cup \mathbf{Id}$ is a strongly hhp-bisimulation. It can be proved similarly to the proof of Static laws about parallel $\parallel$ for strongly hhp-bisimulation in CTC, we omit it;

2. It is sufficient to prove the relation $R = \{(P_1 \parallel P_2, P_2 \parallel P_1)\} \cup \mathbf{Id}$ is a strongly hhp-bisimulation. It can be proved similarly to the proof of Static laws about parallel $\parallel$ for strongly hhp-bisimulation in CTC, we omit it;

3. It is sufficient to prove the relation $R = \{((y)P_1 \parallel P_2, (y)(P_1 \parallel P_2))\} \cup \mathbf{Id}$ is a strongly hhp-bisimulation. It can be proved similarly to the proof of Static laws about parallel $\parallel$ for strongly hhp-bisimulation in CTC, we omit it;

4. It is sufficient to prove the relation $R = \{((P_1 \parallel P_2) \parallel P_3, P_1 \parallel (P_2 \parallel P_3))\} \cup \mathbf{Id}$ is a strongly hhp-bisimulation. It can be proved similarly to the proof of Static laws about parallel $\parallel$ for strongly hhp-bisimulation in CTC, we omit it;

5. It is sufficient to prove the relation $R = \{(y)(P_1 \parallel P_2), (y)P_1 \parallel (y)P_2)\mid$ if $y \notin fn(P_1) \cap fn(P_2)\} \cup \mathbf{Id}$ is a strongly hhp-bisimulation. It can be proved similarly to the proof of Static laws about parallel $\parallel$ for strongly hhp-bisimulation in CTC, we omit it. $\square$

Theorem 5.28 (Expansion law for truly concurrent bisimilarities). *Let $P \equiv \sum_i \alpha_i.P_i$ and $Q \equiv \sum_j \beta_j.Q_j$, where $bn(\alpha_i) \cap fn(Q) = \emptyset$ for all i, and $bn(\beta_j) \cap fn(P) = \emptyset$ for all j. Then*

1. $P \parallel Q \sim_p \sum_i \sum_j (\alpha_i \parallel \beta_j).(P_i \parallel Q_j) + \sum_{\alpha_i \; comp \; \beta_j} \tau.R_{ij}$;
2. $P \parallel Q \sim_s \sum_i \sum_j (\alpha_i \parallel \beta_j).(P_i \parallel Q_j) + \sum_{\alpha_i \; comp \; \beta_j} \tau.R_{ij}$;
3. $P \parallel Q \sim_{hp} \sum_i \sum_j (\alpha_i \parallel \beta_j).(P_i \parallel Q_j) + \sum_{\alpha_i \; comp \; \beta_j} \tau.R_{ij}$;
4. $P \parallel Q \approx_{hhp} \sum_i \sum_j (\alpha_i \parallel \beta_j).(P_i \parallel Q_j) + \sum_{\alpha_i \; comp \; \beta_j} \tau.R_{ij}$.

Where $\alpha_i \; comp \; \beta_j$ and R_{ij} are defined as follows:

1. α_i *is* $\overline{x}u$ *and* β_j *is* $x(v)$, *then* $R_{ij} = P_i \parallel Q_j\{u/v\}$;
2. α_i *is* $\overline{x}(u)$ *and* β_j *is* $x(v)$, *then* $R_{ij} = (w)(P_i\{w/u\} \parallel Q_j\{w/v\})$, *if* $w \notin fn((u)P_i) \cup fn((v)Q_j)$;
3. α_i *is* $x(v)$ *and* β_j *is* $\overline{x}u$, *then* $R_{ij} = P_i\{u/v\} \parallel Q_j$;
4. α_i *is* $x(v)$ *and* β_j *is* $\overline{x}(u)$, *then* $R_{ij} = (w)(P_i\{w/v\} \parallel Q_j\{w/u\})$, *if* $w \notin fn((v)P_i) \cup fn((u)Q_j)$.

Proof. **1.** It is sufficient to prove the relation $R = \{(P \parallel Q, \sum_i \sum_j (\alpha_i \parallel \beta_j).(P_i \parallel Q_j) + \sum_{\alpha_i \; comp \; \beta_j} \tau.R_{ij})\mid$ if $y \notin fn(P)\} \cup \mathbf{Id}$ is a strong pomset bisimulation. It can be proved similarly to the proof of Expansion law for strong pomset bisimulation in CTC, we omit it;

2. It is sufficient to prove the relation $R = \{(P \parallel Q, \sum_i \sum_j (\alpha_i \parallel \beta_j).(P_i \parallel Q_j) + \sum_{\alpha_i \; comp \; \beta_j} \tau.R_{ij})\mid$ if $y \notin fn(P)\} \cup \mathbf{Id}$ is a strong step bisimulation. It can be proved similarly to the proof of Expansion law for strong step bisimulation in CTC, we omit it;

3. It is sufficient to prove the relation $R = \{(P \parallel Q, \sum_i \sum_j (\alpha_i \parallel \beta_j).(P_i \parallel Q_j) + \sum_{\alpha_i \; comp \; \beta_j} \tau.R_{ij})\mid$ if $y \notin fn(P)\} \cup \mathbf{Id}$ is a strong hp-bisimulation. It can be proved similarly to the proof of Expansion law for strong hp-bisimulation in CTC, we omit it;

4. We just prove that for free actions a, b, c, let $s_1 = (a + b) \parallel c$, $t_1 = (a \parallel c) + (b \parallel c)$, and $s_2 = a \parallel (b + c)$, $t_2 = (a \parallel b) + (a \parallel c)$. We know that $s_1 \sim_{hp} t_1$ and $s_2 \sim_{hp} t_2$, we prove that $s_1 \approx_{hhp} t_1$ and $s_2 \approx_{hhp} t_2$. Let $(C(s_1), f_1, C(t_1))$ and $(C(s_2), f_2, C(t_2))$ be the corresponding posetal products.

- $s_1 \approx_{hhp} t_1$. $s_1 \xrightarrow{\{a,c\}} \sqrt{}(s_1')$ $(C(s_1) \xrightarrow{\{a,c\}} C(s_1'))$, then $t_1 \xrightarrow{\{a,c\}} \sqrt{}(t_1')$ $(C(t_1) \xrightarrow{\{a,c\}} C(t_1'))$, we define $f_1' = f_1[a \mapsto a, c \mapsto c]$, obviously, $(C(s_1), f_1, C(t_1)) \in \sim_{hp}$ and $(C(s_1'), f_1', C(t_1')) \in \sim_{hp}$. But, $(C(s_1), f_1, C(t_1)) \in \sim_{hhp}$ and $(C(s_1'), f_1', C(t_1')) \in \approx_{hhp}$, just because they are not downward closed. Let $(C(s_1''), f_1'', C(t_1''))$, and $f_1'' = f_1[c \mapsto c]$, $s_1 \xrightarrow{c} s_1''$ $(C(s_1) \xrightarrow{c} C(s_1''))$, $t_1 \xrightarrow{c} t_1''$ $(C(t_1) \xrightarrow{c} C(t_1''))$, it is easy to see that $(C(s_1''), f_1'', C(t_1'')) \subseteq (C(s_1'), f_1', C(t_1'))$ pointwise, while $(C(s_1''), f_1'', C(t_1'')) \notin \sim_{hp}$, because s_1'' and $C(s_1'')$ exist, but t_1'' and $C(t_1'')$ do not exist.
- $s_2 \approx_{hhp} t_2$. $s_2 \xrightarrow{\{a,c\}} \sqrt{}(s_2')$ $(C(s_2) \xrightarrow{\{a,c\}} C(s_2'))$, then $t_2 \xrightarrow{\{a,c\}} \sqrt{}(t_2')$ $(C(t_2) \xrightarrow{\{a,c\}} C(t_2'))$, we define $f_2' = f_2[a \mapsto a, c \mapsto c]$, obviously, $(C(s_2), f_2, C(t_2)) \in \sim_{hp}$ and $(C(s_2'), f_2', C(t_2')) \in \sim_{hp}$. But, $(C(s_2), f_2, C(t_2)) \in \sim_{hhp}$ and $(C(s_2'), f_2', C(t_2')) \in \approx_{hhp}$, just because they are not downward closed. Let $(C(s_2''), f_2'', C(t_2''))$, and $f_2'' = f_2[a \mapsto a]$, $s_2 \xrightarrow{a} s_2''$ $(C(s_2) \xrightarrow{a} C(s_2''))$, $t_2 \xrightarrow{a} t_2''$ $(C(t_2) \xrightarrow{a} C(t_2''))$, it is easy to see that $(C(s_2''), f_2'', C(t_2'')) \subseteq (C(s_2'), f_2', C(t_2'))$ pointwise, while $(C(s_2''), f_2'', C(t_2'')) \notin \sim_{hp}$, because s_2'' and $C(s_2'')$ exist, but t_2'' and $C(t_2'')$ do not exist. □

Theorem 5.29 (Equivalence and congruence for strong pomset bisimilarity). **1.** $\sim_p$ *is an equivalence relation;*
2. *If* $P \sim_p Q$ *then*
  a. $\alpha.P \sim_p \alpha.Q$, α *is a free action;*
  b. $P + R \sim_p Q + R$;
  c. $P \parallel R \sim_p Q \parallel R$;
  d. $(w)P \sim_p (w)Q$;
  e. $x(y).P \sim_p x(y).Q$.

Proof. **1.** It is sufficient to prove that $\sim_p$ is reflexivity, symmetry, and transitivity, we omit it.
2. If $P \sim_p Q$, then
  a. it is sufficient to prove the relation $R = \{(\alpha.P, \alpha.Q) | \alpha$ is a free action$\} \cup \mathbf{Id}$ is a strong pomset bisimulation. It can be proved similarly to the proof of congruence for strong pomset bisimulation in CTC, we omit it;
  b. it is sufficient to prove the relation $R = \{(P + R, Q + R)\} \cup \mathbf{Id}$ is a strong pomset bisimulation. It can be proved similarly to the proof of congruence for strong pomset bisimulation in CTC, we omit it;
  c. it is sufficient to prove the relation $R = \{(P \parallel R, Q \parallel R)\} \cup \mathbf{Id}$ is a strong pomset bisimulation. It can be proved similarly to the proof of congruence for strong pomset bisimulation in CTC, we omit it;
  d. it is sufficient to prove the relation $R = \{((w)P, (w).Q)\} \cup \mathbf{Id}$ is a strong pomset bisimulation. It can be proved similarly to the proof of congruence for strong pomset bisimulation in CTC, we omit it;

e. it is sufficient to prove the relation $R = \{(x(y).P, x(y).Q)\} \cup \mathbf{Id}$ is a strong pomset bisimulation. It can be proved similarly to the proof of congruence for strong pomset bisimulation in CTC, we omit it. $\qquad\square$

Theorem 5.30 (Equivalence and congruence for strong step bisimilarity). **1.** $\sim_s$ *is an equivalence relation;*

2. *If* $P \sim_s Q$ *then*
   a. $\alpha.P \sim_s \alpha.Q$, α *is a free action;*
   b. $P + R \sim_s Q + R$;
   c. $P \parallel R \sim_s Q \parallel R$;
   d. $(w)P \sim_s (w)Q$;
   e. $x(y).P \sim_s x(y).Q$.

Proof. **1.** It is sufficient to prove that $\sim_s$ is reflexivity, symmetry, and transitivity, we omit it.

2. If $P \sim_s Q$, then
   a. it is sufficient to prove the relation $R = \{(\alpha.P, \alpha.Q)|\alpha$ is a free action$\} \cup \mathbf{Id}$ is a strong step bisimulation. It can be proved similarly to the proof of congruence for strong step bisimulation in CTC, we omit it;
   b. it is sufficient to prove the relation $R = \{(P + R, Q + R)\} \cup \mathbf{Id}$ is a strong step bisimulation. It can be proved similarly to the proof of congruence for strong step bisimulation in CTC, we omit it;
   c. it is sufficient to prove the relation $R = \{(P \parallel R, Q \parallel R)\} \cup \mathbf{Id}$ is a strong step bisimulation. It can be proved similarly to the proof of congruence for strong step bisimulation in CTC, we omit it;
   d. it is sufficient to prove the relation $R = \{((w)P, (w).Q)\} \cup \mathbf{Id}$ is a strong step bisimulation. It can be proved similarly to the proof of congruence for strong step bisimulation in CTC, we omit it;
   e. it is sufficient to prove the relation $R = \{(x(y).P, x(y).Q)\} \cup \mathbf{Id}$ is a strong step bisimulation. It can be proved similarly to the proof of congruence for strong step bisimulation in CTC, we omit it. $\qquad\square$

Theorem 5.31 (Equivalence and congruence for strong hp-bisimilarity). **1.** $\sim_{hp}$ *is an equivalence relation;*

2. *If* $P \sim_{hp} Q$ *then*
   a. $\alpha.P \sim_{hp} \alpha.Q$, α *is a free action;*
   b. $P + R \sim_{hp} Q + R$;
   c. $P \parallel R \sim_{hp} Q \parallel R$;
   d. $(w)P \sim_{hp} (w)Q$;
   e. $x(y).P \sim_{hp} x(y).Q$.

Proof. **1.** It is sufficient to prove that $\sim_{hp}$ is reflexivity, symmetry, and transitivity, we omit it.

2. If $P \sim_{hp} Q$, then

a. it is sufficient to prove the relation $R = \{(\alpha.P, \alpha.Q)|\alpha$ is a free action$\} \cup$ **Id** is a strong hp-bisimulation. It can be proved similarly to the proof of congruence for strong hp-bisimulation in CTC, we omit it;

b. it is sufficient to prove the relation $R = \{(P + R, Q + R)\} \cup$ **Id** is a strong hp-bisimulation. It can be proved similarly to the proof of congruence for strong hp-bisimulation in CTC, we omit it;

c. it is sufficient to prove the relation $R = \{(P \parallel R, Q \parallel R)\} \cup$ **Id** is a strong hp-bisimulation. It can be proved similarly to the proof of congruence for strong hp-bisimulation in CTC, we omit it;

d. it is sufficient to prove the relation $R = \{((w)P, (w).Q)\} \cup$ **Id** is a strong hp-bisimulation. It can be proved similarly to the proof of congruence for strong hp-bisimulation in CTC, we omit it;

e. it is sufficient to prove the relation $R = \{(x(y).P, x(y).Q)\} \cup$ **Id** is a strong hp-bisimulation. It can be proved similarly to the proof of congruence for strong hp-bisimulation in CTC, we omit it.   □

Theorem 5.32 (Equivalence and congruence for strongly hhp-bisimilarity). **1.** $\sim_{hhp}$ *is an equivalence relation;*

2. *If* $P \sim_{hhp} Q$ *then*

  a. $\alpha.P \sim_{hhp} \alpha.Q$, α *is a free action;*

  b. $P + R \sim_{hhp} Q + R;$

  c. $P \parallel R \sim_{hhp} Q \parallel R;$

  d. $(w)P \sim_{hhp} (w)Q;$

  e. $x(y).P \sim_{hhp} x(y).Q.$

Proof. **1.** It is sufficient to prove that $\sim_{hhp}$ is reflexivity, symmetry, and transitivity, we omit it.

2. If $P \sim_p Q$, then

a. it is sufficient to prove the relation $R = \{(\alpha.P, \alpha.Q)|\alpha$ is a free action$\} \cup$ **Id** is a strongly hhp-bisimulation. It can be proved similarly to the proof of congruence for strongly hhp-bisimulation in CTC, we omit it;

b. it is sufficient to prove the relation $R = \{(P + R, Q + R)\} \cup$ **Id** is a strongly hhp-bisimulation. It can be proved similarly to the proof of congruence for strongly hhp-bisimulation in CTC, we omit it;

c. it is sufficient to prove the relation $R = \{(P \parallel R, Q \parallel R)\} \cup$ **Id** is a strongly hhp-bisimulation. It can be proved similarly to the proof of congruence for strongly hhp-bisimulation in CTC, we omit it;

d. it is sufficient to prove the relation $R = \{((w)P, (w).Q)\} \cup$ **Id** is a strongly hhp-bisimulation. It can be proved similarly to the proof of congruence for strongly hhp-bisimulation in CTC, we omit it;

e. it is sufficient to prove the relation $R = \{(x(y).P, x(y).Q)\} \cup$ **Id** is a strongly hhp-bisimulation. It can be proved similarly to the proof of congruence for strongly hhp-bisimulation in CTC, we omit it.   □

5.2.3 Recursion

Definition 5.33. Let X have arity n, and let $\tilde{x} = x_1, \cdots, x_n$ be distinct names, and $fn(P) \subseteq \{x_1, \cdots, x_n\}$. The replacement of $X(\tilde{x})$ by P in E, written $E\{X(\tilde{x}) := P\}$, means the result of replacing each subterm $X(\tilde{y})$ in E by $P\{\tilde{y}/\tilde{x}\}$.

Definition 5.34. Let E and F be two process expressions containing only $X_1, \cdots, X_m$ with associated name sequences $\tilde{x}_1, \cdots, \tilde{x}_m$. Then,

1. $E \sim_p F$ means $E(\tilde{P}) \sim_p F(\tilde{P})$;
2. $E \sim_s F$ means $E(\tilde{P}) \sim_s F(\tilde{P})$;
3. $E \sim_{hp} F$ means $E(\tilde{P}) \sim_{hp} F(\tilde{P})$;
4. $E \sim_{hhp} F$ means $E(\tilde{P}) \sim_{hhp} F(\tilde{P})$;

for all $\tilde{P}$ such that $fn(P_i) \subseteq \tilde{x}_i$ for each i.

Definition 5.35. A term or identifier is weakly guarded in P if it lies within some subterm $\alpha.Q$ or $(\alpha_1 \parallel \cdots \parallel \alpha_n).Q$ of P.

Theorem 5.36. Assume that $\tilde{E}$ and $\tilde{F}$ are expressions containing only X_i with $\tilde{x}_i$, and $\tilde{A}$ and $\tilde{B}$ are identifiers with A_i, B_i. Then, for all i,

1. $E_i \sim_s F_i$, $A_i(\tilde{x}_i) \stackrel{def}{=} E_i(\tilde{A})$, $B_i(\tilde{x}_i) \stackrel{def}{=} F_i(\tilde{B})$, then $A_i(\tilde{x}_i) \sim_s B_i(\tilde{x}_i)$;
2. $E_i \sim_p F_i$, $A_i(\tilde{x}_i) \stackrel{def}{=} E_i(\tilde{A})$, $B_i(\tilde{x}_i) \stackrel{def}{=} F_i(\tilde{B})$, then $A_i(\tilde{x}_i) \sim_p B_i(\tilde{x}_i)$;
3. $E_i \sim_{hp} F_i$, $A_i(\tilde{x}_i) \stackrel{def}{=} E_i(\tilde{A})$, $B_i(\tilde{x}_i) \stackrel{def}{=} F_i(\tilde{B})$, then $A_i(\tilde{x}_i) \sim_{hp} B_i(\tilde{x}_i)$;
4. $E_i \sim_{hhp} F_i$, $A_i(\tilde{x}_i) \stackrel{def}{=} E_i(\tilde{A})$, $B_i(\tilde{x}_i) \stackrel{def}{=} F_i(\tilde{B})$, then $A_i(\tilde{x}_i) \sim_{hhp} B_i(\tilde{x}_i)$.

Proof. **1.** $E_i \sim_s F_i$, $A_i(\tilde{x}_i) \stackrel{def}{=} E_i(\tilde{A})$, $B_i(\tilde{x}_i) \stackrel{def}{=} F_i(\tilde{B})$, then $A_i(\tilde{x}_i) \sim_s B_i(\tilde{x}_i)$.

We will consider the case $I = \{1\}$ with loss of generality, and show the following relation R is a strong step bisimulation.

$$R = \{(G(A), G(B)) : G \text{ has only identifier } X\}.$$

By choosing $G \equiv X(\tilde{y})$, it follows that $A(\tilde{y}) \sim_s B(\tilde{y})$. It is sufficient to prove the following:

a. If $G(A) \xrightarrow{\{\alpha_1, \cdots, \alpha_n\}} P'$, where $\alpha_i (1 \leq i \leq n)$ is a free action or bound output action with $bn(\alpha_1) \cap \cdots \cap bn(\alpha_n) \cap n(G(A), G(B)) = \emptyset$, then $G(B) \xrightarrow{\{\alpha_1, \cdots, \alpha_n\}} Q''$ such that $P' \sim_s Q''$;

b. If $G(A) \xrightarrow{x(y)} P'$ with $x \notin n(G(A), G(B))$, then $G(B) \xrightarrow{x(y)} Q''$, such that for all u, $P'\{u/y\} \sim_s Q''\{u/y\}$.

To prove the above properties, it is sufficient to induct on the depth of inference and quite routine, we omit it.

2. $E_i \sim_p F_i$, $A_i(\tilde{x}_i) \stackrel{def}{=} E_i(\tilde{A})$, $B_i(\tilde{x}_i) \stackrel{def}{=} F_i(\tilde{B})$, then $A_i(\tilde{x}_i) \sim_p B_i(\tilde{x}_i)$. It can be proven similarly to the above case.

3. $E_i \sim_{hp} F_i$, $A_i(\tilde{x}_i) \stackrel{def}{=} E_i(\tilde{A})$, $B_i(\tilde{x}_i) \stackrel{def}{=} F_i(\tilde{B})$, then $A_i(\tilde{x}_i) \sim_{hp} B_i(\tilde{x}_i)$. It can be proven similarly to the above case.

4. $E_i \sim_{hhp} F_i$, $A_i(\tilde{x}_i) \stackrel{\text{def}}{=} E_i(\tilde{A})$, $B_i(\tilde{x}_i) \stackrel{\text{def}}{=} F_i(\tilde{B})$, then $A_i(\tilde{x}_i) \sim_{hhp} B_i(\tilde{x}_i)$. It can be proven similarly to the above case. $\square$

Theorem 5.37 (Unique solution of equations). *Assume $\tilde{E}$ are expressions containing only X_i with $\tilde{x}_i$, and each X_i is weakly guarded in each E_j. Assume that $\tilde{P}$ and $\tilde{Q}$ are processes such that $fn(P_i) \subseteq \tilde{x}_i$ and $fn(Q_i) \subseteq \tilde{x}_i$. Then, for all i,*

1. *if $P_i \sim_p E_i(\tilde{P})$, $Q_i \sim_p E_i(\tilde{Q})$, then $P_i \sim_p Q_i$;*
2. *if $P_i \sim_s E_i(\tilde{P})$, $Q_i \sim_s E_i(\tilde{Q})$, then $P_i \sim_s Q_i$;*
3. *if $P_i \sim_{hp} E_i(\tilde{P})$, $Q_i \sim_{hp} E_i(\tilde{Q})$, then $P_i \sim_{hp} Q_i$;*
4. *if $P_i \sim_{hhp} E_i(\tilde{P})$, $Q_i \sim_{hhp} E_i(\tilde{Q})$, then $P_i \sim_{hhp} Q_i$.*

Proof. **1.** It is similar to the proof of unique solution of equations for strong pomset bisimulation in CTC, we omit it;

2. It is similar to the proof of unique solution of equations for strong step bisimulation in CTC, we omit it;
3. It is similar to the proof of unique solution of equations for strong hp-bisimulation in CTC, we omit it;
4. It is similar to the proof of unique solution of equations for strong hhp-bisimulation in CTC, we omit it. $\square$

5.3 Algebraic theory

In this section, we will try to axiomatize π_{tc}, the theory is **STC** (for strongly truly concurrency).

Definition 5.38 (STC). The theory **STC** is consisted of the following axioms and inference rules:

1. Alpha-conversion **A**.

$$\text{if } P \equiv Q, \text{ then } P = Q$$

2. Congruence **C**. If $P = Q$, then,

$$\tau.P = \tau.Q \quad \overline{x}y.P = \overline{x}y.Q$$

$$P + R = Q + R \quad P \parallel R = Q \parallel R$$

$$(x)P = (x)Q \quad x(y).P = x(y).Q$$

3. Summation **S**.

$$\textbf{S0} \quad P + \textbf{nil} = P$$

$$\textbf{S1} \quad P + P = P$$

$$\textbf{S2} \quad P + Q = Q + P$$

$$\textbf{S3} \quad P + (Q + R) = (P + Q) + R$$

4. Restriction **R.**

$$\textbf{R0} \quad (x)P = P \quad \text{if } x \notin fn(P)$$

$$\textbf{R1} \quad (x)(y)P = (y)(x)P$$

$$\textbf{R2} \quad (x)(P + Q) = (x)P + (x)Q$$

$$\textbf{R3} \quad (x)\alpha.P = \alpha.(x)P \quad \text{if } x \notin n(\alpha)$$

$$\textbf{R4} \quad (x)\alpha.P = \textbf{nil} \quad \text{if } x \text{ is the subject of } \alpha$$

5. Expansion **E.** Let $P \equiv \sum_i \alpha_i.P_i$ and $Q \equiv \sum_j \beta_j.Q_j$, where $bn(\alpha_i) \cap fn(Q) = \emptyset$ for all i, and $bn(\beta_j) \cap fn(P) = \emptyset$ for all j. Then $P \parallel Q = \sum_i \sum_j (\alpha_i \parallel \beta_j).(P_i \parallel Q_j) + \sum_{\alpha_i \text{ comp } \beta_j} \tau.R_{ij}$. Where α_i comp β_j and R_{ij} are defined as follows:

   a. α_i is $\bar{x}u$ and β_j is $x(v)$, then $R_{ij} = P_i \parallel Q_j\{u/v\}$;

   b. α_i is $\bar{x}(u)$ and β_j is $x(v)$, then $R_{ij} = (w)(P_i\{w/u\} \parallel Q_j\{w/v\})$, if $w \notin fn((u)P_i) \cup fn((v)Q_j)$;

   c. α_i is $x(v)$ and β_j is $\bar{x}u$, then $R_{ij} = P_i\{u/v\} \parallel Q_j$;

   d. α_i is $x(v)$ and β_j is $\bar{x}(u)$, then $R_{ij} = (w)(P_i\{w/v\} \parallel Q_j\{w/u\})$, if $w \notin fn((v)P_i) \cup fn((u)Q_j)$.

6. Identifier **I.**

$$\text{If } A(\tilde{x}) \overset{\text{def}}{=} P, \text{ then } A(\tilde{y}) = P\{\tilde{y}/\tilde{x}\}.$$

Theorem 5.39 (Soundness). *If* $STC \vdash P = Q$ *then*

1. $P \sim_p Q$;

2. $P \sim_s Q$;

3. $P \sim_{hp} Q$.

Proof. The soundness of these laws modulo strongly truly concurrent bisimilarities is already proven in Section 5.2. $\qquad\qquad\qquad\qquad\qquad\qquad\qquad\qquad\qquad\qquad \square$

Definition 5.40. The agent identifier A is weakly guardedly defined if every agent identifier is weakly guarded in the right-hand side of the definition of A.

Definition 5.41 (Head normal form). A Process P is in head normal form if it is a sum of the prefixes:

$$P \equiv \sum_i (\alpha_{i1} \parallel \cdots \parallel \alpha_{in}).P_i.$$

Proposition 5.42. *If every agent identifier is weakly guardedly defined, then for any process P, there is a head normal form H such that*

$$STC \vdash P = H.$$

Proof. It is sufficient to induct on the structure of P and quite obvious. $\qquad\square$

Theorem 5.43 (Completeness). *For all processes P and Q,*

1. *if $P \sim_p Q$, then $STC \vdash P = Q$;*
2. *if $P \sim_s Q$, then $STC \vdash P = Q$;*
3. *if $P \sim_{hp} Q$, then $STC \vdash P = Q$.*

Proof. **1.** if $P \sim_s Q$, then $STC \vdash P = Q$. Since P and Q all have head normal forms, let $P \equiv \sum_{i=1}^{k} \alpha_i.P_i$ and $Q \equiv \sum_{i=1}^{k} \beta_i.Q_i$. Then the depth of P, denoted as $d(P) = 0$, if $k = 0$; $d(P) = 1 + max\{d(P_i)\}$ for $1 \le i \le k$. The depth $d(Q)$ can be defined similarly. It is sufficient to induct on $d = d(P) + d(Q)$. When $d = 0$, $P \equiv \textbf{nil}$ and $Q \equiv \textbf{nil}$, $P = Q$, as desired.
Suppose $d > 0$.

- If $(\alpha_1 \| \cdots \| \alpha_n).M$ with $\alpha_i (1 \le i \le n)$ free actions is a summand of P, then $P \xrightarrow{\{\alpha_1, \cdots, \alpha_n\}} M$. Since Q is in head normal form and has a summand $(\alpha_1 \| \cdots \| \alpha_n).N$ such that $M \sim_s N$, by the induction hypothesis $STC \vdash M = N$, $STC \vdash (\alpha_1 \| \cdots \| \alpha_n).M = (\alpha_1 \| \cdots \| \alpha_n).N$;

- If $x(y).M$ is a summand of P, then for $z \notin n(P, Q)$, $P \xrightarrow{x(z)} M' \equiv M\{z/y\}$. Since Q is in head normal form and has a summand $x(w).N$ such that for all v, $M'\{v/z\} \sim_s N'\{v/z\}$ where $N' \equiv N\{z/w\}$, by the induction hypothesis $STC \vdash M'\{v/z\} = N'\{v/z\}$, by the axioms **C** and **A**, $STC \vdash x(y).M = x(w).N$;

- If $\overline{x}(y).M$ is a summand of P, then for $z \notin n(P, Q)$, $P \xrightarrow{\overline{x}(z)} M' \equiv M\{z/y\}$. Since Q is in head normal form and has a summand $\overline{x}(w).N$ such that $M' \sim_s N'$ where $N' \equiv N\{z/w\}$, by the induction hypothesis $STC \vdash M' = N'$, by the axioms **A** and **C**, $STC \vdash \overline{x}(y).M = \overline{x}(w).N$;

2. if $P \sim_p Q$, then $STC \vdash P = Q$. It can be proven similarly to the above case.
3. if $P \sim_{hp} Q$, then $STC \vdash P = Q$. It can be proven similarly to the above case. $\qquad\square$

5.4 Conclusions

This work is a mixture of mobile processes and true concurrency called π_{tc}, which makes truly concurrent process algebra have the ability to model and verify mobile systems in a flavor of true concurrency, and can be used as a formal tool.

Note that, for true concurrency, an empty action playing the role of placeholder is needed, just like APTC with the shadow constant in subsection 4.6.1. We do not do this work and leave it as an exercise to the readers.

6

Guards

This chapter is organized as follows. We introduce the operational semantics of guards in section 6.1, $BATC$ with Guards in section 6.2, $APTC$ with Guards 6.3, recursion in section 6.4, abstraction in section 6.5, Hoare Logic for $APTC_G$ in section 6.6. Finally, in section 6.7, we conclude this chapter.

6.1 Operational semantics

In this section, we extend truly concurrent bisimilarities to the ones containing data states.

Definition 6.1 (Prime event structure with silent event and empty event). Let Λ be a fixed set of labels, ranged over $a, b, c, \cdots$ and τ, ϵ. A (Λ-labeled) prime event structure with silent event τ and empty event ϵ is a tuple $\mathcal{E} = \langle \mathbb{E}, \leq, \sharp, \lambda \rangle$, where $\mathbb{E}$ is a denumerable set of events, including the silent event τ and empty event ϵ. Let $\hat{\mathbb{E}} = \mathbb{E} \backslash \{\tau, \epsilon\}$, exactly excluding τ and ϵ, it is obvious that $\hat{\tau^*} = \epsilon$. Let $\lambda : \mathbb{E} \to \Lambda$ be a labeling function and let $\lambda(\tau) = \tau$ and $\lambda(\epsilon) = \epsilon$. And $\leq, \sharp$ are binary relations on $\mathbb{E}$, called causality and conflict respectively, such that:

1. $\leq$ is a partial order and $\lceil e \rceil = \{e' \in \mathbb{E} | e' \leq e\}$ is finite for all $e \in \mathbb{E}$. It is easy to see that $e \leq \tau^* \leq e' = e \leq \tau \leq \cdots \leq \tau \leq e'$, then $e \leq e'$.
2. $\sharp$ is irreflexive, symmetric and hereditary with respect to $\leq$, that is, for all $e, e', e'' \in \mathbb{E}$, if $e \sharp e' \leq e''$, then $e \sharp e''$.

Then, the concepts of consistency and concurrency can be drawn from the above definition:

1. $e, e' \in \mathbb{E}$ are consistent, denoted as $e \frown e'$, if $\neg(e \sharp e')$. A subset $X \subseteq \mathbb{E}$ is called consistent, if $e \frown e'$ for all $e, e' \in X$.
2. $e, e' \in \mathbb{E}$ are concurrent, denoted as $e \parallel e'$, if $\neg(e \leq e')$, $\neg(e' \leq e)$, and $\neg(e \sharp e')$.

Definition 6.2 (Configuration). Let $\mathcal{E}$ be a PES. A (finite) configuration in $\mathcal{E}$ is a (finite) consistent subset of events $C \subseteq \mathcal{E}$, closed with respect to causality (i.e. $\lceil C \rceil = C$), and a data state $s \in S$ with S the set of all data states, denoted $\langle C, s \rangle$. The set of finite configurations of $\mathcal{E}$ is denoted by $\langle \mathcal{C}(\mathcal{E}), S \rangle$. We let $\hat{C} = C \backslash \{\tau\} \cup \{\epsilon\}$.

A consistent subset of $X \subseteq \mathbb{E}$ of events can be seen as a pomset. Given $X, Y \subseteq \mathbb{E}$, $\hat{X} \sim \hat{Y}$ if $\hat{X}$ and $\hat{Y}$ are isomorphic as pomsets. In the following of the paper, we say $C_1 \sim C_2$, we mean $\hat{C}_1 \sim \hat{C}_2$.

Definition 6.3 (Pomset transitions and step). Let $\mathcal{E}$ be a PES and let $C \in \mathcal{C}(\mathcal{E})$, and $\emptyset \neq X \subseteq \mathbb{E}$, if $C \cap X = \emptyset$ and $C' = C \cup X \in \mathcal{C}(\mathcal{E})$, then $\langle C, s \rangle \xrightarrow{X} \langle C', s' \rangle$ is called a pomset transition from

$\langle C, s \rangle$ to $\langle C', s' \rangle$. When the events in X are pairwise concurrent, we say that $\langle C, s \rangle \xrightarrow{X} \langle C', s' \rangle$ is a step. It is obvious that $\rightarrow^* \xrightarrow{X} \rightarrow^* = \xrightarrow{X}$ and $\rightarrow^* \xrightarrow{e} \rightarrow^* = \xrightarrow{e}$ for any $e \in \mathbb{E}$ and $X \subseteq \mathbb{E}$.

Definition 6.4 (Weak pomset transitions and weak step). Let $\mathcal{E}$ be a PES and let $C \in \mathcal{C}(\mathcal{E})$, and $\emptyset \neq X \subseteq \hat{\mathbb{E}}$, if $C \cap X = \emptyset$ and $\hat{C}' = \hat{C} \cup X \in \mathcal{C}(\mathcal{E})$, then $\langle C, s \rangle \xRightarrow{X} \langle C', s' \rangle$ is called a weak pomset transition from $\langle C, s \rangle$ to $\langle C', s' \rangle$, where we define $\xRightarrow{e} \triangleq \xrightarrow{\tau^*} \xrightarrow{e} \xrightarrow{\tau^*}$. And $\xRightarrow{X} \triangleq \xrightarrow{\tau^*} \xrightarrow{e} \xrightarrow{\tau^*}$, for every $e \in X$. When the events in X are pairwise concurrent, we say that $\langle C, s \rangle \xRightarrow{X} \langle C', s' \rangle$ is a weak step.

We will also suppose that all the PESs in this paper are image finite, that is, for any PES $\mathcal{E}$ and $C \in \mathcal{C}(\mathcal{E})$ and $a \in \Lambda$, $\{e \in \mathbb{E} | \langle C, s \rangle \xrightarrow{e} \langle C', s' \rangle \wedge \lambda(e) = a\}$ and $\{e \in \hat{\mathbb{E}} | \langle C, s \rangle \xRightarrow{e} \langle C', s' \rangle \wedge \lambda(e) = a\}$ is finite.

Definition 6.5 (Pomset, step bisimulation). Let $\mathcal{E}_1$, $\mathcal{E}_2$ be PESs. A pomset bisimulation is a relation $R \subseteq \langle \mathcal{C}(\mathcal{E}_1), S \rangle \times \langle \mathcal{C}(\mathcal{E}_2), S \rangle$, such that if $(\langle C_1, s \rangle, \langle C_2, s \rangle) \in R$, and $\langle C_1, s \rangle \xrightarrow{X_1} \langle C_1', s' \rangle$ then $\langle C_2, s \rangle \xrightarrow{X_2} \langle C_2', s' \rangle$, with $X_1 \subseteq \mathbb{E}_1$, $X_2 \subseteq \mathbb{E}_2$, $X_1 \sim X_2$ and $(\langle C_1', s' \rangle, \langle C_2', s' \rangle) \in R$ for all $s, s' \in S$, and vice-versa. We say that $\mathcal{E}_1$, $\mathcal{E}_2$ are pomset bisimilar, written $\mathcal{E}_1 \sim_p \mathcal{E}_2$, if there exists a pomset bisimulation R, such that $(\langle \emptyset, \emptyset \rangle, \langle \emptyset, \emptyset \rangle) \in R$. By replacing pomset transitions with steps, we can get the definition of step bisimulation. When PESs $\mathcal{E}_1$ and $\mathcal{E}_2$ are step bisimilar, we write $\mathcal{E}_1 \sim_s \mathcal{E}_2$.

Definition 6.6 (Weak pomset, step bisimulation). Let $\mathcal{E}_1$, $\mathcal{E}_2$ be PESs. A weak pomset bisimulation is a relation $R \subseteq \langle \mathcal{C}(\mathcal{E}_1), S \rangle \times \langle \mathcal{C}(\mathcal{E}_2), S \rangle$, such that if $(\langle C_1, s \rangle, \langle C_2, s \rangle) \in R$, and $\langle C_1, s \rangle \xRightarrow{X_1} \langle C_1', s' \rangle$ then $\langle C_2, s \rangle \xRightarrow{X_2} \langle C_2', s' \rangle$, with $X_1 \subseteq \hat{\mathbb{E}}_1$, $X_2 \subseteq \hat{\mathbb{E}}_2$, $X_1 \sim X_2$ and $(\langle C_1', s' \rangle, \langle C_2', s' \rangle) \in R$ for all $s, s' \in S$, and vice-versa. We say that $\mathcal{E}_1$, $\mathcal{E}_2$ are weak pomset bisimilar, written $\mathcal{E}_1 \approx_p \mathcal{E}_2$, if there exists a weak pomset bisimulation R, such that $(\langle \emptyset, \emptyset \rangle, \langle \emptyset, \emptyset \rangle) \in R$. By replacing weak pomset transitions with weak steps, we can get the definition of weak step bisimulation. When PESs $\mathcal{E}_1$ and $\mathcal{E}_2$ are weak step bisimilar, we write $\mathcal{E}_1 \approx_s \mathcal{E}_2$.

Definition 6.7 (Posetal product). Given two PESs $\mathcal{E}_1$, $\mathcal{E}_2$, the posetal product of their configurations, denoted $\langle \mathcal{C}(\mathcal{E}_1), S \rangle \overline{\times} \langle \mathcal{C}(\mathcal{E}_2), S \rangle$, is defined as

$$\{(\langle C_1, s \rangle, f, \langle C_2, s \rangle) | C_1 \in \mathcal{C}(\mathcal{E}_1), C_2 \in \mathcal{C}(\mathcal{E}_2), f : C_1 \rightarrow C_2 \text{ isomorphism}\}.$$

A subset $R \subseteq \langle \mathcal{C}(\mathcal{E}_1), S \rangle \overline{\times} \langle \mathcal{C}(\mathcal{E}_2), S \rangle$ is called a posetal relation. We say that R is downward closed when for any $(\langle C_1, s \rangle, f, \langle C_2, s \rangle), (\langle C_1', s' \rangle, f', \langle C_2', s' \rangle) \in \langle \mathcal{C}(\mathcal{E}_1), S \rangle \overline{\times} \langle \mathcal{C}(\mathcal{E}_2), S \rangle$, if $(\langle C_1, s \rangle, f, \langle C_2, s \rangle) \subseteq (\langle C_1', s' \rangle, f', \langle C_2', s' \rangle)$ pointwise and $(\langle C_1', s' \rangle, f', \langle C_2', s' \rangle) \in R$, then $(\langle C_1, s \rangle, f, \langle C_2, s \rangle) \in R$.

For $f : X_1 \rightarrow X_2$, we define $f[x_1 \mapsto x_2] : X_1 \cup \{x_1\} \rightarrow X_2 \cup \{x_2\}$, $z \in X_1 \cup \{x_1\}$, (1) $f[x_1 \mapsto x_2](z) = x_2$, if $z = x_1$; (2) $f[x_1 \mapsto x_2](z) = f(z)$, otherwise. Where $X_1 \subseteq \mathbb{E}_1$, $X_2 \subseteq \mathbb{E}_2$, $x_1 \in \mathbb{E}_1$, $x_2 \in \mathbb{E}_2$.

Definition 6.8 (Weakly posetal product). Given two PESs $\mathcal{E}_1, \mathcal{E}_2$, the weakly posetal product of their configurations, denoted $\langle \mathcal{C}(\mathcal{E}_1), S \rangle \overline{\times} \langle \mathcal{C}(\mathcal{E}_2), S \rangle$, is defined as

$$\{((\langle C_1, s \rangle, f, \langle C_2, s \rangle)|C_1 \in \mathcal{C}(\mathcal{E}_1), C_2 \in \mathcal{C}(\mathcal{E}_2), f : \hat{C}_1 \to \hat{C}_2 \text{ isomorphism}\}.$$

A subset $R \subseteq \langle \mathcal{C}(\mathcal{E}_1), S \rangle \overline{\times} \langle \mathcal{C}(\mathcal{E}_2), S \rangle$ is called a weakly posetal relation. We say that R is downward closed when for any $((\langle C_1, s \rangle, f, \langle C_2, s \rangle), (\langle C_1', s' \rangle, f, \langle C_2', s' \rangle) \in \langle \mathcal{C}(\mathcal{E}_1), S \rangle \overline{\times} \langle \mathcal{C}(\mathcal{E}_2), S \rangle$, if $(\langle C_1, s \rangle, f, \langle C_2, s \rangle) \subseteq (\langle C_1', s' \rangle, f', \langle C_2', s' \rangle)$ pointwise and $(\langle C_1', s' \rangle, f', \langle C_2', s' \rangle) \in R$, then $(\langle C_1, s \rangle, f, \langle C_2, s \rangle) \in R$.

For $f : X_1 \to X_2$, we define $f[x_1 \mapsto x_2] : X_1 \cup \{x_1\} \to X_2 \cup \{x_2\}$, $z \in X_1 \cup \{x_1\}$, (1) $f[x_1 \mapsto x_2](z) = x_2$, if $z = x_1$; (2) $f[x_1 \mapsto x_2](z) = f(z)$, otherwise. Where $X_1 \subseteq \hat{\mathbb{E}}_1$, $X_2 \subseteq \hat{\mathbb{E}}_2$, $x_1 \in \hat{\mathbb{E}}_1$, $x_2 \in \hat{\mathbb{E}}_2$. Also, we define $f(\tau^*) = f(\tau^*)$.

Definition 6.9 ((Hereditary) history-preserving bisimulation). A history-preserving (hp-) bisimulation is a posetal relation $R \subseteq \langle \mathcal{C}(\mathcal{E}_1), S \rangle \overline{\times} \langle \mathcal{C}(\mathcal{E}_2), S \rangle$ such that if $(\langle C_1, s \rangle, f, \langle C_2, s \rangle) \in R$, and $\langle C_1, s \rangle \xrightarrow{e_1} \langle C_1', s' \rangle$, then $\langle C_2, s \rangle \xrightarrow{e_2} \langle C_2', s' \rangle$, with $(\langle C_1', s' \rangle, f[e_1 \mapsto e_2], \langle C_2', s' \rangle) \in R$ for all $s, s' \in S$, and vice-versa. $\mathcal{E}_1, \mathcal{E}_2$ are history-preserving (hp-)bisimilar and are written $\mathcal{E}_1 \sim_{hp} \mathcal{E}_2$ if there exists a hp-bisimulation R such that $(\langle \emptyset, \emptyset \rangle, \emptyset, \langle \emptyset, \emptyset \rangle) \in R$.

A hereditary history-preserving (hhp-)bisimulation is a downward closed hp-bisimulation. $\mathcal{E}_1, \mathcal{E}_2$ are hereditary history-preserving (hhp-)bisimilar and are written $\mathcal{E}_1 \sim_{hhp} \mathcal{E}_2$.

Definition 6.10 (Weak (hereditary) history-preserving bisimulation). A weak history-preserving (hp-) bisimulation is a weakly posetal relation $R \subseteq \langle \mathcal{C}(\mathcal{E}_1), S \rangle \overline{\times} \langle \mathcal{C}(\mathcal{E}_2), S \rangle$ such that if $(\langle C_1, s \rangle, f, \langle C_2, s \rangle) \in R$, and $\langle C_1, s \rangle \xRightarrow{e_1} \langle C_1', s' \rangle$, then $\langle C_2, s \rangle \xRightarrow{e_2} \langle C_2', s' \rangle$, with $(\langle C_1', s' \rangle, f[e_1 \mapsto e_2], \langle C_2', s' \rangle) \in R$ for all $s, s' \in S$, and vice-versa. $\mathcal{E}_1, \mathcal{E}_2$ are weak history-preserving (hp-)bisimilar and are written $\mathcal{E}_1 \approx_{hp} \mathcal{E}_2$ if there exists a weak hp-bisimulation R such that $(\langle \emptyset, \emptyset \rangle, \emptyset, \langle \emptyset, \emptyset \rangle) \in R$.

A weakly hereditary history-preserving (hhp-)bisimulation is a downward closed weak hp-bisimulation. $\mathcal{E}_1, \mathcal{E}_2$ are weakly hereditary history-preserving (hhp-)bisimilar and are written $\mathcal{E}_1 \approx_{hhp} \mathcal{E}_2$.

6.2 *BATC* with guards

In this subsection, we will discuss the guards for $BATC$, which is denoted as $BATC_G$. Let $\mathbb{E}$ be the set of atomic events (actions), G_{at} be the set of atomic guards, δ be the deadlock constant, and ϵ be the empty event. We extend G_{at} to the set of basic guards G with element $\phi, \psi, \cdots$, which is generated by the following formation rules:

$$\phi ::= \delta|\epsilon|\neg\phi|\psi \in G_{at}|\phi + \psi|\phi \cdot \psi$$

In the following, let $e_1, e_2, e_1', e_2' \in \mathbb{E}$, $\phi, \psi \in G$ and let variables x, y, z range over the set of terms for true concurrency, p, q, s range over the set of closed terms. The predicate $test(\phi, s)$ represents that ϕ holds in the state s, and $test(\epsilon, s)$ holds and $test(\delta, s)$ does not

Table 6.1   Axioms of $BATC_G$.

No.	Axiom
$A1$	$x + y = y + x$
$A2$	$(x + y) + z = x + (y + z)$
$A3$	$x + x = x$
$A4$	$(x + y) \cdot z = x \cdot z + y \cdot z$
$A5$	$(x \cdot y) \cdot z = x \cdot (y \cdot z)$
$A6$	$x + \delta = x$
$A7$	$\delta \cdot x = \delta$
$A8$	$\epsilon \cdot x = x$
$A9$	$x \cdot \epsilon = x$
$G1$	$\phi \cdot \neg\phi = \delta$
$G2$	$\phi + \neg\phi = \epsilon$
$G3$	$\phi\delta = \delta$
$G4$	$\phi(x + y) = \phi x + \phi y$
$G5$	$\phi(x \cdot y) = \phi x \cdot y$
$G6$	$(\phi + \psi)x = \phi x + \psi x$
$G7$	$(\phi \cdot \psi) \cdot x = \phi \cdot (\psi \cdot x)$
$G8$	$\phi = \epsilon$ if $\forall s \in S.test(\phi, s)$
$G9$	$\phi_0 \cdots \cdots \phi_n = \delta$ if $\forall s \in S, \exists i \leq n.test(\neg\phi_i, s)$
$G10$	$wp(e, \phi)e\phi = wp(e, \phi)e$
$G11$	$\neg wp(e, \phi)e\neg\phi = \neg wp(e, \phi)e$

hold. $effect(e, s) \in S$ denotes s' in $s \xrightarrow{e} s'$. The predicate weakest precondition $wp(e, \phi)$ denotes that $\forall s, s' \in S, test(\phi, effect(e, s))$ holds.

The set of axioms of $BATC_G$ consists of the laws given in Table 6.1.

Note that, by eliminating atomic event from the process terms, the axioms in Table 6.1 will lead to a Boolean Algebra. And $G9$ is a precondition of e and ϕ, $G10$ is the weakest precondition of e and ϕ. A data environment with $effect$ function is sufficiently deterministic, and it is obvious that if the weakest precondition is expressible and $G9$, $G10$ are sound, then the related data environment is sufficiently deterministic.

Definition 6.11 (Basic terms of $BATC_G$). The set of basic terms of $BATC_G$, $\mathcal{B}(BATC_G)$, is inductively defined as follows:

1. $\mathbb{E} \subset \mathcal{B}(BATC_G)$;
2. $G \subset \mathcal{B}(BATC_G)$;
3. if $e \in \mathbb{E}, t \in \mathcal{B}(BATC_G)$ then $e \cdot t \in \mathcal{B}(BATC_G)$;
4. if $\phi \in G, t \in \mathcal{B}(BATC_G)$ then $\phi \cdot t \in \mathcal{B}(BATC_G)$;
5. if $t, s \in \mathcal{B}(BATC_G)$ then $t + s \in \mathcal{B}(BATC_G)$.

Table 6.2 Term rewrite system of $BATC_G$.

No.	Rewriting rule
$RA3$	$x + x \to x$
$RA4$	$(x + y) \cdot z \to x \cdot z + y \cdot z$
$RA5$	$(x \cdot y) \cdot z \to x \cdot (y \cdot z)$
$RA6$	$x + \delta \to x$
$RA7$	$\delta \cdot x \to \delta$
$RA8$	$\epsilon \cdot x \to x$
$RA9$	$x \cdot \epsilon \to x$
$RG1$	$\phi \cdot \neg\phi \to \delta$
$RG2$	$\phi + \neg\phi \to \epsilon$
$RG3$	$\phi\delta \to \delta$
$RG4$	$\phi(x + y) \to \phi x + \phi y$
$RG5$	$\phi(x \cdot y) \to \phi x \cdot y$
$RG6$	$(\phi + \psi)x \to \phi x + \psi x$
$RG7$	$(\phi \cdot \psi) \cdot x \to \phi \cdot (\psi \cdot x)$
$RG8$	$\phi \to \epsilon$ if $\forall s \in S.test(\phi, s)$
$RG9$	$\phi_0 \cdot \cdots \cdot \phi_n \to \delta$ if $\forall s \in S, \exists i \leq n.test(\neg\phi_i, s)$
$RG10$	$wp(e, \phi)e\phi \to wp(e, \phi)e$
$RG11$	$\neg wp(e, \phi)e\neg\phi \to \neg wp(e, \phi)e$

Theorem 6.12 (Elimination theorem of $BATC_G$). *Let p be a closed $BATC_G$ term. Then there is a basic $BATC_G$ term q such that $BATC_G \vdash p = q$.*

Proof. (1) Firstly, suppose that the following ordering on the signature of $BATC_G$ is defined: $\cdot > +$ and the symbol $\cdot$ is given the lexicographical status for the first argument, then for each rewrite rule $p \to q$ in Table 6.2 relation $p >_{lpo} q$ can easily be proved. We obtain that the term rewrite system shown in Table 6.2 is strongly normalizing, for it has finitely many rewriting rules, and $>$ is a well-founded ordering on the signature of $BATC_G$, and if $s >_{lpo} t$, for each rewriting rule $s \to t$ is in Table 6.2 (see Theorem 2.12).

(2) Then we prove that the normal forms of closed $BATC_G$ terms are basic $BATC_G$ terms.

Suppose that p is a normal form of some closed $BATC_G$ term and suppose that p is not a basic term. Let p' denote the smallest sub-term of p which is not a basic term. It implies that each sub-term of p' is a basic term. Then we prove that p is not a term in normal form. It is sufficient to induct on the structure of p':

- Case $p' \equiv e, e \in \mathbb{E}$. p' is a basic term, which contradicts the assumption that p' is not a basic term, so this case should not occur.
- Case $p' \equiv \phi, \phi \in G$. p' is a basic term, which contradicts the assumption that p' is not a basic term, so this case should not occur.

Table 6.3 Single event transition rules of $BATC_G$.

$$\overline{\langle \epsilon, s \rangle \rightarrow \langle \sqrt{}, s \rangle}$$

$$\frac{}{\langle e, s \rangle \xrightarrow{e} \langle \sqrt{}, s' \rangle} \quad \text{if } s' \in effect(e, s)$$

$$\frac{}{\langle \phi, s \rangle \rightarrow \langle \sqrt{}, s \rangle} \quad \text{if } test(\phi, s)$$

$$\frac{\langle x, s \rangle \xrightarrow{e} \langle \sqrt{}, s' \rangle}{\langle x + y, s \rangle \xrightarrow{e} \langle \sqrt{}, s' \rangle} \qquad \frac{\langle x, s \rangle \xrightarrow{e} \langle x', s' \rangle}{\langle x + y, s \rangle \xrightarrow{e} \langle x', s' \rangle}$$

$$\frac{\langle y, s \rangle \xrightarrow{e} \langle \sqrt{}, s' \rangle}{\langle x + y, s \rangle \xrightarrow{e} \langle \sqrt{}, s' \rangle} \qquad \frac{\langle y, s \rangle \xrightarrow{e} \langle y', s' \rangle}{\langle x + y, s \rangle \xrightarrow{e} \langle y', s' \rangle}$$

$$\frac{\langle x, s \rangle \xrightarrow{e} \langle \sqrt{}, s' \rangle}{\langle x \cdot y, s \rangle \xrightarrow{e} \langle y, s' \rangle} \qquad \frac{\langle x, s \rangle \xrightarrow{e} \langle x', s' \rangle}{\langle x \cdot y, s \rangle \xrightarrow{e} \langle x' \cdot y, s' \rangle}$$

- Case $p' \equiv p_1 \cdot p_2$. By induction on the structure of the basic term p_1:
  - Subcase $p_1 \in \mathbb{E}$. p' would be a basic term, which contradicts the assumption that p' is not a basic term;
  - Subcase $p_1 \in G$. p' would be a basic term, which contradicts the assumption that p' is not a basic term;
  - Subcase $p_1 \equiv e \cdot p'_1$. $RA5$ or $RA9$ rewriting rule can be applied. So p is not a normal form;
  - Subcase $p_1 \equiv \phi \cdot p'_1$. $RG1$, $RG3$, $RG4$, $RG5$, $RG7$, or $RG8 - 9$ rewriting rules can be applied. So p is not a normal form;
  - Subcase $p_1 \equiv p'_1 + p''_1$. $RA4$, $RA6$, $RG2$, or $RG6$ rewriting rules can be applied. So p is not a normal form.
- Case $p' \equiv p_1 + p_2$. By induction on the structure of the basic terms both p_1 and p_2, all subcases will lead to that p' would be a basic term, which contradicts the assumption that p' is not a basic term. $\square$

We will define a term-deduction system which gives the operational semantics of $BATC_G$. We give the operational transition rules for ϵ, atomic guard $\phi \in G_{at}$, atomic event $e \in \mathbb{E}$, operators $\cdot$ and $+$ as Table 6.3 shows. And the predicate $\xrightarrow{e} \sqrt{}$ represents successful termination after execution of the event e.

Note that, we replace the single atomic event $e \in \mathbb{E}$ by $X \subseteq \mathbb{E}$, we can obtain the pomset transition rules of $BATC_G$, and omit them.

Theorem 6.13 (Congruence of $BATC_G$ with respect to truly concurrent bisimulation equivalences). *(1) Pomset bisimulation equivalence $\sim_p$ is a congruence with respect to $BATC_G$.*

(2) Step bisimulation equivalence $\sim_s$ is a congruence with respect to $BATC_G$.

(3) Hp-bisimulation equivalence $\sim_{hp}$ is a congruence with respect to $BATC_G$.

(4) Hhp-bisimulation equivalence $\sim_{hhp}$ is a congruence with respect to $BATC_G$.

Proof. (1) It is easy to see that pomset bisimulation is an equivalent relation on $BATC_G$ terms, we only need to prove that $\sim_p$ is preserved by the operators $\cdot$ and $+$. It is trivial and we leave the proof as an exercise for the readers.

(2) It is easy to see that step bisimulation is an equivalent relation on $BATC_G$ terms, we only need to prove that $\sim_s$ is preserved by the operators $\cdot$ and $+$. It is trivial and we leave the proof as an exercise for the readers.

(3) It is easy to see that hp-bisimulation is an equivalent relation on $BATC_G$ terms, we only need to prove that $\sim_{hp}$ is preserved by the operators $\cdot$ and $+$. It is trivial and we leave the proof as an exercise for the readers.

(4) It is easy to see that hhp-bisimulation is an equivalent relation on $BATC_G$ terms, we only need to prove that $\sim_{hhp}$ is preserved by the operators $\cdot$ and $+$. It is trivial and we leave the proof as an exercise for the readers. □

Theorem 6.14 (Soundness of $BATC_G$ modulo truly concurrent bisimulation equivalences). *(1) Let x and y be $BATC_G$ terms. If $BATC \vdash x = y$, then $x \sim_p y$.*

(2) Let x and y be $BATC_G$ terms. If $BATC \vdash x = y$, then $x \sim_s y$.

(3) Let x and y be $BATC_G$ terms. If $BATC \vdash x = y$, then $x \sim_{hp} y$.

(4) Let x and y be $BATC_G$ terms. If $BATC \vdash x = y$, then $x \sim_{hhp} y$.

Proof. (1) Since pomset bisimulation $\sim_p$ is both an equivalent and a congruent relation, we only need to check if each axiom in Table 6.1 is sound modulo pomset bisimulation equivalence. We leave the proof as an exercise for the readers.

(2) Since step bisimulation $\sim_s$ is both an equivalent and a congruent relation, we only need to check if each axiom in Table 6.1 is sound modulo step bisimulation equivalence. We leave the proof as an exercise for the readers.

(3) Since hp-bisimulation $\sim_{hp}$ is both an equivalent and a congruent relation, we only need to check if each axiom in Table 6.1 is sound modulo hp-bisimulation equivalence. We leave the proof as an exercise for the readers.

(4) Since hhp-bisimulation $\sim_{hhp}$ is both an equivalent and a congruent relation, we only need to check if each axiom in Table 6.1 is sound modulo hhp-bisimulation equivalence. We leave the proof as an exercise for the readers. □

Theorem 6.15 (Completeness of $BATC_G$ modulo truly concurrent bisimulation equivalences). *(1) Let p and q be closed $BATC_G$ terms, if $p \sim_p q$ then $p = q$.*

(2) Let p and q be closed $BATC_G$ terms, if $p \sim_s q$ then $p = q$.

(3) Let p and q be closed $BATC_G$ terms, if $p \sim_{hp} q$ then $p = q$.

(4) Let p and q be closed $BATC_G$ terms, if $p \sim_{hhp} q$ then $p = q$.

Proof. (1) Firstly, by the elimination theorem of $BATC_G$, we know that for each closed $BATC_G$ term p, there exists a closed basic $BATC_G$ term p', such that $BATC_G \vdash p = p'$, so, we only need to consider closed basic $BATC_G$ terms.

The basic terms (see Definition 6.11) modulo associativity and commutativity (AC) of conflict $+$ (defined by axioms $A1$ and $A2$ in Table 6.1), and this equivalence is denoted by $=_{AC}$. Then, each equivalence class s modulo AC of $+$ has the following normal form

$$s_1 + \cdots + s_k$$

with each s_i either an atomic event, or an atomic guard, or of the form $t_1 \cdot t_2$, and each s_i is called the summand of s.

Now, we prove that for normal forms n and n', if $n \sim_p n'$ then $n =_{AC} n'$. It is sufficient to induct on the sizes of n and n'.

- Consider a summand e of n. Then $\langle n, s \rangle \xrightarrow{e} \langle \sqrt{}, s' \rangle$, so $n \sim_p n'$ implies $\langle n', s \rangle \xrightarrow{e} \langle \sqrt{}, s \rangle$, meaning that n' also contains the summand e.
- Consider a summand ϕ of n. Then $\langle n, s \rangle \rightarrow \langle \sqrt{}, s \rangle$, if $test(\phi, s)$ holds, so $n \sim_p n'$ implies $\langle n', s \rangle \rightarrow \langle \sqrt{}, s \rangle$, if $test(\phi, s)$ holds, meaning that n' also contains the summand ϕ.
- Consider a summand $t_1 \cdot t_2$ of n. Then $\langle n, s \rangle \xrightarrow{t_1} \langle t_2, s' \rangle$, so $n \sim_p n'$ implies $\langle n', s \rangle \xrightarrow{t_1} \langle t'_2, s' \rangle$ with $t_2 \sim_p t'_2$, meaning that n' contains a summand $t_1 \cdot t'_2$. Since t_2 and t'_2 are normal forms and have sizes smaller than n and n', by the induction hypotheses $t_2 \sim_p t'_2$ implies $t_2 =_{AC} t'_2$.

So, we get $n =_{AC} n'$.

Finally, let s and t be basic terms, and $s \sim_p t$, there are normal forms n and n', such that $s = n$ and $t = n'$. The soundness theorem of $BATC_G$ modulo pomset bisimulation equivalence (see Theorem 6.14) yields $s \sim_p n$ and $t \sim_p n'$, so $n \sim_p s \sim_p t \sim_p n'$. Since if $n \sim_p n'$ then $n =_{AC} n'$, $s = n =_{AC} n' = t$, as desired.

(2) It can be proven similarly as (1).

(3) It can be proven similarly as (1).

(4) It can be proven similarly as (1). $\qquad\square$

Theorem 6.16 (Sufficient determinacy). *All related data environments with respect to $BATC_G$ can be sufficiently deterministic.*

Proof. It only needs to check $effect(t, s)$ function is deterministic, and is sufficient to induct on the structure of term t. The only matter is the case $t = t_1 + t_2$, with the help of guards, we can make $t_1 = \phi_1 \cdot t'_1$ and $t_2 = \phi_2 \cdot t'_2$, and $effect(t)$ is sufficiently deterministic. $\qquad\square$

6.3 *APTC* with guards

In this subsection, we will extend $APTC$ with guards, which is abbreviated $APTC_G$. The set of basic guards G with element $\phi, \psi, \cdots$, which is extended by the following formation rules:

$$\phi ::= \delta \,|\, \epsilon \,|\, \neg\phi \,|\, \psi \in G_{at} \,|\, \phi + \psi \,|\, \phi \cdot \psi \,|\, \phi \,\|\, \psi$$

The set of axioms of $APTC_G$ including axioms of $BATC_G$ in Table 6.1 and the axioms are shown in Table 6.4.

Definition 6.17 (Basic terms of $APTC_G$). The set of basic terms of $APTC_G$, $\mathcal{B}(APTC_G)$, is inductively defined as follows:

1. $\mathbb{E} \subset \mathcal{B}(APTC_G)$;
2. $G \subset \mathcal{B}(APTC_G)$;
3. if $e \in \mathbb{E}, t \in \mathcal{B}(APTC_G)$ then $e \cdot t \in \mathcal{B}(APTC_G)$;
4. if $\phi \in G, t \in \mathcal{B}(APTC_G)$ then $\phi \cdot t \in \mathcal{B}(APTC_G)$;
5. if $t, s \in \mathcal{B}(APTC_G)$ then $t + s \in \mathcal{B}(APTC_G)$.
6. if $t, s \in \mathcal{B}(APTC_G)$ then $t \,\|\, s \in \mathcal{B}(APTC_G)$.

Based on the definition of basic terms for $APTC_G$ (see Definition 6.17) and axioms of $APTC_G$, we can prove the elimination theorem of $APTC_G$.

Theorem 6.18 (Elimination theorem of $APTC_G$). *Let p be a closed $APTC_G$ term. Then there is a basic $APTC_G$ term q such that $APTC_G \vdash p = q$.*

Proof. (1) Firstly, suppose that the following ordering on the signature of $APTC_G$ is defined: $\| > \cdot > +$ and the symbol $\|$ is given the lexicographical status for the first argument, then for each rewrite rule $p \to q$ in Table 6.5 relation $p >_{lpo} q$ can easily be proved. We obtain that the term rewrite system shown in Table 6.5 is strongly normalizing, for it has finitely many rewriting rules, and $>$ is a well-founded ordering on the signature of $APTC_G$, and if $s >_{lpo} t$, for each rewriting rule $s \to t$ is in Table 6.5 (see Theorem 2.12).

(2) Then we prove that the normal forms of closed $APTC_G$ terms are basic $APTC_G$ terms.

Suppose that p is a normal form of some closed $APTC_G$ term and suppose that p is not a basic $APTC_G$ term. Let p' denote the smallest sub-term of p which is not a basic $APTC_G$ term. It implies that each sub-term of p' is a basic $APTC_G$ term. Then we prove that p is not a term in normal form. It is sufficient to induct on the structure of p':

- Case $p' \equiv e, e \in \mathbb{E}$. p' is a basic $APTC_G$ term, which contradicts the assumption that p' is not a basic $APTC_G$ term, so this case should not occur.
- Case $p' \equiv \phi, \phi \in G$. p' is a basic term, which contradicts the assumption that p' is not a basic term, so this case should not occur.
- Case $p' \equiv p_1 \cdot p_2$. By induction on the structure of the basic $APTC_G$ term p_1:
  - Subcase $p_1 \in \mathbb{E}$. p' would be a basic $APTC_G$ term, which contradicts the assumption that p' is not a basic $APTC_G$ term;
  - Subcase $p_1 \in G$. p' would be a basic term, which contradicts the assumption that p' is not a basic term;

Table 6.4 Axioms of $APTC_G$.

No.	Axiom
$P1$	$x \between y = x \parallel y + x \mid y$
$P2$	$e_1 \parallel (e_2 \cdot y) = (e_1 \parallel e_2) \cdot y$
$P3$	$(e_1 \cdot x) \parallel e_2 = (e_1 \parallel e_2) \cdot x$
$P4$	$(e_1 \cdot x) \parallel (e_2 \cdot y) = (e_1 \parallel e_2) \cdot (x \between y)$
$P5$	$(x + y) \parallel z = (x \parallel z) + (y \parallel z)$
$P6$	$x \parallel (y + z) = (x \parallel y) + (x \parallel z)$
$P7$	$\delta \parallel x = \delta$
$P8$	$x \parallel \delta = \delta$
$P9$	$\epsilon \parallel x = x$
$P10$	$x \parallel \epsilon = x$
$C1$	$e_1 \mid e_2 = \gamma(e_1, e_2)$
$C2$	$e_1 \mid (e_2 \cdot y) = \gamma(e_1, e_2) \cdot y$
$C3$	$(e_1 \cdot x) \mid e_2 = \gamma(e_1, e_2) \cdot x$
$C4$	$(e_1 \cdot x) \mid (e_2 \cdot y) = \gamma(e_1, e_2) \cdot (x \between y)$
$C5$	$(x + y) \mid z = (x \mid z) + (y \mid z)$
$C6$	$x \mid (y + z) = (x \mid y) + (x \mid z)$
$C7$	$\delta \mid x = \delta$
$C8$	$x \mid \delta = \delta$
$C9$	$\epsilon \mid x = \delta$
$C10$	$x \mid \epsilon = \delta$
$CE1$	$\Theta(e) = e$
$CE2$	$\Theta(\delta) = \delta$
$CE3$	$\Theta(\epsilon) = \epsilon$
$CE4$	$\Theta(x + y) = \Theta(x) + \Theta(y)$
$CE5$	$\Theta(x \cdot y) = \Theta(x) \cdot \Theta(y)$
$CE6$	$\Theta(x \parallel y) = ((\Theta(x) \triangleleft y) \parallel y) + ((\Theta(y) \triangleleft x) \parallel x)$
$CE7$	$\Theta(x \mid y) = ((\Theta(x) \triangleleft y) \mid y) + ((\Theta(y) \triangleleft x) \mid x)$
$U1$	$(\sharp(e_1, e_2))$ $e_1 \triangleleft e_2 = \tau$
$U2$	$(\sharp(e_1, e_2), e_2 \leq e_3)$ $e_1 \triangleleft e_3 = \tau$
$U3$	$(\sharp(e_1, e_2), e_2 \leq e_3)$ $e3 \triangleleft e_1 = \tau$
$U4$	$e \triangleleft \delta = e$
$U5$	$\delta \triangleleft e = \delta$
$U6$	$e \triangleleft \epsilon = e$
$U7$	$\epsilon \triangleleft e = e$
$U8$	$(x + y) \triangleleft z = (x \triangleleft z) + (y \triangleleft z)$
$U9$	$(x \cdot y) \triangleleft z = (x \triangleleft z) \cdot (y \triangleleft z)$

continued on next page

Table 6.4 (continued)

No.	Axiom
$U10$	$(x \parallel y) \triangleleft z = (x \triangleleft z) \parallel (y \triangleleft z)$
$U11$	$(x \mid y) \triangleleft z = (x \triangleleft z) \mid (y \triangleleft z)$
$U12$	$x \triangleleft (y + z) = (x \triangleleft y) \triangleleft z$
$U13$	$x \triangleleft (y \cdot z) = (x \triangleleft y) \triangleleft z$
$U14$	$x \triangleleft (y \parallel z) = (x \triangleleft y) \triangleleft z$
$U15$	$x \triangleleft (y \mid z) = (x \triangleleft y) \triangleleft z$
$D1$	$e \notin H \quad \partial_H(e) = e$
$D2$	$e \in H \quad \partial_H(e) = \delta$
$D3$	$\partial_H(\delta) = \delta$
$D4$	$\partial_H(x + y) = \partial_H(x) + \partial_H(y)$
$D5$	$\partial_H(x \cdot y) = \partial_H(x) \cdot \partial_H(y)$
$D6$	$\partial_H(x \parallel y) = \partial_H(x) \parallel \partial_H(y)$
$G12$	$\phi(x \parallel y) = \phi x \parallel \phi y$
$G13$	$\phi(x \mid y) = \phi x \mid \phi y$
$G14$	$\phi \parallel \delta = \delta$
$G15$	$\delta \parallel \phi = \delta$
$G16$	$\phi \mid \delta = \delta$
$G17$	$\delta \mid \phi = \delta$
$G18$	$\phi \parallel \epsilon = \phi$
$G19$	$\epsilon \parallel \phi = \phi$
$G20$	$\phi \mid \epsilon = \delta$
$G21$	$\epsilon \mid \phi = \delta$
$G22$	$\phi \parallel \neg\phi = \delta$
$G23$	$\Theta(\phi) = \phi$
$G24$	$\partial_H(\phi) = \phi$
$G25$	$\phi_0 \parallel \cdots \parallel \phi_n = \delta$ if $\forall s_0, \cdots, s_n \in S, \exists i \leq n.test(\neg\phi_i, s_0 \cup \cdots \cup s_n)$

- Subcase $p_1 \equiv e \cdot p_1'$. $RA5$ or $RA9$ rewriting rules in Table 6.2 can be applied. So p is not a normal form;
- Subcase $p_1 \equiv \phi \cdot p_1'$. $RG1$, $RG3$, $RG4$, $RG5$, $RG7$, or $RG8 - 9$ rewriting rules can be applied. So p is not a normal form;
- Subcase $p_1 \equiv p_1' + p_1''$. $RA4$, $RA6$, $RG2$, or $RG6$ rewriting rules in Table 6.2 can be applied. So p is not a normal form;
- Subcase $p_1 \equiv p_1' \parallel p_1''$. $RP2$-$RP10$ rewrite rules in Table 6.5 can be applied. So p is not a normal form;

Table 6.5   Term rewrite system of $APTC_G$.

No.	Rewriting rule
$RP1$	$x \between y \to x \parallel y + x \mid y$
$RP2$	$e_1 \parallel (e_2 \cdot y) \to (e_1 \parallel e_2) \cdot y$
$RP3$	$(e_1 \cdot x) \parallel e_2 \to (e_1 \parallel e_2) \cdot x$
$RP4$	$(e_1 \cdot x) \parallel (e_2 \cdot y) \to (e_1 \parallel e_2) \cdot (x \between y)$
$RP5$	$(x + y) \parallel z \to (x \parallel z) + (y \parallel z)$
$RP6$	$x \parallel (y + z) \to (x \parallel y) + (x \parallel z)$
$RP7$	$\delta \parallel x \to \delta$
$RP8$	$x \parallel \delta \to \delta$
$RP9$	$\epsilon \parallel x \to x$
$RP10$	$x \parallel \epsilon \to x$
$RC1$	$e_1 \mid e_2 \to \gamma(e_1, e_2)$
$RC2$	$e_1 \mid (e_2 \cdot y) \to \gamma(e_1, e_2) \cdot y$
$RC3$	$(e_1 \cdot x) \mid e_2 \to \gamma(e_1, e_2) \cdot x$
$RC4$	$(e_1 \cdot x) \mid (e_2 \cdot y) \to \gamma(e_1, e_2) \cdot (x \between y)$
$RC5$	$(x + y) \mid z \to (x \mid z) + (y \mid z)$
$RC6$	$x \mid (y + z) \to (x \mid y) + (x \mid z)$
$RC7$	$\delta \mid x \to \delta$
$RC8$	$x \mid \delta \to \delta$
$RC9$	$\epsilon \mid x \to \delta$
$RC10$	$x \mid \epsilon \to \delta$
$RCE1$	$\Theta(e) \to e$
$RCE2$	$\Theta(\delta) \to \delta$
$RCE3$	$\Theta(\epsilon) \to \epsilon$
$RCE4$	$\Theta(x + y) \to \Theta(x) + \Theta(y)$
$RCE5$	$\Theta(x \cdot y) \to \Theta(x) \cdot \Theta(y)$
$RCE6$	$\Theta(x \parallel y) \to ((\Theta(x) \triangleleft y) \parallel y) + ((\Theta(y) \triangleleft x) \parallel x)$
$RCE7$	$\Theta(x \mid y) \to ((\Theta(x) \triangleleft y) \mid y) + ((\Theta(y) \triangleleft x) \mid x)$
$RU1$	$(\sharp(e_1, e_2)) \quad e_1 \triangleleft e_2 \to \tau$
$RU2$	$(\sharp(e_1, e_2), e_2 \leq e_3) \quad e_1 \triangleleft e_3 \to \tau$
$RU3$	$(\sharp(e_1, e_2), e_2 \leq e_3) \quad e3 \triangleleft e_1 \to \tau$
$RU4$	$e \triangleleft \delta \to e$
$RU5$	$\delta \triangleleft e \to \delta$
$RU6$	$e \triangleleft \epsilon \to e$
$RU7$	$\epsilon \triangleleft e \to e$
$RU8$	$(x + y) \triangleleft z \to (x \triangleleft z) + (y \triangleleft z)$
$RU9$	$(x \cdot y) \triangleleft z \to (x \triangleleft z) \cdot (y \triangleleft z)$

continued on next page

Table 6.5 (continued)

No.	Rewriting rule
$RU10$	$(x \parallel y) \triangleleft z \to (x \triangleleft z) \parallel (y \triangleleft z)$
$RU11$	$(x \mid y) \triangleleft z \to (x \triangleleft z) \mid (y \triangleleft z)$
$RU12$	$x \triangleleft (y + z) \to (x \triangleleft y) \triangleleft z$
$RU13$	$x \triangleleft (y \cdot z) \to (x \triangleleft y) \triangleleft z$
$RU14$	$x \triangleleft (y \parallel z) \to (x \triangleleft y) \triangleleft z$
$RU15$	$x \triangleleft (y \mid z) \to (x \triangleleft y) \triangleleft z$
$RD1$	$e \notin H \quad \partial_H(e) \to e$
$RD2$	$e \in H \quad \partial_H(e) \to \delta$
$RD3$	$\partial_H(\delta) \to \delta$
$RD4$	$\partial_H(x + y) \to \partial_H(x) + \partial_H(y)$
$RD5$	$\partial_H(x \cdot y) \to \partial_H(x) \cdot \partial_H(y)$
$RD6$	$\partial_H(x \parallel y) \to \partial_H(x) \parallel \partial_H(y)$
$RG12$	$\phi(x \parallel y) \to \phi x \parallel \phi y$
$RG13$	$\phi(x \mid y) \to \phi x \mid \phi y$
$RG14$	$\phi \parallel \delta \to \delta$
$RG15$	$\delta \parallel \phi \to \delta$
$RG16$	$\phi \mid \delta \to \delta$
$RG17$	$\delta \mid \phi \to \delta$
$RG18$	$\phi \parallel \epsilon \to \phi$
$RG19$	$\epsilon \parallel \phi \to \phi$
$RG20$	$\phi \mid \epsilon \to \delta$
$RG21$	$\epsilon \mid \phi \to \delta$
$RG22$	$\phi \parallel \neg\phi \to \delta$
$RG23$	$\Theta(\phi) \to \phi$
$RG24$	$\partial_H(\phi) \to \phi$
$RG25$	$\phi_0 \parallel \cdots \parallel \phi_n \to \delta$ if $\forall s_0, \cdots, s_n \in S, \exists i \le n.test(\neg\phi_i, s_0 \cup \cdots \cup s_n)$

- Subcase $p_1 \equiv p_1' \mid p_1''$. $RC1$-$RC10$ rewrite rules in Table 6.5 can be applied. So p is not a normal form;
- Subcase $p_1 \equiv \Theta(p_1')$. $RCE1$-$RCE7$ rewrite rules in Table 6.5 can be applied. So p is not a normal form;
- Subcase $p_1 \equiv \partial_H(p_1')$. $RD1$-$RD6$ rewrite rules in Table 6.5 can be applied. So p is not a normal form.
- Case $p' \equiv p_1 + p_2$. By induction on the structure of the basic $APTC_G$ terms both p_1 and p_2, all subcases will lead to that p' would be a basic $APTC_G$ term, which contradicts the assumption that p' is not a basic $APTC_G$ term.

- Case $p' \equiv p_1 \parallel p_2$. By induction on the structure of the basic $APTC_G$ terms both p_1 and p_2, all subcases will lead to that p' would be a basic $APTC_G$ term, which contradicts the assumption that p' is not a basic $APTC_G$ term.
- Case $p' \equiv p_1 \mid p_2$. By induction on the structure of the basic $APTC_G$ terms both p_1 and p_2, all subcases will lead to that p' would be a basic $APTC_G$ term, which contradicts the assumption that p' is not a basic $APTC_G$ term.
- Case $p' \equiv \Theta(p_1)$. By induction on the structure of the basic $APTC_G$ term p_1, $RCE1 - RCE7$ rewrite rules in Table 6.5 can be applied. So p is not a normal form.
- Case $p' \equiv p_1 \triangleleft p_2$. By induction on the structure of the basic $APTC_G$ terms both p_1 and p_2, all subcases will lead to that p' would be a basic $APTC_G$ term, which contradicts the assumption that p' is not a basic $APTC_G$ term.
- Case $p' \equiv \partial_H(p_1)$. By induction on the structure of the basic $APTC_G$ terms of p_1, all subcases will lead to that p' would be a basic $APTC_G$ term, which contradicts the assumption that p' is not a basic $APTC_G$ term.   □

We will define a term-deduction system which gives the operational semantics of $APTC_G$. Two atomic events e_1 and e_2 are in race condition, which are denoted $e_1 \% e_2$.

Theorem 6.19 (Generalization of $APTC_G$ with respect to $BATC_G$). *$APTC_G$ is a generalization of $BATC_G$.*

Proof. It follows from the following three facts.

1. The transition rules of $BATC_G$ in section 6.2 are all source-dependent;
2. The sources of the transition rules $APTC_G$ contain an occurrence of $\emptyset$, or $\parallel$, or $\mid$, or Θ, or $\triangleleft$;
3. The transition rules of $APTC_G$ (Table 6.6) are all source-dependent.

So, $APTC_G$ is a generalization of $BATC_G$, that is, $BATC_G$ is an embedding of $APTC_G$, as desired.   □

Theorem 6.20 (Congruence of $APTC_G$ with respect to truly concurrent bisimulation equivalences). *(1) Pomset bisimulation equivalence $\sim_p$ is a congruence with respect to $APTC_G$.*

(2) Step bisimulation equivalence $\sim_s$ is a congruence with respect to $APTC_G$.

(3) Hp-bisimulation equivalence $\sim_{hp}$ is a congruence with respect to $APTC_G$.

(4) Hhp-bisimulation equivalence $\sim_{hhp}$ is a congruence with respect to $APTC_G$.

Proof. (1) It is easy to see that pomset bisimulation is an equivalent relation on $APTC_G$ terms, we only need to prove that $\sim_p$ is preserved by the operators $\parallel$, $\mid$, Θ, $\triangleleft$, ∂_H. It is trivial and we leave the proof as an exercise for the readers.

(2) It is easy to see that step bisimulation is an equivalent relation on $APTC_G$ terms, we only need to prove that $\sim_s$ is preserved by the operators $\parallel$, $\mid$, Θ, $\triangleleft$, ∂_H. It is trivial and we leave the proof as an exercise for the readers.

Table 6.6 Transition rules of $APTC_G$.

$$\frac{}{\langle e_1 \parallel \cdots \parallel e_n, s\rangle \xrightarrow{\{e_1,\cdots,e_n\}} \langle \surd, s'\rangle} \quad \text{if } s' \in effect(e_1, s) \cup \cdots \cup effect(e_n, s)$$

$$\frac{}{\langle \phi_1 \parallel \cdots \parallel \phi_n, s\rangle \to \langle \surd, s\rangle} \quad \text{if } test(\phi_1, s), \cdots, test(\phi_n, s)$$

$$\frac{\langle x, s\rangle \xrightarrow{e_1} \langle \surd, s'\rangle \quad \langle y, s\rangle \xrightarrow{e_2} \langle \surd, s''\rangle}{\langle x \parallel y, s\rangle \xrightarrow{\{e_1, e_2\}} \langle \surd, s' \cup s''\rangle} \qquad \frac{\langle x, s\rangle \xrightarrow{e_1} \langle x', s'\rangle \quad \langle y, s\rangle \xrightarrow{e_2} \langle \surd, s''\rangle}{\langle x \parallel y, s\rangle \xrightarrow{\{e_1, e_2\}} \langle x', s' \cup s''\rangle}$$

$$\frac{\langle x, s\rangle \xrightarrow{e_1} \langle \surd, s'\rangle \quad \langle y, s\rangle \xrightarrow{e_2} \langle y', s''\rangle}{\langle x \parallel y, s\rangle \xrightarrow{\{e_1, e_2\}} \langle y', s' \cup s''\rangle} \qquad \frac{\langle x, s\rangle \xrightarrow{e_1} \langle x', s'\rangle \quad \langle y, s\rangle \xrightarrow{e_2} \langle y', s''\rangle}{\langle x \parallel y, s\rangle \xrightarrow{\{e_1, e_2\}} \langle x' \between y', s' \cup s''\rangle}$$

$$\frac{\langle x, s\rangle \xrightarrow{e_1} \langle \surd, s'\rangle \quad \langle y, s\rangle \xnrightarrow{e_2} \quad (e_1 \% e_2)}{\langle x \parallel y, s\rangle \xrightarrow{e_1} \langle y, s'\rangle} \qquad \frac{\langle x, s\rangle \xrightarrow{e_1} \langle x', s'\rangle \quad \langle y, s\rangle \xnrightarrow{e_2} \quad (e_1 \% e_2)}{\langle x \parallel y, s\rangle \xrightarrow{e_1} \langle x' \between y, s'\rangle}$$

$$\frac{\langle x, s\rangle \xnrightarrow{e_1} \quad \langle y, s\rangle \xrightarrow{e_2} \langle \surd, s''\rangle \quad (e_1 \% e_2)}{\langle x \parallel y, s\rangle \xrightarrow{e_2} \langle x, s''\rangle} \qquad \frac{\langle x, s\rangle \xnrightarrow{e_1} \quad \langle y, s\rangle \xrightarrow{e_2} \langle y', s''\rangle \quad (e_1 \% e_2)}{\langle x \parallel y, s\rangle \xrightarrow{e_2} \langle x \between y', s''\rangle}$$

$$\frac{\langle x, s\rangle \xrightarrow{e_1} \langle \surd, s'\rangle \quad \langle y, s\rangle \xrightarrow{e_2} \langle \surd, s''\rangle}{\langle x \mid y, s\rangle \xrightarrow{\gamma(e_1, e_2)} \langle \surd, effect(\gamma(e_1, e_2), s)\rangle} \qquad \frac{\langle x, s\rangle \xrightarrow{e_1} \langle x', s'\rangle \quad \langle y, s\rangle \xrightarrow{e_2} \langle \surd, s''\rangle}{\langle x \mid y, s\rangle \xrightarrow{\gamma(e_1, e_2)} \langle x', effect(\gamma(e_1, e_2), s)\rangle}$$

$$\frac{\langle x, s\rangle \xrightarrow{e_1} \langle \surd, s'\rangle \quad \langle y, s\rangle \xrightarrow{e_2} \langle y', s''\rangle}{\langle x \mid y, s\rangle \xrightarrow{\gamma(e_1, e_2)} \langle y', effect(\gamma(e_1, e_2), s)\rangle} \qquad \frac{\langle x, s\rangle \xrightarrow{e_1} \langle x', s'\rangle \quad \langle y, s\rangle \xrightarrow{e_2} \langle y', s''\rangle}{\langle x \mid y, s\rangle \xrightarrow{\gamma(e_1, e_2)} \langle x' \between y', effect(\gamma(e_1, e_2), s)\rangle}$$

$$\frac{\langle x, s\rangle \xrightarrow{e_1} \langle \surd, s'\rangle \quad (\sharp(e_1, e_2))}{\langle \Theta(x), s\rangle \xrightarrow{e_1} \langle \surd, s'\rangle} \qquad \frac{\langle x, s\rangle \xrightarrow{e_2} \langle \surd, s''\rangle \quad (\sharp(e_1, e_2))}{\langle \Theta(x), s\rangle \xrightarrow{e_2} \langle \surd, s''\rangle}$$

$$\frac{\langle x, s\rangle \xrightarrow{e_1} \langle x', s'\rangle \quad (\sharp(e_1, e_2))}{\langle \Theta(x), s\rangle \xrightarrow{e_1} \langle \Theta(x'), s'\rangle} \qquad \frac{\langle x, s\rangle \xrightarrow{e_2} \langle x'', s''\rangle \quad (\sharp(e_1, e_2))}{\langle \Theta(x), s\rangle \xrightarrow{e_2} \langle \Theta(x''), s''\rangle}$$

$$\frac{\langle x, s\rangle \xrightarrow{e_1} \langle \surd, s'\rangle \quad \langle y, s\rangle \xnrightarrow{e_2} \quad (\sharp(e_1, e_2))}{\langle x \triangleleft y, s\rangle \xrightarrow{\tau} \langle \surd, s'\rangle} \qquad \frac{\langle x, s\rangle \xrightarrow{e_1} \langle x', s'\rangle \quad \langle y, s\rangle \xnrightarrow{e_2} \quad (\sharp(e_1, e_2))}{\langle x \triangleleft y, s\rangle \xrightarrow{\tau} \langle x', s'\rangle}$$

$$\frac{\langle x, s\rangle \xrightarrow{e_1} \langle \surd, s\rangle \quad \langle y, s\rangle \xnrightarrow{e_3} \quad (\sharp(e_1, e_2), e_2 \leq e_3)}{\langle x \triangleleft y, s\rangle \xrightarrow{\tau} \langle \surd, s'\rangle} \qquad \frac{\langle x, s\rangle \xrightarrow{e_1} \langle x', s'\rangle \quad \langle y, s\rangle \xnrightarrow{e_3} \quad (\sharp(e_1, e_2), e_2 \leq e_3)}{\langle x \triangleleft y, s\rangle \xrightarrow{\tau} \langle x', s'\rangle}$$

continued on next page

Table 6.6 (continued)

$$\frac{\langle x,s\rangle \overset{e_3}{\to} \langle \sqrt{},s'\rangle \quad \langle y,s\rangle \overset{e_2}{\nrightarrow} \quad (\sharp(e_1,e_2),e_1 \leq e_3)}{\langle x \vartriangleleft y,s\rangle \overset{\tau}{\to} \langle \sqrt{},s'\rangle} \qquad \frac{\langle x,s\rangle \overset{e_3}{\to} \langle x',s'\rangle \quad \langle y,s\rangle \overset{e_2}{\nrightarrow} \quad (\sharp(e_1,e_2),e_1 \leq e_3)}{\langle x \vartriangleleft y,s\rangle \overset{\tau}{\to} \langle x',s'\rangle}$$

$$\frac{\langle x,s\rangle \overset{e}{\to} \langle \sqrt{},s'\rangle}{\langle \partial_H(x),s\rangle \overset{e}{\to} \langle \sqrt{},s'\rangle} \quad (e \notin H) \qquad \frac{\langle x,s\rangle \overset{e}{\to} \langle x',s'\rangle}{\langle \partial_H(x),s\rangle \overset{e}{\to} \langle \partial_H(x'),s'\rangle} \quad (e \notin H)$$

(3) It is easy to see that hp-bisimulation is an equivalent relation on $APTC_G$ terms, we only need to prove that $\sim_{hp}$ is preserved by the operators $\parallel$, $\mid$, Θ, $\vartriangleleft$, ∂_H. It is trivial and we leave the proof as an exercise for the readers.

(4) It is easy to see that hhp-bisimulation is an equivalent relation on $APTC_G$ terms, we only need to prove that $\sim_{hhp}$ is preserved by the operators $\parallel$, $\mid$, Θ, $\vartriangleleft$, ∂_H. It is trivial and we leave the proof as an exercise for the readers.   □

Theorem 6.21 (Soundness of $APTC_G$ modulo truly concurrent bisimulation equivalences). *(1) Let x and y be $APTC_G$ terms. If $APTC \vdash x = y$, then $x \sim_p y$.*

(2) Let x and y be $APTC_G$ terms. If $APTC \vdash x = y$, then $x \sim_s y$.

(3) Let x and y be $APTC_G$ terms. If $APTC \vdash x = y$, then $x \sim_{hp} y$.

Proof. (1) Since pomset bisimulation $\sim_p$ is both an equivalent and a congruent relation, we only need to check if each axiom in Table 6.4 is sound modulo pomset bisimulation equivalence. We leave the proof as an exercise for the readers.

(2) Since step bisimulation $\sim_s$ is both an equivalent and a congruent relation, we only need to check if each axiom in Table 6.4 is sound modulo step bisimulation equivalence. We leave the proof as an exercise for the readers.

(3) Since hp-bisimulation $\sim_{hp}$ is both an equivalent and a congruent relation, we only need to check if each axiom in Table 6.4 is sound modulo hp-bisimulation equivalence. We leave the proof as an exercise for the readers.   □

Theorem 6.22 (Completeness of $APTC_G$ modulo truly concurrent bisimulation equivalences). *(1) Let p and q be closed $APTC_G$ terms, if $p \sim_p q$ then $p = q$.*

(2) Let p and q be closed $APTC_G$ terms, if $p \sim_s q$ then $p = q$.

(3) Let p and q be closed $APTC_G$ terms, if $p \sim_{hp} q$ then $p = q$.

Proof. (1) Firstly, by the elimination theorem of $APTC_G$ (see Theorem 6.18), we know that for each closed $APTC_G$ term p, there exists a closed basic $APTC_G$ term p', such that $APTC \vdash p = p'$, so, we only need to consider closed basic $APTC_G$ terms.

The basic terms (see Definition 6.17) modulo associativity and commutativity (AC) of conflict $+$ (defined by axioms $A1$ and $A2$ in Table 6.1), and these equivalences are denoted by $=_{AC}$. Then, each equivalence class s modulo AC of $+$ has the following normal form

$$s_1 + \cdots + s_k$$

with each s_i either an atomic event, or an atomic guard, or of the form

$$t_1 \cdot \cdots \cdot t_m$$

with each t_j either an atomic event, or an atomic guard, or of the form

$$u_1 \parallel \cdots \parallel u_l$$

with each u_l an atomic event, or an atomic guard, and each s_i is called the summand of s.

Now, we prove that for normal forms n and n', if $n \sim_p n'$ then $n =_{AC} n'$. It is sufficient to induct on the sizes of n and n'.

- Consider a summand e of n. Then $\langle n, s \rangle \xrightarrow{e} \langle \surd, s' \rangle$, so $n \sim_p n'$ implies $\langle n', s \rangle \xrightarrow{e} \langle \surd, s \rangle$, meaning that n' also contains the summand e.
- Consider a summand ϕ of n. Then $\langle n, s \rangle \rightarrow \langle \surd, s \rangle$, if $test(\phi, s)$ holds, so $n \sim_p n'$ implies $\langle n', s \rangle \rightarrow \langle \surd, s \rangle$, if $test(\phi, s)$ holds, meaning that n' also contains the summand ϕ.
- Consider a summand $t_1 \cdot t_2$ of n,
  - if $t_1 \equiv e'$, then $\langle n, s \rangle \xrightarrow{e'} \langle t_2, s' \rangle$, so $n \sim_p n'$ implies $\langle n', s \rangle \xrightarrow{e'} \langle t'_2, s' \rangle$ with $t_2 \sim_p t'_2$, meaning that n' contains a summand $e' \cdot t'_2$. Since t_2 and t'_2 are normal forms and have sizes smaller than n and n', by the induction hypotheses if $t_2 \sim_p t'_2$ then $t_2 =_{AC} t'_2$;
  - if $t_1 \equiv \phi'$, then $\langle n, s \rangle \rightarrow \langle t_2, s \rangle$, if $test(\phi', s)$ holds, so $n \sim_p n'$ implies $\langle n', s \rangle \rightarrow \langle t'_2, s \rangle$ with $t_2 \sim_p t'_2$, if $test(\phi', s)$ holds, meaning that n' contains a summand $\phi' \cdot t'_2$. Since t_2 and t'_2 are normal forms and have sizes smaller than n and n', by the induction hypotheses if $t_2 \sim_p t'_2$ then $t_2 =_{AC} t'_2$;
  - if $t_1 \equiv e_1 \parallel \cdots \parallel e_l$, then $\langle n, s \rangle \xrightarrow{\{e_1, \cdots, e_l\}} \langle t_2, s' \rangle$, so $n \sim_p n'$ implies $\langle n', s \rangle \xrightarrow{\{e_1, \cdots, e_l\}} \langle t'_2, s' \rangle$ with $t_2 \sim_p t'_2$, meaning that n' contains a summand $(e_1 \parallel \cdots \parallel e_l) \cdot t'_2$. Since t_2 and t'_2 are normal forms and have sizes smaller than n and n', by the induction hypotheses if $t_2 \sim_p t'_2$ then $t_2 =_{AC} t'_2$;
  - if $t_1 \equiv \phi_1 \parallel \cdots \parallel \phi_l$, then $\langle n, s \rangle \rightarrow \langle t_2, s \rangle$, if $test(\phi_1, s), \cdots, test(\phi_l, s)$ hold, so $n \sim_p n'$ implies $\langle n', s \rangle \rightarrow \langle t'_2, s \rangle$ with $t_2 \sim_p t'_2$, if $test(\phi_1, s), \cdots, test(\phi_l, s)$ hold, meaning that n' contains a summand $(\phi_1 \parallel \cdots \parallel \phi_l) \cdot t'_2$. Since t_2 and t'_2 are normal forms and have sizes smaller than n and n', by the induction hypotheses if $t_2 \sim_p t'_2$ then $t_2 =_{AC} t'_2$.

So, we get $n =_{AC} n'$.

Finally, let s and t be basic $APTC_G$ terms, and $s \sim_p t$, there are normal forms n and n', such that $s = n$ and $t = n'$. The soundness theorem of $APTC_G$ modulo pomset bisimulation equivalence (see Theorem 6.21) yields $s \sim_p n$ and $t \sim_p n'$, so $n \sim_p s \sim_p t \sim_p n'$. Since if $n \sim_p n'$ then $n =_{AC} n'$, $s = n =_{AC} n' = t$, as desired.

(2) It can be proven similarly as (1).

(3) It can be proven similarly as (1).   □

Theorem 6.23 (Sufficient determinacy). *All related data environments with respect to $APTC_G$ can be sufficiently deterministic.*

Proof. It only needs to check $effect(t, s)$ function is deterministic, and is sufficient to induct on the structure of term t. The new matter is the case $t = t_1 \between t_2$, because t_1 and t_2 may be in race condition, the whole thing is $t_1 \between t_2 = t_1 \cdot t_2 + t_2 \cdot t_1 + t_1 \parallel t_2 + t_1 \mid t_2$. We can make $effect(t)$ be sufficiently deterministic: eliminating non-determinacy caused by race condition during modeling time by use of empty event ϵ. We can make $t = t_1 \parallel t_2$ $(t_1 \% t_2)$ be $t = (\epsilon \cdot t_1) \parallel t_2$ or $t = t_1 \parallel (\epsilon \cdot t_2)$ during modeling phase, and then $effect(t, s)$ becomes sufficiently deterministic. $\square$

6.4 Recursion

In this subsection, we introduce recursion to capture infinite processes based on $APTC_G$. In the following, E, F, G are recursion specifications, X, Y, Z are recursive variables.

Definition 6.24 (Guarded recursive specification). A recursive specification

$$X_1 = t_1(X_1, \cdots, X_n)$$

$$\cdots$$

$$X_n = t_n(X_1, \cdots, X_n)$$

is guarded if the right-hand sides of its recursive equations can be adapted to the form by applications of the axioms in $APTC$ and replacing recursion variables by the right-hand sides of their recursive equations,

$$(a_{11} \parallel \cdots \parallel a_{1i_1}) \cdot s_1(X_1, \cdots, X_n) + \cdots + (a_{k1} \parallel \cdots \parallel a_{ki_k}) \cdot s_k(X_1, \cdots, X_n) + (b_{11} \parallel \cdots \parallel b_{1j_1}) + \cdots$$
$$+ (b_{1j_1} \parallel \cdots \parallel b_{lj_l})$$

where $a_{11}, \cdots, a_{1i_1}, a_{k1}, \cdots, a_{ki_k}, b_{11}, \cdots, b_{1j_1}, b_{1j_1}, \cdots, b_{lj_l} \in \mathbb{E}$, and the sum above is allowed to be empty, in which case it represents the deadlock δ. And there does not exist an infinite sequence of ϵ-transitions $\langle X|E \rangle \rightarrow \langle X'|E \rangle \rightarrow \langle X''|E \rangle \rightarrow \cdots$.

Theorem 6.25 (Conservativity of $APTC_G$ with guarded recursion). *$APTC_G$ with guarded recursion is a conservative extension of $APTC_G$.*

Proof. Since the transition rules of $APTC_G$ are source-dependent, and the transition rules for guarded recursion in Table 6.7 contain only a fresh constant in their source, so the transition rules of $APTC_G$ with guarded recursion are conservative extensions of those of $APTC_G$. $\square$

Theorem 6.26 (Congruence theorem of $APTC_G$ with guarded recursion). *Truly concurrent bisimulation equivalences $\sim_p$, $\sim_s$ and $\sim_{hp}$ are all congruences with respect to $APTC_G$ with guarded recursion.*

Proof. It follows the following two facts:

Table 6.7 Transition rules of guarded recursion.

$$\frac{\langle t_i(\langle X_1|E\rangle, \cdots, \langle X_n|E\rangle), s\rangle \xrightarrow{\{e_1,\cdots,e_k\}} \langle \sqrt{}, s'\rangle}{\langle \langle X_i|E\rangle, s\rangle \xrightarrow{\{e_1,\cdots,e_k\}} \langle \sqrt{}, s'\rangle}$$

$$\frac{\langle t_i(\langle X_1|E\rangle, \cdots, \langle X_n|E\rangle), s\rangle \xrightarrow{\{e_1,\cdots,e_k\}} \langle y, s'\rangle}{\langle \langle X_i|E\rangle, s\rangle \xrightarrow{\{e_1,\cdots,e_k\}} \langle y, s'\rangle}$$

1. in a guarded recursive specification, right-hand sides of its recursive equations can be adapted to the form by applications of the axioms in $APTC_G$ and replacing recursion variables by the right-hand sides of their recursive equations;
2. truly concurrent bisimulation equivalences $\sim_p$, $\sim_s$ and $\sim_{hp}$ are all congruences with respect to all operators of $APTC_G$. □

Theorem 6.27 (Elimination theorem of $APTC_G$ with linear recursion). *Each process term in $APTC_G$ with linear recursion is equal to a process term $\langle X_1|E\rangle$ with E a linear recursive specification.*

Proof. By applying structural induction with respect to term size, each process term t_1 in $APTC_G$ with linear recursion generates a process can be expressed in the form of equations

$$t_i = (a_{i11} \parallel \cdots \parallel a_{i1i_1})t_{i1} + \cdots + (a_{ik_i1} \parallel \cdots \parallel a_{ik_ii_k})t_{ik_i} + (b_{i11} \parallel \cdots \parallel b_{i1i_1}) + \cdots + (b_{il_i1} \parallel \cdots \parallel b_{il_ii_l})$$

for $i \in \{1, \cdots, n\}$. Let the linear recursive specification E consist of the recursive equations

$$X_i = (a_{i11} \parallel \cdots \parallel a_{i1i_1})X_{i1} + \cdots + (a_{ik_i1} \parallel \cdots \parallel a_{ik_ii_k})X_{ik_i} + (b_{i11} \parallel \cdots \parallel b_{i1i_1}) + \cdots$$
$$+ (b_{il_i1} \parallel \cdots \parallel b_{il_ii_l})$$

for $i \in \{1, \cdots, n\}$. Replacing X_i by t_i for $i \in \{1, \cdots, n\}$ is a solution for E, RSP yields $t_1 = \langle X_1|E\rangle$. □

Theorem 6.28 (Soundness of $APTC_G$ with guarded recursion). *Let x and y be $APTC_G$ with guarded recursion terms. If $APTC_G$ with guarded recursion $\vdash x = y$, then*
   (1) $x \sim_s y$.
   (2) $x \sim_p y$.
   (3) $x \sim_{hp} y$.

Proof. (1) Since step bisimulation $\sim_s$ is both an equivalent and a congruent relation with respect to $APTC_G$ with guarded recursion, we only need to check if each axiom in Table 4.16 is sound modulo step bisimulation equivalence. We leave them as exercises to the readers.

(2) Since pomset bisimulation $\sim_p$ is both an equivalent and a congruent relation with respect to the guarded recursion, we only need to check if each axiom in Table 4.16 is sound modulo pomset bisimulation equivalence. We leave them as exercises to the readers.

(3) Since hp-bisimulation $\sim_{hp}$ is both an equivalent and a congruent relation with respect to guarded recursion, we only need to check if each axiom in Table 4.16 is sound modulo hp-bisimulation equivalence. We leave them as exercises to the readers. $\square$

Theorem 6.29 (Completeness of $APTC_G$ with linear recursion). *Let p and q be closed $APTC_G$ with linear recursion terms, then,*

(1) if $p \sim_s q$ then $p = q$.
(2) if $p \sim_p q$ then $p = q$.
(3) if $p \sim_{hp} q$ then $p = q$.

Proof. Firstly, by the elimination theorem of $APTC_G$ with guarded recursion (see Theorem 6.27), we know that each process term in $APTC_G$ with linear recursion is equal to a process term $\langle X_1 | E \rangle$ with E a linear recursive specification. And for the simplicity, without loss of generalization, we do not consider empty event ϵ, just because recursion with ϵ is similar to that with silent event τ, please refer to the proof of Theorem 4.64 for details.

It remains to prove the following cases.

(1) If $\langle X_1 | E_1 \rangle \sim_s \langle Y_1 | E_2 \rangle$ for linear recursive specification E_1 and E_2, then $\langle X_1 | E_1 \rangle = \langle Y_1 | E_2 \rangle$.

Let E_1 consist of recursive equations $X = t_X$ for $X \in \mathcal{X}$ and E_2 consists of recursion equations $Y = t_Y$ for $Y \in \mathcal{Y}$. Let the linear recursive specification E consist of recursion equations $Z_{XY} = t_{XY}$, and $\langle X | E_1 \rangle \sim_s \langle Y | E_2 \rangle$, and t_{XY} consists of the following summands:

1. t_{XY} contains a summand $(a_1 \| \cdots \| a_m)Z_{X'Y'}$ iff t_X contains the summand $(a_1 \| \cdots \| a_m)X'$ and t_Y contains the summand $(a_1 \| \cdots \| a_m)Y'$ such that $\langle X' | E_1 \rangle \sim_s \langle Y' | E_2 \rangle$;
2. t_{XY} contains a summand $b_1 \| \cdots \| b_n$ iff t_X contains the summand $b_1 \| \cdots \| b_n$ and t_Y contains the summand $b_1 \| \cdots \| b_n$.

Let σ map recursion variable X in E_1 to $\langle X | E_1 \rangle$, and let π map recursion variable Z_{XY} in E to $\langle X | E_1 \rangle$. So, $\sigma((a_1 \| \cdots \| a_m)X') \equiv (a_1 \| \cdots \| a_m)\langle X' | E_1 \rangle \equiv \pi((a_1 \| \cdots \| a_m)Z_{X'Y'})$, so by RDP, we get $\langle X | E_1 \rangle = \sigma(t_X) = \pi(t_{XY})$. Then by RSP, $\langle X | E_1 \rangle = \langle Z_{XY} | E \rangle$, particularly, $\langle X_1 | E_1 \rangle = \langle Z_{X_1 Y_1} | E \rangle$. Similarly, we can obtain $\langle Y_1 | E_2 \rangle = \langle Z_{X_1 Y_1} | E \rangle$. Finally, $\langle X_1 | E_1 \rangle = \langle Z_{X_1 Y_1} | E \rangle = \langle Y_1 | E_2 \rangle$, as desired.

(2) If $\langle X_1 | E_1 \rangle \sim_p \langle Y_1 | E_2 \rangle$ for linear recursive specification E_1 and E_2, then $\langle X_1 | E_1 \rangle = \langle Y_1 | E_2 \rangle$.

It can be proven similarly to (1), we omit it.

(3) If $\langle X_1 | E_1 \rangle \sim_{hp} \langle Y_1 | E_2 \rangle$ for linear recursive specification E_1 and E_2, then $\langle X_1 | E_1 \rangle = \langle Y_1 | E_2 \rangle$.

It can be proven similarly to (1), we omit it. $\square$

Table 6.8 Transition rule of the silent step.

$$\frac{}{\langle \tau_\phi, s \rangle \to \langle \sqrt{}, s \rangle} \quad \text{if } test(\tau_\phi, s)$$

$$\frac{}{\langle \tau, s \rangle \xrightarrow{\tau} \langle \sqrt{}, \tau(s) \rangle}$$

6.5 Abstraction

To abstract away from the internal implementations of a program, and verify that the program exhibits the desired external behaviors, the silent step τ and abstraction operator τ_I are introduced, where $I \subseteq \mathbb{E} \cup G_{at}$ denotes the internal events or guards. The silent step τ represents the internal events, and τ_ϕ denotes the internal guards, when we consider the external behaviors of a process, τ steps can be removed, that is, τ steps must keep silent. The transition rule of τ is shown in Table 6.8. In the following, let the atomic event e range over $\mathbb{E} \cup \{\epsilon\} \cup \{\delta\} \cup \{\tau\}$, and ϕ range over $G \cup \{\tau\}$, and let the communication function $\gamma : \mathbb{E} \cup \{\tau\} \times \mathbb{E} \cup \{\tau\} \to \mathbb{E} \cup \{\delta\}$, with each communication involved τ resulting in δ. We use $\tau(s)$ to denote $effect(\tau, s)$, for the fact that τ only change the state of internal data environment, that is, for the external data environments, $s = \tau(s)$.

In section 6.1, we introduce τ into event structure, and also give the concept of weakly true concurrency. In this subsection, we give the concepts of rooted branching truly concurrent bisimulation equivalences, based on these concepts, we can design the axiom system of the silent step τ and the abstraction operator τ_I.

Definition 6.30 (Branching pomset, step bisimulation). Assume a special termination predicate $\downarrow$, and let $\sqrt{}$ represent a state with $\sqrt{} \downarrow$. Let $\mathcal{E}_1$, $\mathcal{E}_2$ be PESs. A branching pomset bisimulation is a relation $R \subseteq \langle \mathcal{C}(\mathcal{E}_1), S \rangle \times \langle \mathcal{C}(\mathcal{E}_2), S \rangle$, such that:

1. if $(\langle C_1, s \rangle, \langle C_2, s \rangle) \in R$, and $\langle C_1, s \rangle \xrightarrow{X} \langle C_1', s' \rangle$ then

   - either $X \equiv \tau^*$, and $(\langle C_1', s' \rangle, \langle C_2, s \rangle) \in R$ with $s' \in \tau(s)$;
   - or there is a sequence of (zero or more) τ-transitions $\langle C_2, s \rangle \xrightarrow{\tau^*} \langle C_2^0, s^0 \rangle$, such that $(\langle C_1, s \rangle, \langle C_2^0, s^0 \rangle) \in R$ and $\langle C_2^0, s^0 \rangle \xRightarrow{X} \langle C_2', s' \rangle$ with $(\langle C_1', s' \rangle, \langle C_2', s' \rangle) \in R$;

2. if $(\langle C_1, s \rangle, \langle C_2, s \rangle) \in R$, and $\langle C_2, s \rangle \xrightarrow{X} \langle C_2', s' \rangle$ then

   - either $X \equiv \tau^*$, and $(\langle C_1, s \rangle, \langle C_2', s' \rangle) \in R$;
   - or there is a sequence of (zero or more) τ-transitions $\langle C_1, s \rangle \xrightarrow{\tau^*} \langle C_1^0, s^0 \rangle$, such that $(\langle C_1^0, s^0 \rangle, \langle C_2, s \rangle) \in R$ and $\langle C_1^0, s^0 \rangle \xRightarrow{X} \langle C_1', s' \rangle$ with $(\langle C_1', s' \rangle, \langle C_2', s' \rangle) \in R$;

3. if $(\langle C_1, s \rangle, \langle C_2, s \rangle) \in R$ and $\langle C_1, s \rangle \downarrow$, then there is a sequence of (zero or more) τ-transitions $\langle C_2, s \rangle \xrightarrow{\tau^*} \langle C_2^0, s^0 \rangle$ such that $(\langle C_1, s \rangle, \langle C_2^0, s^0 \rangle) \in R$ and $\langle C_2^0, s^0 \rangle \downarrow$;

4. if $(\langle C_1, s \rangle, \langle C_2, s \rangle) \in R$ and $\langle C_2, s \rangle \downarrow$, then there is a sequence of (zero or more) τ-transitions $\langle C_1, s \rangle \xrightarrow{\tau^*} \langle C_1^0, s^0 \rangle$ such that $(\langle C_1^0, s^0 \rangle, \langle C_2, s \rangle) \in R$ and $\langle C_1^0, s^0 \rangle \downarrow$.

We say that $\mathcal{E}_1, \mathcal{E}_2$ are branching pomset bisimilar, written $\mathcal{E}_1 \approx_{bp} \mathcal{E}_2$, if there exists a branching pomset bisimulation R, such that $(\langle \emptyset, \emptyset \rangle, \langle \emptyset, \emptyset \rangle) \in R$.

By replacing pomset transitions with steps, we can get the definition of branching step bisimulation. When PESs $\mathcal{E}_1$ and $\mathcal{E}_2$ are branching step bisimilar, we write $\mathcal{E}_1 \approx_{bs} \mathcal{E}_2$.

Definition 6.31 (Rooted branching pomset, step bisimulation). Assume a special termination predicate $\downarrow$, and let $\sqrt{}$ represent a state with $\sqrt{} \downarrow$. Let $\mathcal{E}_1, \mathcal{E}_2$ be PESs. A rooted branching pomset bisimulation is a relation $R \subseteq \langle \mathcal{C}(\mathcal{E}_1), S \rangle \times \langle \mathcal{C}(\mathcal{E}_2), S \rangle$, such that:

1. if $(\langle C_1, s \rangle, \langle C_2, s \rangle) \in R$, and $\langle C_1, s \rangle \xrightarrow{X} \langle C_1', s' \rangle$ then $\langle C_2, s \rangle \xrightarrow{X} \langle C_2', s' \rangle$ with $\langle C_1', s' \rangle \approx_{bp} \langle C_2', s' \rangle$;
2. if $(\langle C_1, s \rangle, \langle C_2, s \rangle) \in R$, and $\langle C_2, s \rangle \xrightarrow{X} \langle C_2', s' \rangle$ then $\langle C_1, s \rangle \xrightarrow{X} \langle C_1', s' \rangle$ with $\langle C_1', s' \rangle \approx_{bp} \langle C_2', s' \rangle$;
3. if $(\langle C_1, s \rangle, \langle C_2, s \rangle) \in R$ and $\langle C_1, s \rangle \downarrow$, then $\langle C_2, s \rangle \downarrow$;
4. if $(\langle C_1, s \rangle, \langle C_2, s \rangle) \in R$ and $\langle C_2, s \rangle \downarrow$, then $\langle C_1, s \rangle \downarrow$.

We say that $\mathcal{E}_1, \mathcal{E}_2$ are rooted branching pomset bisimilar, written $\mathcal{E}_1 \approx_{rbp} \mathcal{E}_2$, if there exists a rooted branching pomset bisimulation R, such that $(\langle \emptyset, \emptyset \rangle, \langle \emptyset, \emptyset \rangle) \in R$.

By replacing pomset transitions with steps, we can get the definition of rooted branching step bisimulation. When PESs $\mathcal{E}_1$ and $\mathcal{E}_2$ are rooted branching step bisimilar, we write $\mathcal{E}_1 \approx_{rbs} \mathcal{E}_2$.

Definition 6.32 (Branching (hereditary) history-preserving bisimulation). Assume a special termination predicate $\downarrow$, and let $\sqrt{}$ represent a state with $\sqrt{} \downarrow$. A branching history-preserving (hp-) bisimulation is a weakly posetal relation $R \subseteq \langle \mathcal{C}(\mathcal{E}_1), S \rangle \overline{\times} \langle \mathcal{C}(\mathcal{E}_2), S \rangle$ such that:

1. if $(\langle C_1, s \rangle, f, \langle C_2, s \rangle) \in R$, and $\langle C_1, s \rangle \xrightarrow{e_1} \langle C_1', s' \rangle$ then

   - either $e_1 \equiv \tau$, and $(\langle C_1', s' \rangle, f[e_1 \mapsto \tau^{e_1}], \langle C_2, s \rangle) \in R$;
   - or there is a sequence of (zero or more) τ-transitions $\langle C_2, s \rangle \xrightarrow{\tau^*} \langle C_2^0, s^0 \rangle$, such that $(\langle C_1, s \rangle, f, \langle C_2^0, s^0 \rangle) \in R$ and $\langle C_2^0, s^0 \rangle \xrightarrow{e_2} \langle C_2', s' \rangle$ with $(\langle C_1', s' \rangle, f[e_1 \mapsto e_2], \langle C_2', s' \rangle) \in R$;

2. if $(\langle C_1, s \rangle, f, \langle C_2, s \rangle) \in R$, and $\langle C_2, s \rangle \xrightarrow{e_2} \langle C_2', s' \rangle$ then

   - either $e_2 \equiv \tau$, and $(\langle C_1, s \rangle, f[e_2 \mapsto \tau^{e_2}], \langle C_2', s' \rangle) \in R$;
   - or there is a sequence of (zero or more) τ-transitions $\langle C_1, s \rangle \xrightarrow{\tau^*} \langle C_1^0, s^0 \rangle$, such that $(\langle C_1^0, s^0 \rangle, f, \langle C_2, s \rangle) \in R$ and $\langle C_1^0, s^0 \rangle \xrightarrow{e_1} \langle C_1', s' \rangle$ with $(\langle C_1', s' \rangle, f[e_2 \mapsto e_1], \langle C_2', s' \rangle) \in R$;

3. if $(\langle C_1, s \rangle, f, \langle C_2, s \rangle) \in R$ and $\langle C_1, s \rangle \downarrow$, then there is a sequence of (zero or more) τ-transitions $\langle C_2, s \rangle \xrightarrow{\tau^*} \langle C_2^0, s^0 \rangle$ such that $(\langle C_1, s \rangle, f, \langle C_2^0, s^0 \rangle) \in R$ and $\langle C_2^0, s^0 \rangle \downarrow$;

4. if $((\langle C_1, s \rangle, f, \langle C_2, s \rangle) \in R$ and $\langle C_2, s \rangle \downarrow$, then there is a sequence of (zero or more) τ-transitions $\langle C_1, s \rangle \xrightarrow{\tau^*} \langle C_1^0, s^0 \rangle$ such that $((\langle C_1^0, s^0 \rangle, f, \langle C_2, s \rangle) \in R$ and $\langle C_1^0, s^0 \rangle \downarrow$.

$\mathcal{E}_1, \mathcal{E}_2$ are branching history-preserving (hp-)bisimilar and are written $\mathcal{E}_1 \approx_{bhp} \mathcal{E}_2$ if there exists a branching hp-bisimulation R such that $((\emptyset, \emptyset), \emptyset, (\emptyset, \emptyset)) \in R$.

A branching hereditary history-preserving (hhp-)bisimulation is a downward closed branching hp-bisimulation. $\mathcal{E}_1, \mathcal{E}_2$ are branching hereditary history-preserving (hhp-)bisimilar and are written $\mathcal{E}_1 \approx_{bhhp} \mathcal{E}_2$.

Definition 6.33 (Rooted branching (hereditary) history-preserving bisimulation). Assume a special termination predicate $\downarrow$, and let $\sqrt{}$ represent a state with $\sqrt{} \downarrow$. A rooted branching history-preserving (hp-) bisimulation is a weakly posetal relation $R \subseteq \langle \mathcal{C}(\mathcal{E}_1), S \rangle \overline{\times} \langle \mathcal{C}(\mathcal{E}_2), S \rangle$ such that:

1. if $((\langle C_1, s \rangle, f, \langle C_2, s \rangle) \in R$, and $\langle C_1, s \rangle \xrightarrow{e_1} \langle C_1', s' \rangle$, then $\langle C_2, s \rangle \xrightarrow{e_2} \langle C_2', s' \rangle$ with $\langle C_1', s' \rangle \approx_{bhp} \langle C_2', s' \rangle$;
2. if $((\langle C_1, s \rangle, f, \langle C_2, s \rangle) \in R$, and $\langle C_2, s \rangle \xrightarrow{e_2} \langle C_2', s' \rangle$, then $\langle C_1, s \rangle \xrightarrow{e_1} \langle C_1', s' \rangle$ with $\langle C_1', s' \rangle \approx_{bhp} \langle C_2', s' \rangle$;
3. if $((\langle C_1, s \rangle, f, \langle C_2, s \rangle) \in R$ and $\langle C_1, s \rangle \downarrow$, then $\langle C_2, s \rangle \downarrow$;
4. if $((\langle C_1, s \rangle, f, \langle C_2, s \rangle) \in R$ and $\langle C_2, s \rangle \downarrow$, then $\langle C_1, s \rangle \downarrow$.

$\mathcal{E}_1, \mathcal{E}_2$ are rooted branching history-preserving (hp-)bisimilar and are written $\mathcal{E}_1 \approx_{rbhp} \mathcal{E}_2$ if there exists a rooted branching hp-bisimulation R such that $((\emptyset, \emptyset), \emptyset, (\emptyset, \emptyset)) \in R$.

A rooted branching hereditary history-preserving (hhp-)bisimulation is a downward closed rooted branching hp-bisimulation. $\mathcal{E}_1, \mathcal{E}_2$ are rooted branching hereditary history-preserving (hhp-)bisimilar and are written $\mathcal{E}_1 \approx_{rbhhp} \mathcal{E}_2$.

Definition 6.34 (Guarded linear recursive specification). A linear recursive specification E is guarded if there does not exist an infinite sequence of τ-transitions $\langle X|E \rangle \xrightarrow{\tau} \langle X'|E \rangle \xrightarrow{\tau} \langle X''|E \rangle \xrightarrow{\tau} \cdots$, and there does not exist an infinite sequence of ϵ-transitions $\langle X|E \rangle \to \langle X'|E \rangle \to \langle X''|E \rangle \to \cdots$.

Theorem 6.35 (Conservativity of $APTC_G$ with silent step and guarded linear recursion). *$APTC_G$ with silent step and guarded linear recursion is a conservative extension of $APTC_G$ with linear recursion.*

Proof. Since the transition rules of $APTC_G$ with linear recursion are source-dependent, and the transition rules for silent step in Table 6.8 contain only a fresh constant τ in their source, so the transition rules of $APTC_G$ with silent step and guarded linear recursion are conservative extensions of those of $APTC_G$ with linear recursion. □

Theorem 6.36 (Congruence theorem of $APTC_G$ with silent step and guarded linear recursion). *Rooted branching truly concurrent bisimulation equivalences $\approx_{rbp}, \approx_{rbs}$ and $\approx_{rbhp}$ are all congruences with respect to $APTC_G$ with silent step and guarded linear recursion.*

Proof. It follows the following three facts:

Table 6.9 Axioms of silent step.

No.	Axiom
$B1$	$e \cdot \tau = e$
$B2$	$e \cdot (\tau \cdot (x + y) + x) = e \cdot (x + y)$
$B3$	$x \parallel \tau = x$
$G26$	$\tau_\phi \cdot x = x$
$G27$	$x \cdot \tau_\phi = x$
$G28$	$x \parallel \tau_\phi = x$

1. in a guarded linear recursive specification, right-hand sides of its recursive equations can be adapted to the form by applications of the axioms in $APTC_G$ and replacing recursion variables by the right-hand sides of their recursive equations;

2. truly concurrent bisimulation equivalences $\sim_p$, $\sim_s$ and $\sim_{hp}$ are all congruences with respect to all operators of $APTC_G$, while truly concurrent bisimulation equivalences $\sim_p$, $\sim_s$ and $\sim_{hp}$ imply the corresponding rooted branching truly concurrent bisimulations $\approx_{rbp}$, $\approx_{rbs}$ and $\approx_{rbhp}$ (see Proposition 2.23), so rooted branching truly concurrent bisimulations $\approx_{rbp}$, $\approx_{rbs}$ and $\approx_{rbhp}$ are all congruences with respect to all operators of $APTC_G$;

3. While $\mathbb{E}$ is extended to $\mathbb{E} \cup \{\tau\}$, and G is extended to $G \cup \{\tau\}$, it can be proved that rooted branching truly concurrent bisimulations $\approx_{rbp}$, $\approx_{rbs}$ and $\approx_{rbhp}$ are all congruences with respect to all operators of $APTC_G$, we omit it. $\qquad\square$

We design the axioms for the silent step τ in Table 6.9.

Theorem 6.37 (Elimination theorem of $APTC_G$ with silent step and guarded linear recursion). *Each process term in $APTC_G$ with silent step and guarded linear recursion is equal to a process term $\langle X_1 | E \rangle$ with E a guarded linear recursive specification.*

Proof. By applying structural induction with respect to term size, each process term t_1 in $APTC_G$ with silent step and guarded linear recursion generates a process can be expressed in the form of equations

$$t_i = (a_{i11} \parallel \cdots \parallel a_{i1l_i})t_{i1} + \cdots + (a_{ik_i1} \parallel \cdots \parallel a_{ik_ii_k})t_{ik_i} + (b_{i11} \parallel \cdots \parallel b_{i1l_i}) + \cdots + (b_{il_i1} \parallel \cdots \parallel b_{il_ii_l})$$

for $i \in \{1, \cdots, n\}$. Let the linear recursive specification E consist of the recursive equations

$$X_i = (a_{i11} \parallel \cdots \parallel a_{i1l_i})X_{i1} + \cdots + (a_{ik_i1} \parallel \cdots \parallel a_{ik_ii_k})X_{ik_i} + (b_{i11} \parallel \cdots \parallel b_{i1l_i}) + \cdots$$
$$+ (b_{il_i1} \parallel \cdots \parallel b_{il_ii_l})$$

for $i \in \{1, \cdots, n\}$. Replacing X_i by t_i for $i \in \{1, \cdots, n\}$ is a solution for E, RSP yields $t_1 = \langle X_1 | E \rangle$. $\qquad\square$

Theorem 6.38 (Soundness of $APTC_G$ with silent step and guarded linear recursion). *Let x and y be $APTC_G$ with silent step and guarded linear recursion terms. If $APTC_G$ with silent step and guarded linear recursion $\vdash x = y$, then*

   (1) $x \approx_{rbs} y$.
   (2) $x \approx_{rbp} y$.
   (3) $x \approx_{rbhp} y$.

Proof. (1) Since rooted branching step bisimulation $\approx_{rbs}$ is both an equivalent and a congruent relation with respect to $APTC_G$ with silent step and guarded linear recursion, we only need to check if each axiom in Table 6.9 is sound modulo rooted branching step bisimulation equivalence. We leave them as exercises to the readers.

(2) Since rooted branching pomset bisimulation $\approx_{rbp}$ is both an equivalent and a congruent relation with respect to $APTC_G$ with silent step and guarded linear recursion, we only need to check if each axiom in Table 6.9 is sound modulo rooted branching pomset bisimulation $\approx_{rbp}$. We leave them as exercises to the readers.

(3) Since rooted branching hp-bisimulation $\approx_{rbhp}$ is both an equivalent and a congruent relation with respect to $APTC_G$ with silent step and guarded linear recursion, we only need to check if each axiom in Table 6.9 is sound modulo rooted branching hp-bisimulation $\approx_{rbhp}$. We leave them as exercises to the readers. $\qquad\square$

Theorem 6.39 (Completeness of $APTC_G$ with silent step and guarded linear recursion). *Let p and q be closed $APTC_G$ with silent step and guarded linear recursion terms, then,*

   (1) if $p \approx_{rbs} q$ then $p = q$.
   (2) if $p \approx_{rbp} q$ then $p = q$.
   (3) if $p \approx_{rbhp} q$ then $p = q$.

Proof. Firstly, by the elimination theorem of $APTC_G$ with silent step and guarded linear recursion (see Theorem 6.37), we know that each process term in $APTC_G$ with silent step and guarded linear recursion is equal to a process term $\langle X_1|E\rangle$ with E a guarded linear recursive specification.

It remains to prove the following cases.

(1) If $\langle X_1|E_1\rangle \approx_{rbs} \langle Y_1|E_2\rangle$ for guarded linear recursive specification E_1 and E_2, then $\langle X_1|E_1\rangle = \langle Y_1|E_2\rangle$.

Firstly, the recursive equation $W = \tau + \cdots + \tau$ with $W \not\equiv X_1$ in E_1 and E_2, can be removed, and the corresponding summands aW are replaced by a, to get E_1' and E_2', by use of the axioms RDP, $A3$ and $B1$, and $\langle X|E_1\rangle = \langle X|E_1'\rangle$, $\langle Y|E_2\rangle = \langle Y|E_2'\rangle$.

Let E_1 consist of recursive equations $X = t_X$ for $X \in \mathcal{X}$ and E_2 consist of recursion equations $Y = t_Y$ for $Y \in \mathcal{Y}$, and are not the form $\tau + \cdots + \tau$. Let the guarded linear recursive specification E consists of recursion equations $Z_{XY} = t_{XY}$, and $\langle X|E_1\rangle \approx_{rbs} \langle Y|E_2\rangle$, and t_{XY} consists of the following summands:

1. t_{XY} contains a summand $(a_1 \parallel \cdots \parallel a_m)Z_{X'Y'}$ iff t_X contains the summand $(a_1 \parallel \cdots \parallel a_m)X'$ and t_Y contains the summand $(a_1 \parallel \cdots \parallel a_m)Y'$ such that $\langle X'|E_1\rangle \approx_{rbs} \langle Y'|E_2\rangle$;

2. t_{XY} contains a summand $b_1 \parallel \cdots \parallel b_n$ iff t_X contains the summand $b_1 \parallel \cdots \parallel b_n$ and t_Y contains the summand $b_1 \parallel \cdots \parallel b_n$;

3. t_{XY} contains a summand $\tau Z_{X'Y}$ iff $XY \not\equiv X_1 Y_1$, t_X contains the summand $\tau X'$, and $\langle X'|E_1 \rangle \approx_{rbs} \langle Y|E_2 \rangle$;

4. t_{XY} contains a summand $\tau Z_{XY'}$ iff $XY \not\equiv X_1 Y_1$, t_Y contains the summand $\tau Y'$, and $\langle X|E_1 \rangle \approx_{rbs} \langle Y'|E_2 \rangle$.

Since E_1 and E_2 are guarded, E is guarded. Constructing the process term u_{XY} consist of the following summands:

1. u_{XY} contains a summand $(a_1 \parallel \cdots \parallel a_m)\langle X'|E_1 \rangle$ iff t_X contains the summand $(a_1 \parallel \cdots \parallel a_m)X'$ and t_Y contains the summand $(a_1 \parallel \cdots \parallel a_m)Y'$ such that $\langle X'|E_1 \rangle \approx_{rbs} \langle Y'|E_2 \rangle$;

2. u_{XY} contains a summand $b_1 \parallel \cdots \parallel b_n$ iff t_X contains the summand $b_1 \parallel \cdots \parallel b_n$ and t_Y contains the summand $b_1 \parallel \cdots \parallel b_n$;

3. u_{XY} contains a summand $\tau \langle X'|E_1 \rangle$ iff $XY \not\equiv X_1 Y_1$, t_X contains the summand $\tau X'$, and $\langle X'|E_1 \rangle \approx_{rbs} \langle Y|E_2 \rangle$.

Let the process term s_{XY} be defined as follows:

1. $s_{XY} \triangleq \tau \langle X|E_1 \rangle + u_{XY}$ iff $XY \not\equiv X_1 Y_1$, t_Y contains the summand $\tau Y'$, and $\langle X|E_1 \rangle \approx_{rbs} \langle Y'|E_2 \rangle$;

2. $s_{XY} \triangleq \langle X|E_1 \rangle$, otherwise.

So, $\langle X|E_1 \rangle = \langle X|E_1 \rangle + u_{XY}$, and $(a_1 \parallel \cdots \parallel a_m)(\tau \langle X|E_1 \rangle + u_{XY}) = (a_1 \parallel \cdots \parallel a_m)((\tau \langle X|E_1 \rangle + u_{XY}) + u_{XY}) = (a_1 \parallel \cdots \parallel a_m)(\langle X|E_1 \rangle + u_{XY}) = (a_1 \parallel \cdots \parallel a_m)\langle X|E_1 \rangle$, hence, $(a_1 \parallel \cdots \parallel a_m)s_{XY} = (a_1 \parallel \cdots \parallel a_m)\langle X|E_1 \rangle$.

Let σ map recursion variable X in E_1 to $\langle X|E_1 \rangle$, and let π map recursion variable Z_{XY} in E to s_{XY}. It is sufficient to prove $s_{XY} = \pi(t_{XY})$ for recursion variables Z_{XY} in E. Either $XY \equiv X_1 Y_1$ or $XY \not\equiv X_1 Y_1$, we all can get $s_{XY} = \pi(t_{XY})$. So, $s_{XY} = \langle Z_{XY}|E \rangle$ for recursive variables Z_{XY} in E is a solution for E. Then by RSP, particularly, $\langle X_1|E_1 \rangle = \langle Z_{X_1 Y_1}|E \rangle$. Similarly, we can obtain $\langle Y_1|E_2 \rangle = \langle Z_{X_1 Y_1}|E \rangle$. Finally, $\langle X_1|E_1 \rangle = \langle Z_{X_1 Y_1}|E \rangle = \langle Y_1|E_2 \rangle$, as desired.

(2) If $\langle X_1|E_1 \rangle \approx_{rbp} \langle Y_1|E_2 \rangle$ for guarded linear recursive specification E_1 and E_2, then $\langle X_1|E_1 \rangle = \langle Y_1|E_2 \rangle$.

It can be proven similarly to (1), we omit it.

(3) If $\langle X_1|E_1 \rangle \approx_{rbhb} \langle Y_1|E_2 \rangle$ for guarded linear recursive specification E_1 and E_2, then $\langle X_1|E_1 \rangle = \langle Y_1|E_2 \rangle$.

It can be proven similarly to (1), we omit it.   □

The unary abstraction operator τ_I ($I \subseteq \mathbb{E} \cup G_{at}$) renames all atomic events or atomic guards in I into τ. $APTC_G$ with silent step and abstraction operator is called $APTC_{G_\tau}$. The transition rules of operator τ_I are shown in Table 6.10.

Theorem 6.40 (Conservativity of $APTC_{G_\tau}$ with guarded linear recursion). *$APTC_{G_\tau}$ with guarded linear recursion is a conservative extension of $APTC_G$ with silent step and guarded linear recursion.*

Table 6.10   Transition rule of the abstraction operator.

$$\frac{\langle x, s\rangle \xrightarrow{e} \langle \surd, s'\rangle}{\langle \tau_I(x), s\rangle \xrightarrow{e} \langle \surd, s'\rangle} \quad e \notin I \qquad \frac{\langle x, s\rangle \xrightarrow{e} \langle x', s'\rangle}{\langle \tau_I(x), s\rangle \xrightarrow{e} \langle \tau_I(x'), s'\rangle} \quad e \notin I$$

$$\frac{\langle x, s\rangle \xrightarrow{e} \langle \surd, s'\rangle}{\langle \tau_I(x), s\rangle \xrightarrow{\tau} \langle \surd, \tau(s)\rangle} \quad e \in I \qquad \frac{\langle x, s\rangle \xrightarrow{e} \langle x', s'\rangle}{\langle \tau_I(x), s\rangle \xrightarrow{\tau} \langle \tau_I(x'), \tau(s)\rangle} \quad e \in I$$

Table 6.11   Axioms of abstraction operator.

No.	Axiom	
$TI1$	$e \notin I$	$\tau_I(e) = e$
$TI2$	$e \in I$	$\tau_I(e) = \tau$
$TI3$	$\tau_I(\delta) = \delta$	
$TI4$	$\tau_I(x + y) = \tau_I(x) + \tau_I(y)$	
$TI5$	$\tau_I(x \cdot y) = \tau_I(x) \cdot \tau_I(y)$	
$TI6$	$\tau_I(x \parallel y) = \tau_I(x) \parallel \tau_I(y)$	
$G29$	$\phi \notin I$	$\tau_I(\phi) = \phi$
$G30$	$\phi \in I$	$\tau_I(\phi) = \tau_\phi$

Proof. Since the transition rules of $APTC_G$ with silent step and guarded linear recursion are source-dependent, and the transition rules for abstraction operator in Table 6.10 contain only a fresh operator τ_I in their source, so the transition rules of $APTC_{G_\tau}$ with guarded linear recursion are conservative extensions of those of $APTC_G$ with silent step and guarded linear recursion. □

Theorem 6.41 (Congruence theorem of $APTC_{G_\tau}$ with guarded linear recursion). *Rooted branching truly concurrent bisimulation equivalences $\approx_{rbp}$, $\approx_{rbs}$ and $\approx_{rbhp}$ are all congruences with respect to $APTC_{G_\tau}$ with guarded linear recursion.*

Proof. (1) It is easy to see that rooted branching pomset bisimulation is an equivalent relation on $APTC_{G_\tau}$ with guarded linear recursion terms, we only need to prove that $\approx_{rbp}$ is preserved by the operators τ_I. It is trivial and we leave the proof as an exercise for the readers.

(2) The cases of rooted branching step bisimulation $\approx_{rbs}$, rooted branching hp-bisimulation $\approx_{rbhp}$ can be proven similarly, we omit them. □

We design the axioms for the abstraction operator τ_I in Table 6.11.

Theorem 6.42 (Soundness of $APTC_{G_\tau}$ with guarded linear recursion). *Let x and y be $APTC_{G_\tau}$ with guarded linear recursion terms. If $APTC_{G_\tau}$ with guarded linear recursion $\vdash x = y$, then*

(1) $x \approx_{rbs} y$.

(2) $x \approx_{rbp} y$.

(3) $x \approx_{rbhp} y$.

Proof. (1) Since rooted branching step bisimulation $\approx_{rbs}$ is both an equivalent and a congruent relation with respect to $APTC_{G_\tau}$ with guarded linear recursion, we only need to check if each axiom in Table 6.11 is sound modulo rooted branching step bisimulation equivalence. We leave them as exercises to the readers.

(2) Since rooted branching pomset bisimulation $\approx_{rbp}$ is both an equivalent and a congruent relation with respect to $APTC_{G_\tau}$ with guarded linear recursion, we only need to check if each axiom in Table 6.11 is sound modulo rooted branching pomset bisimulation $\approx_{rbp}$. We leave them as exercises to the readers.

(3) Since rooted branching hp-bisimulation $\approx_{rbhp}$ is both an equivalent and a congruent relation with respect to $APTC_{G_\tau}$ with guarded linear recursion, we only need to check if each axiom in Table 6.11 is sound modulo rooted branching hp-bisimulation $\approx_{rbhp}$. We leave them as exercises to the readers. $\square$

Though τ-loops are prohibited in guarded linear recursive specifications (see Definition 6.34) in a specifiable way, they can be constructed using the abstraction operator, for example, there exist τ-loops in the process term $\tau_{\{a\}}(\langle X | X = aX\rangle)$. To avoid τ-loops caused by τ_I and ensure fairness, the concept of cluster and $CFAR$ (Cluster Fair Abstraction Rule) [17] are still needed.

Theorem 6.43 (Completeness of $APTC_{G_\tau}$ with guarded linear recursion and $CFAR$). *Let p and q be closed $APTC_{G_\tau}$ with guarded linear recursion and $CFAR$ terms, then,*

(1) *if $p \approx_{rbs} q$ then $p = q$.*

(2) *if $p \approx_{rbp} q$ then $p = q$.*

(3) *if $p \approx_{rbhp} q$ then $p = q$.*

Proof. (1) For the case of rooted branching step bisimulation, the proof is following.

Firstly, in the proof of Theorem 6.39, we know that each process term p in $APTC_G$ with silent step and guarded linear recursion is equal to a process term $\langle X_1 | E\rangle$ with E a guarded linear recursive specification. And we prove if $\langle X_1 | E_1\rangle \approx_{rbs} \langle Y_1 | E_2\rangle$, then $\langle X_1 | E_1\rangle = \langle Y_1 | E_2\rangle$

The only new case is $p \equiv \tau_I(q)$. Let $q = \langle X | E\rangle$ with E a guarded linear recursive specification, so $p = \tau_I(\langle X | E\rangle)$. Then the collection of recursive variables in E can be divided into its clusters $C_1, \cdots, C_N$ for I. Let

$$(a_{1i1} \parallel \cdots \parallel a_{k_i 1 i1})Y_{i1} + \cdots + (a_{1im_i} \parallel \cdots \parallel a_{k_{im_i} im_i})Y_{im_i} + b_{1i1} \parallel \cdots \parallel b_{l_{i1} i1} + \cdots$$

$$+ b_{1im_i} \parallel \cdots \parallel b_{l_{im_i} im_i}$$

be the conflict composition of exits for the cluster C_i, with $i \in \{1, \cdots, N\}$.

For $Z \in C_i$ with $i \in \{1, \cdots, N\}$, we define

$$s_Z \triangleq (a_{\widehat{1i1}} \parallel \cdots \parallel a_{k_{i1}\widehat{i1}})\tau_I(\langle Y_{i1}|E\rangle) + \cdots + (a_{1\widehat{im_i}} \parallel \cdots \parallel a_{k_{im_i}\widehat{im_i}})\tau_I(\langle Y_{im_i}|E\rangle)$$
$$+ b_{\widehat{1i1}} \parallel \cdots \parallel b_{l_{i1}\widehat{i1}} + \cdots + b_{1\widehat{im_i}} \parallel \cdots \parallel b_{l_{im_i}\widehat{im_i}}$$

For $Z \in C_i$ and $a_1, \cdots, a_j \in \mathbb{E} \cup \{\tau\}$ with $j \in \mathbb{N}$, we have

$$(a_1 \parallel \cdots \parallel a_j)\tau_I(\langle Z|E\rangle)$$
$$= (a_1 \parallel \cdots \parallel a_j)\tau_I((a_{1i1} \parallel \cdots \parallel a_{k_{i1}i1})\langle Y_{i1}|E\rangle + \cdots + (a_{1im_i} \parallel \cdots \parallel a_{k_{im_i}im_i})\langle Y_{im_i}|E\rangle$$
$$+ b_{1i1} \parallel \cdots \parallel b_{l_{i1}i1} + \cdots + b_{1im_i} \parallel \cdots \parallel b_{l_{im_i}im_i})$$
$$= (a_1 \parallel \cdots \parallel a_j)s_Z$$

Let the linear recursive specification F contain the same recursive variables as E, for $Z \in C_i$, F contains the following recursive equation

$$Z = (a_{\widehat{1i1}} \parallel \cdots \parallel a_{k_{i1}\widehat{i1}})Y_{i1} + \cdots + (a_{1\widehat{im_i}} \parallel \cdots \parallel a_{k_{im_i}\widehat{im_i}})Y_{im_i} + b_{\widehat{1i1}} \parallel \cdots \parallel b_{l_{i1}\widehat{i1}} + \cdots$$
$$+ b_{1\widehat{im_i}} \parallel \cdots \parallel b_{l_{im_i}\widehat{im_i}}$$

It is easy to see that there is no sequence of one or more τ-transitions from $\langle Z|F\rangle$ to itself, so F is guarded.

For

$$s_Z = (a_{\widehat{1i1}} \parallel \cdots \parallel a_{k_{i1}\widehat{i1}})Y_{i1} + \cdots + (a_{1\widehat{im_i}} \parallel \cdots \parallel a_{k_{im_i}\widehat{im_i}})Y_{im_i} + b_{\widehat{1i1}} \parallel \cdots \parallel b_{l_{i1}\widehat{i1}} + \cdots$$
$$+ b_{1\widehat{im_i}} \parallel \cdots \parallel b_{l_{im_i}\widehat{im_i}}$$

is a solution for F. So, $(a_1 \parallel \cdots \parallel a_j)\tau_I(\langle Z|E\rangle) = (a_1 \parallel \cdots \parallel a_j)s_Z = (a_1 \parallel \cdots \parallel a_j)\langle Z|F\rangle$.

So,

$$\langle Z|F\rangle = (a_{\widehat{1i1}} \parallel \cdots \parallel a_{k_{i1}\widehat{i1}})\langle Y_{i1}|F\rangle + \cdots + (a_{1\widehat{im_i}} \parallel \cdots \parallel a_{k_{im_i}\widehat{im_i}})\langle Y_{im_i}|F\rangle + b_{\widehat{1i1}} \parallel \cdots \parallel b_{l_{i1}\widehat{i1}} + \cdots$$
$$+ b_{1\widehat{im_i}} \parallel \cdots \parallel b_{l_{im_i}\widehat{im_i}}$$

Hence, $\tau_I(\langle X|E\rangle = \langle Z|F\rangle)$, as desired.

(2) For the case of rooted branching pomset bisimulation, it can be proven similarly to (1), we omit it.

(3) For the case of rooted branching hp-bisimulation, it can be proven similarly to (1), we omit it. □

6.6 Hoare logic for $APTC_G$

In this section, we introduce Hoare logic for $APTC_G$. We do not introduce the preliminaries of Hoare logic, please refer to [20] for details.

Table 6.12   The proof system H.

$(H1)$ $\{wp(e, \alpha)\}e\{\alpha\}$ if $e \in \mathbb{E}$

$(H2)$ $\{\alpha\}\phi\{\alpha \cdot \phi\}$ if $\phi \in G$

$(H3)$ $\dfrac{\{\alpha\}t\{\beta\}\quad\{\alpha\}t'\{\beta\}}{\{\alpha\}t+t'\{\beta\}}$

$(H4)$ $\dfrac{\{\alpha\}t\{\alpha'\}\quad\{\alpha'\}t'\{\beta\}}{\{\alpha\}t \cdot t'\{\beta\}}$

$(H5)$ $\dfrac{\{\alpha\}t\{\alpha'\}\quad\{\beta\}t'\{\beta'\}}{\{\alpha\|\beta\}t \between t'\{\alpha'\|\beta'\}}$

$(H6)$ $\dfrac{\{\alpha\}t\{\beta\}}{\{\alpha\}\Theta(t)\{\beta\}}$

$(H7)$ $\dfrac{\{\alpha\}t\{\beta\}}{\{\alpha\}\partial_H(t)\{\beta\}}$

$(H8)$ $\dfrac{\{\alpha\}t\{\beta\}}{\{\alpha\}\tau_I(t)\{\beta\}}$

$(H9)$ $\dfrac{\alpha \to \alpha'\quad\{\alpha'\}t\{\beta'\}\quad\beta' \to \beta}{\{\alpha\}t\{\beta\}}$

$(H10)$ For $E = \{x = t_x | x \in V_E\}$ a guarded linear recursive specification $\forall y \in V_E$ and $z \in V_E$:

$$\frac{\{\alpha_x\}t_x\{\beta_x\}\quad\cdots\quad\{\alpha_y\}t_y\{\beta_y\}}{\{\alpha_z\}\langle z|E\rangle\{\beta_z\}}$$

$(H10')$ For $E = \{x = t_x | x \in V_E\}$ a guarded linear recursive specification $\forall y \in V_E$ and $z \in V_E$:

$$\frac{\{\alpha_{x_1} \| \cdots \| \alpha_{x_{nx}}\}t_x\{\beta_{x_1} \| \cdots \| \beta_{x_{nx}}\}\quad\cdots\quad\{\alpha_{y_1} \| \cdots \| \alpha_{y_{ny}}\}t_y\{\beta_{y_1} \| \cdots \| \beta_{y_{ny}}\}}{\{\alpha_{z_1} \| \cdots \| \alpha_{z_{nz}}\}\langle z|E\rangle\{\beta_{z_1} \| \cdots \| \beta_{z_{nz}}\}}$$

A partial correct formula has the form

$$\{pre\}P\{post\}$$

where *pre* are preconditions, *post* are postconditions, and P are programs. $\{pre\}P\{post\}$ means that *pre* hold, then P are executed and *post* hold. We take the guards G of $APTC_G$ as the language of conditions, and closed terms of $APTC_G$ as programs. For some condition $\alpha \in G$ and some data state $s \in S$, we denote $S \models \alpha[s]$ for $\langle \alpha, s \rangle \to \langle \sqrt{}, s \rangle$, and $S \models \alpha$ for $\forall s \in S, S \models \alpha[s]$, $S \models \{\alpha\}p\{\beta\}$ for all $s \in S$, $\mu \subseteq \mathbb{E} \cup G$, $S \models \alpha[s]$, $\langle p, s \rangle \xrightarrow{\mu} \langle p', s' \rangle$, $S \models \beta[s']$ with $s' \in S$. It is obvious that $S \models \{\alpha\}p\{\beta\} \Leftrightarrow \alpha p \approx_{rbp} (\approx_{rbs}, \approx_{rbhp})\alpha p\beta$.

We design a proof system H to deriving partial correct formulas over terms of $APTC_G$ as Table 6.12 shows. Let Γ be a set of conditions and partial correct formulas, we denote $\Gamma \vdash \{\alpha\}t\{\beta\}$ iff we can derive $\{\alpha\}t\{\beta\}$ in H, note that t does not need to be closed terms. And we write $\alpha \to \beta$ for $S \models \alpha \Rightarrow S \models \beta$.

Theorem 6.44 (Soundness of H). *Let T_{r_S} be the set of conditions that hold in S. Let p be a closed term of $APTC_{G_\tau}$ with guarded linear recursion and $CFAR$, and $\alpha, \beta \in G$ be guards.*

Then

$$Tr_S \vdash \{\alpha\}p\{\beta\} \Rightarrow APTC_{G_\tau} \text{ with guarded linear recursion and } CFAR \vdash \alpha p = \alpha p\beta$$
$$\Leftrightarrow \alpha p \approx_{rbs} (\approx_{rbp}, \approx_{rbhp})\alpha p\beta$$
$$\Leftrightarrow S \models \{\alpha\}p\{\beta\}$$

Proof. We only need to prove

$$Tr_S \vdash \{\alpha\}p\{\beta\} \Rightarrow APTC_{G_\tau} \text{ with guarded linear recursion and } CFAR \vdash \alpha p = \alpha p\beta$$

For $H1$-$H10$, by induction on the length of derivation, the soundness of $H1$-$H10$ is straightforward. We only prove the soundness of $H10'$.

Let $E = \{x_i = t_i(x_1, \cdots, x_n)|i = 1, \cdots, n\}$ be a guarded linear recursive specification. Assume that

$$Tr_S, \{\{\alpha_1 \parallel \cdots \parallel \alpha_{n_i}\}x_i\{\beta_1 \parallel \cdots \parallel \beta_{n_i}\}|i = 1, \cdots, n\} \vdash \{\alpha_1 \parallel \cdots \parallel \alpha_{n_j}\}t_j(x_1, \cdots, x_n)\{\beta_1 \parallel \cdots \parallel \beta_{n_j}\}$$

for $j = 1, \cdots, n$. We would show that $APTC_{G_\tau}$ with guarded linear recursion and $CFAR \vdash$ $(\alpha_1 \parallel \cdots \parallel \alpha_{n_j})X_j = (\alpha_1 \parallel \cdots \parallel \alpha_{n_j})X_j(\beta_1 \parallel \cdots \parallel \beta_{n_j})$.

We write recursive specifications E' and E'' for

$$E' = \{y_i = (\alpha_1 \parallel \cdots \parallel \alpha_{n_i})t_i(y_1, \cdots, y_n)|i = 1, \cdots, n\}$$

$$E'' = \{z_i = (\alpha_1 \parallel \cdots \parallel \alpha_{n_i})t_i(z_1(\beta_1 \parallel \cdots \parallel \beta_{n_1}), \cdots, z_n(\beta_1 \parallel \cdots \parallel \beta_{n_n}))|i = 1, \cdots, n\}$$

and would show that for $j = 1, \cdots, n$,

(1) $(\alpha_1 \parallel \cdots \parallel \alpha_{n_j})X_j = Y_j$;
(2) $Z_j(\beta_1 \parallel \cdots \parallel \beta_{n_j}) = Z_j$;
(3) $Z_j = Y_j$.

For (1), we have

$$(\alpha_1 \parallel \cdots \parallel \alpha_{n_j})X_j = (\alpha_1 \parallel \cdots \parallel \alpha_{n_j})t_j(X_1, \cdots, X_n)$$
$$= (\alpha_1 \parallel \cdots \parallel \alpha_{n_j})t_j((\alpha_1 \parallel \cdots \parallel \alpha_{n_1})X_1, \cdots, (\alpha_1 \parallel \cdots \parallel \alpha_{n_n})X_n)$$

by RDP, we have $(\alpha_1 \parallel \cdots \parallel \alpha_{n_j})X_j = Y_j$.

For (2), we have

$$Z_j(\beta_1 \parallel \cdots \parallel \beta_{n_j}) = (\alpha_1 \parallel \cdots \parallel \alpha_{n_j})t_j(Z_1(\beta_1 \parallel \cdots \parallel \beta_{n_1}), \cdots, Z_n(\beta_1 \parallel \cdots \parallel \beta_{n_n}))(\beta_1 \parallel \cdots \parallel \beta_{n_j})$$
$$= (\alpha_1 \parallel \cdots \parallel \alpha_{n_j})t_j(Z_1(\beta_1 \parallel \cdots \parallel \beta_{n_1}), \cdots, Z_n(\beta_1 \parallel \cdots \parallel \beta_{n_n}))$$
$$= (\alpha_1 \parallel \cdots \parallel \alpha_{n_j})t_j((Z_1(\beta_1 \parallel \cdots \parallel \beta_{n_1}))(\beta_1 \parallel \cdots \parallel \beta_{n_1}), \cdots, (Z_n(\beta_1 \parallel \cdots \parallel \beta_{n_n}))$$
$$(\beta_1 \parallel \cdots \parallel \beta_{n_n}))$$

by RDP, we have $Z_j(\beta_1 \parallel \cdots \parallel \beta_{n_j}) = Z_j$.

For (3), we have

$$Z_j = (\alpha_1 \parallel \cdots \parallel \alpha_{n_j})t_j(Z_1(\beta_1 \parallel \cdots \parallel \beta_{n_1}), \cdots, Z_n(\beta_1 \parallel \cdots \parallel \beta_{n_n}))$$
$$= (\alpha_1 \parallel \cdots \parallel \alpha_{n_j})t_j(Z_1, \cdots, Z_n)$$

by RDP, we have $Z_j = Y_j$. □

6.7 Conclusions

Based on $APTC$, we add guards into it by adopting the solution of [21] in two ways: (1) the same solution to guards to form a Boolean Algebra; (2) the similar operational semantics based on configuration, which has two parts: the processes and the data states. Finally, we get some further results as follows:

1. we design a sound and complete theory of concurrency and parallelism with guards;
2. we design a sound and complete theory of recursion including concurrency with guards;
3. we design a sound and complete theory of abstraction with guards;
4. we design a sound Hoare logic [20] including concurrency and parallelism, recursion, and abstraction.

Note that, for true concurrency, the empty action ϵ can play the role of placeholder, just like APTC with the shadow constant in subsection 4.6.1. We do not do this work and leave it as an exercise to the readers.

And like APTC for hhp-bisimulation equivalence in subsection 4.7, APTC with guards can also have a theoretic correspondence for hhp-bisimulation equivalence to that of APTC.

References

[1] M. Hennessy, R. Milner, Algebraic laws for nondeterminism and concurrency, Journal of the ACM 32 (1985) 137–161.
[2] R. Milner, Communication and Concurrency, Printice Hall, 1989.
[3] R. Milner, A Calculus of Communicating Systems, Lecture Notes in Computer Science, vol. 92, Springer, 1980.
[4] W. Fokkink, Introduction to Process Algebra, 2nd ed., Springer-Verlag, 2007.
[5] M. Nielsen, G.D. Plotkin, G. Winskel, Petri nets, event structures and domains, Part I, Theoretical Computer Science 13 (1981) 85–108.
[6] G. Winskel, Event structures, in: Wilfried Brauer, Wolfgang Reisig, Grzegorz Rozenberg (Eds.), Petri Nets: Applications and Relationships to Other Models of Concurrency, in: Lecture Notes in Computer Science, vol. 255, Springer, Berlin, 1987, pp. 325–392.
[7] G. Winskel, M. Nielsen, Models for concurrency, in: Samson Abramsky, Dov M. Gabbay, Thomas S.E. Maibaum (Eds.), Handbook of logic in Computer Science, vol. 4, Clarendon Press, Oxford, UK, 1995.
[8] M.A. Bednarczyk, Hereditary history preserving bisimulations or what is the power of the future perfect in program logics, Tech. Rep. Polish Academy of Sciences, 1991.
[9] S.B. Fröschle, T.T. Hildebrandt, On plain and hereditary history-preserving bisimulation, in: Miroslaw Kutylowski, Leszek Pacholski, Tomasz Wierzbicki (Eds.), Proceedings of MFCS'99, in: Lecture Notes in Computer Science, vol. 1672, Springer, Berlin, 1999, pp. 354–365.
[10] J. Bradfield, C. Stirling, Modal mu-calculi, in: Patrick Blackburn, Johan van Benthem, Franck Wolter (Eds.), Handbook of Modal Logic, Elsevier, Amsterdam, the Netherlands, 2006, pp. 721–756.
[11] I. Phillips, I. Ulidowski, Reverse bisimulations on stable configuration structures, in: B. Klin, P. Sobociṅski (Eds.), Proceedings of SOS'09, in: Electronic Proceedings in Theoretical Computer Science, vol. 18, Elsevier, Amsterdam, the Netherlands, 2010, pp. 62–76.
[12] I. Phillips, I. Ulidowski, A logic with reverse modalities for history-preserving bisimulations, in: Bas Luttik, Frank Valencia (Eds.), Proceedings of EXPRESS'11, in: Electronic Proceedings in Theoretical Computer Science, vol. 64, Elsevier, Amsterdam, the Netherlands, 2011, pp. 104–118.
[13] J. Gutierrez, On bisimulation and model-checking for concurrent systems with partial order semantics, Ph.D. dissertation, LFCS – University of Edinburgh, 2011.
[14] P. Baldan, S. Crafa, A logic for true concurrency, in: Paul Gastin, François Laroussinie (Eds.), Proceedings of CONCUR'10, in: Lecture Notes in Computer Science, vol. 6269, Springer, Berlin, 2010, pp. 147–161.
[15] P. Baldan, S. Crafa, A logic for true concurrency, Journal of the ACM 61 (4) (2014), 36 pages.
[16] Y. Wang, Weakly true concurrency and its logic, Manuscript, arXiv:1606.06422, 2016.
[17] F.W. Vaandrager, Verification of two communication protocols by means of process algebra, Report CS-R8608, CWI, Amsterdam, 1986.
[18] R. Glabbeek, U. Goltz, Refinement of actions and equivalence notions for concurrent systems, Acta Informatica (2001) 229–327.
[19] K.A. Bartlett, R.A. Scantlebury, P.T. Wilkinson, A note on reliable full-duplex transmission over half-duplex links, Communications of the ACM 12 (5) (1969) 260–261.
[20] C.A.R. Hoare, An axiomatic basis for computer programming, Communications of the ACM 12 (10) (October 1969).
[21] J.F. Groote, A. Ponse, Process algebra with guards: combining hoare logic with process algebra, Formal Aspects of Computing 6 (2) (1994) 115–164.
[22] F. Moller, The importance of the left merge operator in process algebras, in: M.S. Paterson (Ed.), Proceedings 17th Colloquium on Automata, Languages and Programming (ICALP'90), Warwick, in: Lecture Notes in Computer Science, vol. 443, Springer, 1990, pp. 752–764.

[23] R. Milner, J. Parrow, D. Walker, A calculus of mobile processes, Part I, Information and Computation 100 (1) (1992) 1–40.
[24] R. Milner, J. Parrow, D. Walker, A calculus of mobile processes, Part II, Information and Computation 100 (1) (1992) 41–77.

Index